Fungi

Biology and Applications

Second Edition

Editor

Kevin Kavanagh

Department of Biology
National University of Ireland Maynooth
Maynooth
County Kildare
Ireland

A John Wiley & Sons, Ltd., Publication

This edition first published 2011 © 2011 by John Wiley & Sons, Ltd.

Wiley-Blackwell is an imprint of John Wiley & Sons, formed by the merger of Wiley's global Scientific, Technical and Medical business with Blackwell Publishing.

Registered Office: John Wiley & Sons Ltd, The Atrium, Southern Gate, Chichester, West Sussex, PO19 8SQ, UK

Editorial Offices: 9600 Garsington Road, Oxford, OX4 2DQ, UK
The Atrium, Southern Gate, Chichester, West Sussex, PO19 8SQ, UK
111 River Street, Hoboken, NJ 07030-5774, USA

For details of our global editorial offices, for customer services and for information about how to apply for permission to reuse the copyright material in this book please see our website at www.wiley.com/wiley-blackwell.

The right of the author to be identified as the author of this work has been asserted in accordance with the UK Copyright, Designs and Patents Act 1988.

Library of Congress Cataloging-in-Publication Data

Fungi : biology and applications / editor, Kevin Kavanagh. – 2nd ed.
     p. cm.
  Includes bibliographical references and index.
  ISBN 978-0-470-97710-1 (cloth) – ISBN 978-0-470-97709-5 (pbk.)
  1. Fungi–Biotechnology.   2. Fungi.   I. Kavanagh, Kevin.
  TP248.27.F86F875 2011
  579.5–dc22

                                                    2011013563

A catalogue record for this book is available from the British Library.

This book is published in the following electronic formats: ePDF 9781119976967; ePub 9781119977698; Wiley Online Library 9781119976950; Mobi 9781119977704

Set in 10.5/13pt Sabon by Aptara Inc., New Delhi, India.

Printed and bound in Singapore by Markono Print Media Pte Ltd

First Impression 2011

Contents

List of Contributors

Professor Khaled H. Abu-Elteen, Department of Biological Science, Hashemite University, Zarqa 13133, Jordan.

Dr Catherine Bachewich, Biotechnology Research Institute, National Research Council of Canada, 6100 Royalmount Avenue, Montreal, QC, Canada H4P 2R2.

Dr Virginia Bugeja, School of Life Sciences, University of Hertfordshire, College Lane, Hatfield, Hertfordshire AL10 9AB, UK.

Professor David Coleman, Microbiology Research Laboratory, Dublin Dental Hospital, Trinity College, Dublin 2, Ireland.

Dr Brendan Curran, School of Biological and Chemical Science, Queen Mary, University of London, Mile End Road, London E1 4NS, UK.

Dr Fiona Doohan, Department of Plant Pathology, University College Dublin, Belfield, Dublin 4, Ireland.

Professor Sean Doyle, Department of Biology, National University of Ireland Maynooth, Maynooth, Co. Kildare, Ireland.

Dr David Fitzpartick, Department of Biology, National University of Ireland Maynooth, Co. Kildare, Ireland.

Dr Mawieh Hamad, Research and Development Unit, JMS Medicals, Amman, Jordan.

Dr Karina A. Horgan, Alltech Biotechnology Centre, Summerhill Road, Dunboyne, Co. Meath, Ireland.

Dr Kevin Kavanagh, Department of Biology, National University of Ireland Maynooth, Maynooth, Co. Kildare, Ireland.

Dr Shauna M. McKelvey, Alltech Biotechnology Centre, Summerhill Road, Dunboyne, Co. Meath, Ireland.

Dr Gary Moran, Microbiology Research Laboratory, Dublin Dental Hospital, Trinity College, Dublin 2, Ireland.

Dr Richard A. Murphy, Alltech Biotechnology Centre, Summerhill Road, Dunboyne, Co. Meath, Ireland.

Dr Derek Sullivan, Microbiology Research Unit, Dublin Dental School and Hospital, Trinity College, Dublin 2, Ireland.

Dr Edgar Medina Tovar, Mycology and Phytopathology Laboratory (LAMFU), Biological Sciences Department, Universidad de Los Andes, Bogotá, Colombia.

Dr Graeme M. Walker, Biotechnology and Forensic Sciences, School of Contemporary Sciences, University of Abertay Dundee, Kydd Building, Dundee DD1 1HG, Scotland, UK.

Dr Nia A. White, Biotechnology and Forensic Sciences, School of Contemporary Sciences, University of Abertay Dundee, Kydd Building, Dundee DD1 1HG, Scotland, UK.

Dr Malcolm Whiteway, Biotechnology Research Institute, National Research Council of Canada, 6100 Royalmount Avenue, Montreal, QC, Canada H4P 2R2.

1

Introduction to Fungal Physiology

Graeme M. Walker and Nia A. White

1.1 Introduction

Fungal physiology refers to the nutrition, metabolism, growth, reproduction and death of fungal cells. It also generally relates to interaction of fungi with their biotic and abiotic environment, including cellular responses to stress. The physiology of fungal cells impacts significantly on the environment, industrial processes and human health. In relation to ecological aspects, the biogeochemical cycling of carbon in nature would not be possible without the participation of fungi acting as primary decomposers of organic material. Furthermore, in agricultural operations, fungi play important roles as mutualistic symbionts, pathogens and saprophytes, where they mobilize nutrients and affect the physico-chemical environment. Fungal metabolism is also responsible for the detoxification of organic pollutants and for bioremediating heavy metals in the environment. The production of many economically important industrial commodities relies on the exploitation of yeast and fungal metabolism, and these include such diverse products as whole foods, food additives, fermented beverages, antibiotics, probiotics, pigments, pharmaceuticals, biofuels, enzymes, vitamins, organic and fatty acids and sterols. In terms of human health, some yeasts and fungi represent major opportunistic life-threatening pathogens, whilst others are life-savers, as they provide antimicrobial and chemotherapeutic agents. In modern biotechnology, several yeast species are being exploited as ideal hosts for the expression of human therapeutic proteins following recombinant DNA technology. In addition to the direct industrial exploitation of yeasts and fungi, it is important

Fungi: Biology and Applications, Second Edition. Edited by Kevin Kavanagh.
© 2011 John Wiley & Sons, Ltd. Published 2011 by John Wiley & Sons, Ltd.

to note that these organisms, most notably the yeast *Saccharomyces cerevisiae*, play increasingly significant roles as model eukaryotic cells in furthering our fundamental knowledge of biological and biomedical science. This is especially the case now that numerous fungal genomes have been completely sequenced, and the information gleaned from fungal genomics and proteomics is providing valuable insight into human genetics and heritable disorders. However, knowledge of cell physiology is essential if the functions of many of the currently unknown fungal genes are to be fully elucidated.

It is apparent, therefore, that fungi are important organisms for human society, health and well-being and that studies of fungal physiology are very pertinent to our understanding, control and exploitation of this group of microorganisms. This chapter describes some basic aspects of fungal cell physiology, focusing primarily on nutrition, growth and metabolism in unicellular yeasts and filamentous fungi.

1.2 Morphology of Yeasts and Fungi

Most fungi are filamentous, many grow as unicellular yeasts and some primitive fungi, such as the chytridomycetes, grow as individual rounded cells or dichotomous branched chains of cells with root-like rhizoids for attachment to a nutrient resource. Here, we will consider the most common growth forms: the filamentous fungi and unicellular yeasts.

1.2.1 Filamentous Fungi

The gross morphologies of macrofungi and microfungi are very diverse (see Plate 1.1). For example, we can easily recognize a variety of mushrooms and toadstools, the sexual fruiting bodies of certain macrofungi (the higher fungi Asomycotina and Basidiomycotina and related forms), during a walk through pasture or woodland. Microfungi (the moulds) are also diverse and are often observed on decaying foods and detritus, whereas many, including the coloured rusts, smuts and mildews, are common plant pathogens. Closer inspection of these visible structures, however, reveals that all are composed of aggregated long, branching threads termed hyphae (singular: hypha), organized to support spores for reproduction and dissemination. The hyphae of these aerial structures extend and branch within the supporting substratum as a network, termed a mycelium, from which the apically growing hyphae seek out, exploit and translocate available nutrients. Apically growing hyphae usually have a relatively constant diameter ranging from 1 to 30 μm or more, depending on fungal species and growth conditions. Filamentous fungi may be cultivated within the laboratory on a variety of different liquid or solid media. On agar, the radially expanding colonial growth form of the fungal mycelium is most evident, extending from an inoculum, on, within and sometimes above the substrate, forming a near spherical three-dimensional colony. This radiating, circular pattern is also visible during the growth of fairy ring fungi in grassland and as ringworm infections of the skin.

The hyphae of individual fungi may (theoretically) extend endlessly via apical growth, provided they are supported with appropriate nutrients and other environmental conditions. Eucarpic fungi, therefore, are spatially and temporally indeterminate organisms and, unlike animal, plant and other microbial individuals, have no predetermined maximum size or age. The mycelium is not, however, simply a homogeneously extending entity, but displays considerable developmental plasticity. Different interconnected regions of the fungal mycelium may grow, branch, anastomose (fuse), age, die, sporulate and display varying physiological and biochemical activities at different times or even simultaneously, depending on local micro-environmental conditions. Thus, colonies growing on relatively homogeneous media may be pigmented, exhibit different morphological sectors, produce aerial structures, grow as fast-effuse or slow-dense forms and even exhibit rhythmic growth (Plate 1.1). As well as reproductive structures and substrate mycelium, certain higher fungi, most notably the basidiomycetes, when growing within an environment where nutrients are distributed heterogeneously, can differentiate into long string-like structures called rhizomorphs or cords. These linear organs have evolved to rapidly explore for, connect and translocate water and nutrients between patches of resource (e.g. pieces of fallen timber on the forest floor or from tree root to tree root). Accordingly, many, particularly mature, rhizomorphs contain internal vessel hyphae which possess a wide diameter, forming a channel running along the organ. The peripheral hyphae are often closely packed and melanized for insulation.

Filamentous fungi and yeasts are simply different styles of fungal growth suitable for occupation of different habitats and produced by differing cell growth polarities. Many species termed dimorphic fungi can adopt either the hyphal or unicellular yeast forms according to environmental circumstances. For example, certain important human and animal pathogens exist as yeast forms mobilized in body fluids but are able to form hyphae or pseudohyphae for tissue invasion.

1.2.2 Yeasts

Yeasts are unicellular (mostly Ascomycete, Basidiomycete or Deuteromycete) fungi that divide asexually by budding or fission and whose individual cell size can vary widely from 2 to 3 μm to 20–50 μm in length and 1–10 μm in width. *S. cerevisiae* (commonly referred to as brewer's or baker's yeast), is generally ellipsoid in shape with a large diameter of 5–10 μm and a small diameter of 1–7 μm (Figure 1.1).

The morphology of agar-grown yeasts shows great diversity in terms of colour, texture and geometry (peripheries, contours) of giant colonies. Several yeasts

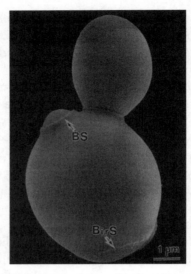

Figure 1.1 Scanning electron micrograph of a typical yeast cell. (×10 000). BS, bud scar; BirS, birth scar. (Reproduced with kind permission of Professor Masako Osumi, Japan Women's University, Tokyo.)

are pigmented, and the following colours may be visualized in surface-grown colonies: cream (e.g. *S. cerevisiae*); white (e.g. *Geotrichum candidum*); black (e.g. *Aureobasidium pullulans*); pink (e.g. *Phaffia rhodozyma*); red (e.g. *Rhodotorula rubra*); orange (e.g. *Rhodosporidium* spp.); and yellow (e.g. *Cryptococcus laurentii*). The pigments of some yeasts have biotechnological uses, including astaxanthin from *P. rhodozyma* in aquacultural feed supplements for farmed salmon (that are unable to synthesize these natural pink compounds) (Table 1.1).

1.3 Ultrastructure and Function of Fungal Cells

1.3.1 The Fungal Cell Surface

The cell envelope in yeasts and fungi is the peripheral structure that encases the cytoplasm and comprises the plasma membrane, the periplasm, the cell wall and additional extracellular structural components (such as fimbriae and capsules). The cell wall represents a dynamically forming exoskeleton that protects the fungal protoplast from the external environment and defines directional growth, cellular strength, shape and interactive properties. In filamentous fungi, cell-wall formation and organization is intimately bound to the process of apical growth. Thus, for example in *Neurospora crassa*, the wall is thin (approximately 50 nm) at the apex but becomes thicker (approximately 125 nm) at 250 μm behind the tip. The plasma membrane component of the fungal cell envelope is a phospholipid bilayer interspersed with globular proteins that dictates entry of nutrients and exit of metabolites and represents a selective barrier for their translocation.

Table 1.1 Diversity of yeast cell shapes.

Cell shape	Description	Examples of yeast genera
Ellipsoid	Ovoid shaped cells	*Saccharomyces*
Cylindrical	Elongated cells with hemispherical ends	*Schizosaccharomyces*
Apiculate	Lemon shaped	*Hanseniaspora, Saccharomycodes*
Ogival	Elongated cell rounded at one end and pointed at other	*Dekkera, Brettanomyces*
Flask-shaped	Cells dividing by bud-fission	*Pityrosporum*
Miscellaneous shapes	Triangular	*Trigonopsis*
	Curved	*Cryptococcus* (e.g. *Cryptococcus cereanus*)
	Spherical	*Debaryomyces*
	Stalked	*Sterigmatomyces*
Pseudohyphal	Chains of budding yeast cells which have elongated without detachment	*Candida* (e.g. *Candida albicans*)
Hyphal	Branched or unbranched filamentous cells which form from germ tubes. Septa may be laid down by the continuously extending hyphal tip. Hyphae may give rise to blastospores	*Candida albicans*
Dimorphic	Yeasts that grow vegetatively in either yeast or filamentous (hyphal or pseudohyphal) form	*Candida albicans* *Saccharomycopsis fibuligera* *Kluyveromyces marxianus* *Malassezia furfur* *Yarrowia lipolytica* *Histoplasma capsulatum*

Ergosterol is the major sterol found in the membranes of fungi, in contrast to the cholesterol found in the membranes of animals and phytosterols in plants. This distinction is exploited during the use of certain antifungal agents used to treat some fungal infections, and can be used as an assay tool to quantify fungal growth. The periplasm, or periplasmic space, is the region external to the plasma membrane and internal to the cell wall. In yeast cells, it comprises secreted proteins (mannoproteins) and enzymes (such as invertase and acid phosphatase) that are unable to traverse the cell wall. In filamentous fungi, the cell membrane and wall may be intimately bound as hyphae and are often resistant to plasmolysis.

Fungal cell surface topological features can be visualized using scanning electron microscopy (SEM), and nanometre resolution is achieved using atomic force microscopy (AFM). The latter is beneficial, as it can be employed with unfixed, living cells and avoids potentially misleading artefacts that may arise when preparing cells for electron microscopy. Figure 1.1 shows SEM micrographs of a typical unicellular yeast cell envelope.

Ultrastructural analysis of fungal cell walls reveals a thick, complex fibrillar network. The cell walls of filamentous fungi are mainly composed of different polysaccharides according to taxonomic group. For example, they may contain either chitin, glucans, mannoproteins, chitosan, polyglucuronic acid or cellulose, together with smaller quantities of proteins and glycoproteins (Table 1.2). Generally, the semi-crystalline microfibrillar components are organized in a network mainly in the central cell wall region and are embedded within an amorphous matrix. Bonding occurs between certain components behind the extending hyphal tip, thereby strengthening the entire wall structure. There is evidence to suggest that the cell wall is a dynamic structure where considerable quantitative and qualitative differences occur not only between different fungal species, but also between different morphological forms of the same species and even in response to environmental stress. For example, a class of hydrophobic proteins called hydrophobins are localized within the aerial growth or appresoria (terminal swellings involved in infection) of certain fungi, whereas pigmented melanins are often found within some fungal cell walls to insulate against biotic and abiotic stresses.

Table 1.2 The major polymers found in different taxonomical groups of fungi, together with the presence of perforate septa in these groups (adapted from Deacon (2005) and Carlile *et al.* (2001)).

Taxonomic grouping	Fibrillar polymers	Matrix polymers	Perforate septa present or absent
Oomycetes	β(1,3)-, β(1,6)-Glucan Cellulose	Glucan	Absent
Chytridomycetes	Chitin; glucan	Glucan	Absent
Zygomycetes	Chitin; chitosan	Polyglucuronic acid; glucuronomannoproteins	Absent
Basidiomycetes	Chitin; β(1,3)-, β(1,6)-glucans	α(1,3)-Glucan; xylomannoproteins	Present (mostly Dolipore)
Ascomycetes/ Deuteromycetes	Chitin; β(1,3)-, β(1,6)-glucans	α(1,3)-Glucan; galactomannoproteins	Present (mostly simple with large central pore)

The hyphae of higher fungi extend via tip growth followed by cross-wall formation or septation, whereas the lower fungi remain aseptate (except when segregating spores or in damaged colony regions). Septa may offer some structural support to hyphae. Significantly, septa serve to compartmentalize hyphae but are typically perforated, thereby permitting passage and communication of cytoplasm or even protoplasm between compartments. However, septal pores can become blocked by Woronin bodies or other materials. This aids morphological and biochemical differentiation and serves to seal off stressed or damaged hyphae from undamaged colony regions. Again, different pore types are representative of different taxonomic groups and species (Table 1.2).

In yeasts, the cell-wall structure comprises polysaccharides (predominantly β-glucans for rigidity), proteins (mainly mannoproteins on the outermost layer for determining porosity), together with some lipid, chitin (e.g. in bud scar tissue) and inorganic phosphate material. Figure 1.2 shows the composition and structure of the *S. cerevisiae* cell wall. Hyphal cell walls generally contain fewer mannans than yeast cell forms, and such changes in composition are even observed during the transition from unicellular to mycelial growth of dimorphic fungi.

Chitin is also found in yeast cell walls and is a major constituent of bud scars (Figure 1.3). These are remnants of previous budding events found on the surface of mother cells following birth of daughter cells (buds). The chitin-rich bud scars of yeast cells can be stained with fluorescent dyes (e.g. calcoflour white), and this can provide useful information regarding cellular age, since the number of scars represents the number of completed cell division cycles. Outside the cell wall in fungi, several extramural layers may exist, including fimbriae and capsules.

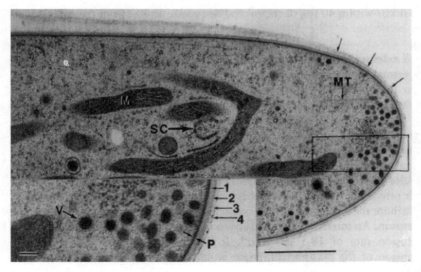

Figure 1.2 Cell envelope structure of the yeast *S. cerevisiae* (from Walker (1998). Permission obtained for First Edition).

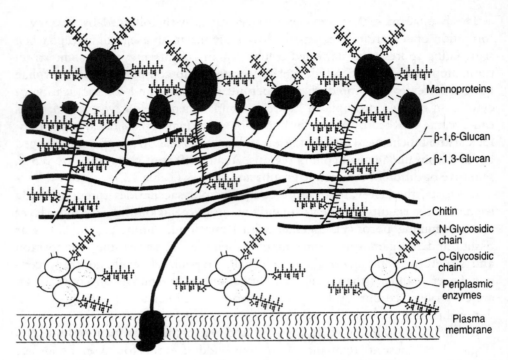

Figure 1.3 Transmission electron microscopy of ultrathin sections of fungal cells reveals intracellular fine structure.

Fungal fimbriae are long, protein-containing protrusions appearing from the cell wall of certain basidiomycetous and ascomycetous fungi that are involved in cell-cell conjugation. Capsules are extracellular polysaccharide-containing structures found in basidiomycetous fungi that are involved in stress protection. In *Cryptococcus neoformans* (the pathogenic yeast state of *Filobasidiella neoformans*) the capsule may determine virulence properties and evasion from macrophages. One extrahyphal substance, the polymer pullulan, is produced commercially from *A. pullulans*.

1.3.2 Subcellular Architecture and Organelle Function

Transmission electron microscopy of ultrathin sections of fungal cells reveals intracellular fine structure (Figures 1.2 and 1.4). Subcellular compartments (organelles) are bathed in an aqueous cytoplasm containing soluble proteins and other macromolecules, together with low-molecular weight metabolites. However, the hyphae of central (and therefore older) colony regions of filamentous fungi may become devoid of protoplasm and organelles, as protoplasmic components are driven forward or are recycled, to support the growth

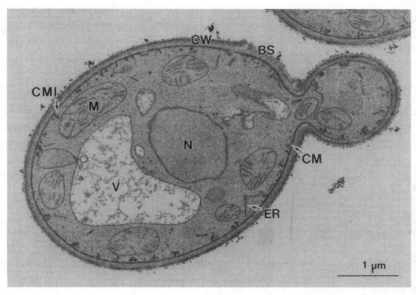

Figure 1.4 Electron micrograph of a typical yeast cell. (CW, cell wall; CM, cell membrane; CMI, cell membrane invagination; BS, bud scar; M, mitochondrion, N, nucleus; V, vacuole; ER, endoplasmic reticulum. (Reproduced with kind permission of Professor Masako Osumi, Japan Women's University, Tokyo.)

of actively growing hyphal tips. Cytoplasmic components additionally comprise microbodies, ribosomes, proteasomes, lipid particles and a cytoskeletal network. The latter confers structural stability to the fungal cytoplasm and consists of microtubules and microfilaments. The following membrane-bound organelles may be found in a typical fungal cell: nucleus: endoplasmic reticulum (ER), mitochondria, Golgi apparatus, secretory vesicles and vacuoles. Several of these organelles form extended membranous systems. For example, the ER is contiguous with the nuclear membrane and secretion of fungal proteins involves intermembrane trafficking in which the ER, Golgi apparatus, plasma membrane and vesicles all participate. The physiological function of the various fungal cell organelles is summarized in Table 1.3.

The nucleus is the structure that defines the eukaryotic nature of fungal cells. It is bound by a double membrane and encases the chromosomes in a nucleoplasm. Most yeast and fungi are haploid, although some (e.g. *S. cerevisiae*) may alternate between haploidy and diploidy. Chromosomes comprise DNA–protein structures that replicate and segregate to newly divided cells or hyphal compartments at mitosis. This, of course, ensures that genetic material is passed onto daughter cells or septated compartments at cell division. Yeasts usually contain a single nucleus per cell. However, the hyphal compartments of filamentous fungi may contain one or more nuclei. Monokaryotic basidiomycetes possess one nucleus per compartment, whereas dikaryons or heterokaryons possess two

Table 1.3 Functional components of an idealized fungal cell.

Organelle or cellular structure	Function
Cell envelope	Comprising: the plasma membrane, which acts as a selectively permeable barrier for transport of hydrophilic molecules in and out of fungal cells; the periplasm, containing proteins and enzymes unable to permeate the cell wall; the cell wall, which provides protection, shape and is involved in cell–cell interactions, signal reception and specialized enzyme activities; fimbriae involved in sexual conjugation; capsules to protect cells from dehydration and immune cell attack.
Nucleus	Relatively small. Containing chromosomes (DNA–protein complexes) that pass genetic information to daughter cells at cell division and the nucleolus, which is the site of ribosomal RNA transcription and processing.
Mitochondria	Site of respiratory metabolism under aerobic conditions and, under anaerobic conditions, for fatty acid, sterol and amino acid metabolism.
Endoplasmic reticulum	Ribosomes on the rough ER are the sites of protein biosynthesis.
Proteasome	Multi-subunit protease complexes involved in regulating protein turnover.
Golgi apparatus and vesicles	Secretory system for import (endocytosis) and export (exocytosis) of proteins.
Vacuole	Intracellular reservoir (amino acids, polyphosphate, metal ions); proteolysis; protein trafficking; control of cellular pH. In filamentous fungi, tubular vacuoles transport materials bidirectionally along hyphae.
Peroxisome	Oxidative utilization of specific carbon and nitrogen sources (contain catalase, oxidases). Glyoxysomes contain enzymes of the glyoxylate cycle.

or more genetically distinct haploid nuclei. The maintenance of multiple nuclei within individual hyphal compartments allows fungi to take advantage of both haploid and diploid lifestyles. This is discussed further in Chapter 2.

In filamentous fungi, a phase-dark near-spherical region, which also stains with iron haemotoxylin, is evident by light microscopy at the apex during hyphal tip growth. The region is termed the Spitzenkörper, the apical vesicle cluster or centre or apical body, and it consists of masses of small membrane-bound vesicles around a vesicle-free core with emergent microfilaments and microtubules. The Spitzenkörper contains differently sized vesicles derived from Golgi bodies, either large vesicles or microvesicles (chitosomes), with varying composition. It orientates to the side as the direction of tip growth changes, and disappears when growth ceases. This vesicle supply centre is involved in wall extension and, hence, tip growth, branching, clamp connection formation (in Basidiomycetes) and germ tube formation.

1.4 Fungal Nutrition and Cellular Biosyntheses

1.4.1 Chemical Requirements for Growth

Yeasts and fungi have relatively simple nutritional needs and most species would be able to survive quite well in aerobic conditions if supplied with glucose, ammonium salts, inorganic ions and a few growth factors. Exceptions to this would include, for example, obligate symbionts such as the vesicular–arbuscular mycorrhizal (VAM) fungi which require growth of a plant partner for cultivation. Macronutrients, supplied at millimolar concentrations, comprise sources of carbon, nitrogen, oxygen, sulfur, phosphorus, potassium and magnesium; and micronutrients, supplied at micromolar concentrations, comprising trace elements like calcium, copper, iron, manganese and zinc, would be required for fungal cell growth (Table 1.4). Some fungi are oligotrophic, apparently growing with very limited nutrient supply, surviving by scavenging minute quantities of volatile organic compounds from the atmosphere.

Being chemoorganotrophs, fungi need fixed forms of organic compounds for their carbon and energy supply. Sugars are widely utilized for fungal growth and can range from simple hexoses, like glucose, to polysaccharides, like starch and cellulose. Some fungi can occasionally utilize aromatic hydrocarbons (e.g. lignin by the white-rot fungi). Table 1.5 outlines the variety of carbon sources which can be utilized by yeasts and filamentous fungi for growth.

Fungi are non-diazotrophic (cannot fix nitrogen) and need to be supplied with nitrogenous compounds, either in inorganic form, such as ammonium salts, or in organic form, such as amino acids. Ammonium sulfate is a commonly used nitrogen source in fungal growth media, since it also provides a source of utilizable sulfur. Many fungi (but not the yeast *S. cerevisiae*) can also grow on nitrate and, if able to do so, may also utilize nitrite. Nitrate reductase followed by nitrite reductase are the enzymes responsible for converting nitrate to ammonia. Most fungi can assimilate amino acids, amines and amides as nitrogen sources. Most fungi (but not many yeasts) are also proteolytic and can hydrolyse proteins (via extracellularly secreted proteases) to liberate utilizable amino acids for growth. Urea utilization is common in fungi, and some basidiomycetous yeasts are classed as urease positive (able to utilize urea) whilst most ascomycetous yeasts are urease negative.

In terms of oxygen requirements, most fungi are aerobes and are often microaerophilic (preferring an oxygen tension below that of normal atmospheric).

Table 1.4 Elemental requirements of fungal cells.

Element	Common sources	Cellular functions
Carbon	Sugars	Structural element of fungal cells in combination with hydrogen, oxygen and nitrogen. Energy source
Hydrogen	Protons from acidic environments	Transmembrane proton motive force vital for fungal nutrition. Intracellular acidic pH (around 5–6) necessary for fungal metabolism
Oxygen	Air, O_2	Substrate for respiratory and other mixed-function oxidative enzymes. Essential for ergosterol and unsaturated fatty acid synthesis
Nitrogen	NH_4^+ salts, urea, amino acids	Structurally and functionally as organic amino nitrogen in proteins and enzymes
Phosphorus	Phosphates	Energy transduction, nucleic acid and membrane structure
Potassium	K^+ salts	Ionic balance, enzyme activity
Magnesium	Mg^{2+} salts	Enzyme activity, cell and organelle structure
Sulfur	Sulfates, methionine	Sulfydryl amino acids and vitamins
Calcium	Ca^{2+} salts	Possible second messenger in signal transduction
Copper	Cupric salts	Redox pigments
Iron	Ferric salts. Fe^{3+} is chelated by siderophores and released as Fe^{2+} within the cell	Haem-proteins, cytochromes
Manganese	Mn^{2+} salts	Enzyme activity
Zinc	Zn^{2+} salts	Enzyme activity
Nickel	Ni^{2+} salts	Urease activity
Molybdenum	Na_2MoO_4	Nitrate metabolism, vitamin B_{12}

Although yeasts like *S. cerevisiae* are sometimes referred to as facultative anaerobes, they cannot actually grow in strictly anaerobic conditions unless supplied with certain fatty acids and sterols (which they cannot synthesize without molecular oxygen). In fact, there are thought to be very few yeast species that are obligately anaerobic. For aerobically respiring yeasts and fungi, oxygen is required as the terminal electron acceptor, where it is finally reduced to water in the electron-transport chain. Different fungal species respond to oxygen

Table 1.5 Diversity of carbon sources for yeast and filamentous fungal growth (adapted from Walker (1998)).

Carbon source	Typical examples	Comments
Hexose sugars	D-Glucose, D-galactose,	Glucose metabolized by majority of yeasts and filamentous fungi
	D-Fructose, D-mannose	If a yeast does not ferment glucose, it will not ferment other sugars. If a yeast ferments glucose, it will also ferment fructose and mannose, but not necessarily galactose
Pentose sugars	L-Arabinose, D-xylose, D-xyulose, L-rhamnose	Some fungi respire pentoses better than glucose. *S. cerevisiae* can utilize xylulose but not xylose
Disaccharides	Maltose, sucrose, lactose, trehalose, melibiose, cellobiose, melezitose	If a yeast ferments maltose, it does not generally ferment lactose and vice versa. Melibiose utilization used to distinguish ale and lager brewing yeasts. Large number of yeasts utilize disaccharides. Few filamentous fungi (e.g. *Rhizopus nigricans*) cannot utilize sucrose
Trisaccharides	Raffinose, maltotriose	Raffinose only partially used by *S. cerevisiae*, but completely used by other *Saccharomyces* spp. (*S. carlsbergensis, S. kluyveri*)
Oligosaccharides	Maltotetraose, maltodextrins	Metabolized by amylolytic yeasts, not by brewing strains
Polysaccharides	Starch, inulin, cellulose, hemicellulose, chitin, pectic substances	Polysaccharide-fermenting yeasts are rare. *Saccharomycopsis* spp. and *S. diastaticus* can utilize soluble starch; *Kluyveromyces* spp. possess inulinase. Many filamentous fungi can utilize these depending on extracellular enzyme activity
Lower aliphatic alcohols	Methanol, ethanol	Respiratory substrates for many fungi. Several methylotrophic yeasts (e.g. *Pichia pastoris, Hansenula polymorpha*) have industrial potential
Sugar alcohols	Glycerol, glucitol	Can be respired by yeasts and a few fungi
Organic acids	Acetate, citrate, lactate, malate, pyruvate, succinate	Many yeasts can respire organic acids, but few can ferment them

(*continued*)

Table 1.5 (*Continued*)

Carbon source	Typical examples	Comments
Fatty acids	Oleate, palmitate	Several species of oleaginous yeasts can assimilate fatty acids as carbon and energy sources
Hydrocarbons	*n*-Alkanes	Many yeast and a few filamentous species grown well on C_{12}-C_{18} *n*-alkanes
Aromatics	Phenol, cresol, quinol, resourcinol, catechol, benzoate	Few yeasts can utilize these compounds. Several *n*-alkane-utilizing yeasts use phenol as carbon source via the β-ketoadipate pathway
Miscellaneous	Adenine, uric acid, butylamine, pentylamine, putrescine	Some mycelial fungi and yeasts, for example, *Arxula adeninivorans* and *A. terestre* can utilize such compounds as sole source of carbon and nitrogen
	Lignin	Can be decayed only by white-rot fungi (basidiomycotina). Little net energy gained directly, but makes available other polysaccharides such as cellulose and hemicellulose
	'Hard' keratin	Keratinophilic fungi

availability in diverse ways, and Table 1.6 categorizes fungi into different groups on this basis.

Sulfur sources for fungal growth include sulfate, sulfite, thiosulfate, methionine and glutathione with inorganic sulfate and the sulfur amino acid methionine being effectively utilized. Virtually all yeasts can synthesize sulfur amino acids from sulfate, the most oxidized form of inorganic sulfur.

Phosphorus is essential for biosynthesis of fungal nucleic acids, phospholipids, ATP and glycophosphates. Hence, the phosphate content of fungi is considerable (e.g. in yeast cells it accounts for around 3–5 % of dry weight; the major part of this is in the form of orthophosphate ($H_2PO_4^-$), which acts as a substrate and enzyme effector). The fungal vacuole can serve as a storage site for phosphate in the form of complexed inorganic polyphosphates (also referred to as volutin granules). Both nitrogen and phosphorus availability may be growth limiting in nature. Filamentous fungi have evolved a number of biochemical and morphological strategies allowing capture of often poorly available phosphorus within the natural environment. Plants exploit such efficiency during symbioses between their roots and certain mycorrhizal fungi. The major storage form of phosphorus in plants is phytic acid (myo-inositol hexa-dihydrogenphosphate), which is poorly utilized by monogastrics (e.g. humans, pigs, poultry), and fungal (and yeast) phytases have applications in reducing phytate content of foods and feeds.

Concerning requirements for minerals, potassium, magnesium and several trace elements are necessary for fungal growth. Potassium and magnesium are

Table 1.6 Yeast and fungal metabolism based on responses to oxygen availability.

Mode of energy metabolism	Examples	Comments
Obligate fermentative	Yeasts: *Candida pintolopesii* (*Saccharomyces telluris*)	Naturally occurring respiratory-deficient yeasts. Only ferment, even in presence of oxygen
	Fungi: facultative and obligate anaerobes	No oxygen requirement for these fungi. Two categories exist with respect to the effects of air: facultative anaerobes (e.g. *Aqualinderella* and *Blastocladia*) and obligate anaerobes (e.g. *Neocallimastix*)
Facultatively fermentative		
Crabtree-positive	*Saccharomyces cerevisiae*	Such yeasts predominantly ferment high sugar-containing media in the presence of oxygen
Crabtree-negative	*Candida utilis*	Such yeasts do not form ethanol under aerobic conditions and cannot grow anaerobically
Non-fermentative	Yeasts: *Rhodotorula rubra*	Such yeasts do not produce ethanol, either in the presence or absence of oxygen
	Fungi: *Phycomyces*	Oxygen essential for such (obligately oxidative) fungi
Obligate aerobes	*Gaemannomyces graminis* (the take-all fungus)	The growth of these is markedly reduced if oxygen partial pressure falls below normal atmospheric

Adapted from Walker (1998), Deacon (2005) and Carlile *et al.* (2001).

macroelements required in millimolar concentrations, primarily as enzyme co-factors, whereas other microelements (trace elements) are generally required in the micromolar range. These include Mn, Ca Fe, Zn, Cu, Ni, Co and Mo. Table 1.7 summarizes the main metals required for fungal growth. Toxic minerals (e.g. Ag, As, Ba, Cs, Cd, Hg, Li, Pb) adversely affect fungal growth generally at concentrations greater than 100 μM.

Fungal growth factors are organic compounds occasionally needed in very low concentrations for specific enzymatic or structural roles, but not as energy sources. These include vitamins (e.g. thiamine, biotin), purines, pyrimidines, nucleosides, nucleotides, amino acids, fatty acids and sterols. For fungi to have a growth factor requirement, this indicates that cells cannot synthesize the particular factor, resulting in the curtailment of growth without its provision

Table 1.7 Metals required for fungal growth and metabolic functions (adapted from Walker (2004)).

Metal ion	Concentration supplied in growth medium[a]	Main cellular functions
Macroelements		
K	2–4 mM	Osmoregulation, enzyme activity
Mg Mg	2–4 mM	Enzyme activity, cell division
Microelements		
Mn	2–4 μM	Enzyme cofactor
Ca	<μM	Second messenger, yeast flocculation
Cu	1.5 μM	Redox pigments
Fe	1–3 μM	Haem-proteins, cytochromes
Zn	4–8 μM	Enzyme activity, protein structure
Ni	~10 μM	Urease activity
Mo	1.5 μM	Nitrate metabolism, vitamin B_{12}
Co	0.1 μM	Cobalamin, coenzymes

[a]Figures relate to yeast (*S. cerevisiae*) growth stimulation and are dependent on the species/strain and conditions of growth, but they would be generally applicable for fungal growth.

in culture media. Some fungi (e.g. *Aspergillus niger*, *Penicillium chrysogenum*) have very simple nutritional needs and are able to synthesize their own growth factors from glucose.

1.4.2 Fungal Cultivation Media

Fungal nutritional requirements are important not only for successful cultivation in the laboratory, but also for the optimization of industrial fermentation processes. In the laboratory, it is relatively easy to grow yeasts and fungi on complex culture media such as malt extract or potato–dextrose agar or broth, which are both carbon rich and in the acidic pH range. Mushrooms are cultivated on various solid substrates, depending on provincial availability. Therefore, *Agaricus bisporus* (common button mushroom) is grown in the UK, USA and France on wheat straw; the padi-straw mushroom (*Volvariella volvacea*) is grown in South-east Asia on damp rice-straw and in Hong Kong on cotton waste; and in Japan, the shiitake mushroom (*Lentinus edodes*) is cultivated on fresh oak logs. In industry, media for fungal fermentation purposes need to be optimized with regard to the specific application and production process. For some industrial processes, growth media may already be relatively complete in a nutritional sense, such as malt wort or molasses for brewing or baker's yeast production respectively (Table 1.8). However, for other processes, supplementation of

Table 1.8 Principal ingredients of selected industrial media for yeasts and fungi.

Components	Molasses	Malt wort	Wine must	Cheese whey	Corn steep liquor
Carbon sources	Sucrose Fructose Glucose Raffinose	Maltose Sucrose Fructose Glucose Maltotriose	Glucose Fructose Sucrose (trace)	Lactose	Glucose, other sugars
Nitrogen sources	Nitrogen compounds as unassimilable proteins. Nitrogen sources need to be supplemented	Low molecular α-amino nitrogen compounds, ammonium ions and a range of amino acids	Variable levels of ammonia nitrogen, which may be limiting. Range of amino acids	Unassimilable globulin and albumin proteins. Low levels of ammonium and urea nitrogen	Amino acids, protein
Minerals	Supply of P, K and S available. High K⁺ levels may be inhibitory	Supply of P, K, Mg and S available	Supply of P, K, Mg and S available. High levels of sulfite often present	Supply of P, K, Mg, S	Supply of P, K, Mg, S
Vitamins	Small, but generally adequate supplies. Biotin is deficient in beet molasses	Supply of vitamins is usually adequate. High adjunct sugar wort may be deficient in biotin	Vitamin supply generally sufficient	Wide range of vitamins present	Biotin, pyridoxine, thiamin
Trace elements	Range of trace metals present, although Mn²⁺ may be limiting	All supplied, although Zn²⁺ may be limiting	Sufficient quantities available	Fe, Zn, Mn, Ca, Cu present	Range of trace elements present
Other components	Unfermentable sugars (2–4 %), organic acids, waxes, pigments, silica, pesticide residues, carmelized compounds, betaine	Unfermentable maltodextrins, pyrazines, hop compounds	Unfermentable pentoses. Tartaric and malic acids. Decanoic and octanoic acids may be inhibitory. May be deficient in sterols and unsaturated fatty acids	Lipids, NaCl. Lactic and citric acids	High levels of lactic acid present. Fat and fibre also present

agriculturally derived substrates, like corn steep liquor, molasses or malt broth, with additional nutrients and growth factors may be necessary. For example, the following may constitute a suitable fermentation medium for penicillin production by *Penicillium* spp.: sucrose (3 g/L), corn steep liquor (100 g/L), KH$_2$PO$_4$ (1 g/L), (NH$_4$)$_2$SO$_4$ (12 g/L), CaCl$_2$·2H$_2$O (0.06 g/L), phenoxyacetic acid (5.7 g/L) – information from Jorgensen *et al.* (1995). However, other industrial processes, such as the growth of *Fusarium graminarium* for production of Quorn^{TM} mycoprotein, require culture on a completely defined medium (see Chapter 5).

1.4.3 Nutrient Uptake and Assimilation

Fungal cells utilize a diverse range of nutrients and employ equally diverse nutrient acquisition strategies. Fungi are nonmotile, saprophytic (and sometimes parasitic), chemo-organotrophic organisms. They exhibit dynamic interactions with their nutritional environment that may be exemplified by certain morphological changes depending on nutrient availability. For example, the filamentous mode of growth observed at the periphery of yeast colonies growing in agar is akin to a foraging for nutrients as observed in certain eucarpic fungi. Metabolic dynamism is also evident in yeasts which, although not avid secretors of hydrolytic enzymes like higher fungi, are nevertheless able to secrete some enzymes to degrade polymers such as starch (as in amylolytic yeasts like *Schwanniomyces occidentalis*).

Several cellular envelope barriers to nutrient uptake by fungal cells exist, namely: the capsule, the cell wall, the periplasm and the cell membrane. Although not considered as a freely porous structures, fungal cell walls are relatively porous to molecules up to an average molecular mass of around 300 Da, and will generally retain molecules greater than around 700 Da. Typically, fungi absorb only small soluble nutrients, such as monosaccharides and amino acids.

The plasma membrane is the major selectively permeable barrier which dictates nutrient entry and metabolite exit from the fungal cell. Membrane transport mechanisms are important in fungal physiology, since they govern the rates at which cells metabolize, grow and divide. Fungi possess different modes of passive and active uptake at the plasma membrane: free diffusion, facilitated diffusion, diffusion channels and active transport (Table 1.9). Active transport of nutrients, such as sugars, amino acids, nitrate, ammonium, sulfate and phosphate, in filamentous fungi involves spatial separation of the ion pumps mostly behind the apex, whereas the symport proteins are active close to the tip. Thus, nutrient uptake occurs at the hyphal tip as it continuously drives into fresh resource, and the mitochondria localized behind the apex supply ATP to support the ion pump and generate proton motive force.

Table 1.9 Modes of nutrient transport in fungi.

Mode of nutrient transport	Description	Examples of nutrients transported
Free diffusion	Passive penetration of lipid-soluble solutes through the plasma membrane following the law of mass action from a high extracellular concentration to a lower intracellular concentration	Organic acids, short-chain alkanes and long-chain fatty acids by fungi and the export of lipophilic metabolites (e.g. ethanol) and gaseous compounds
Facilitated diffusion	Translocates solutes down a transmembrane concentration gradient in an enzyme (permease)-mediated manner. As with passive diffusion, nutrient translocation continues until the intracellular concentration equals that of the extracellular medium	In the yeast *S. cerevisiae*, glucose is transported in this manner
Diffusion channels	These operate as voltage-dependent 'gates' to transiently move certain nutrient ions down concentration gradients. They are normally closed at the negative membrane potential of resting yeast cells but are open when the membrane potential becomes positive	Ions such as potassium may be transported in this fashion
Active transport	The driving force is the membrane potential and the transmembrane electrochemical proton gradient generated by the plasma membrane H^+-ATPase. The latter extrudes protons using the free energy of ATP hydrolysis that enables nutrients to either enter with influxed protons, as in 'symport' mechanisms, or against effluxed protons, as in 'antiport' mechanisms	Many nutrients (sugars, amino acids, ions)

1.4.4 Overview of Fungal Biosynthetic Pathways

Anabolic pathways are energy-consuming, reductive processes which lead to the biosynthesis of new cellular material and are mediated by dehydrogenase enzymes which predominantly use reduced $NADP^+$ as the redox cofactor. NADPH is generated by the hexose monophosphate pathway (or Warburg–Dickens pathway) which accompanies glycolysis (see Section 1.5.1). In *S. cerevisiae*, up to

20 % of total glucose may be degraded via the hexose monophosphate pathway. This pathway generates cytosolic NADPH (following the dehydrogenation of glucose 6-phosphate using glucose 6-phosphate dehydrogenase and $NADP^+$ as hydrogen acceptor) for biosynthetic reactions leading to the production of fatty acids, amino acids, sugar alcohols, structural and storage polysaccharides and secondary metabolites. Besides generating NADPH, the hexose monophosphate pathway also produces ribose sugars for the synthesis of nucleic acids, RNA and DNA and for nucleotide coenzymes, NAD, NADP, FAD and FMN. This is summarized as follows:

$$\text{Glucose 6-phosphate} + 2NADP^+ \rightarrow \text{Ribulose 5-phosphate} + CO_2 + \text{NADPH} + 2H^+$$

And complete oxidation of glucose 6-phosphate would result in

$$\text{Glucose 6-phosphate} + 12NADP^+ \rightarrow 6CO_2 + 12\text{NADPH} + 12H^+ \rightarrow Pi$$

Fungal growth on non-carbohydrate substrates as sole carbon sources (e.g. ethanol, glycerol, succinate and acetate) may lead to gluconeogenesis (conversion of pyruvate to glucose) and polysaccharide biosynthesis. Gluconeogenesis may be regarded as a reversal of glycolysis and requires ATP as energy and NADH as reducing power.

Concerning fungal amino acid biosynthesis, simple nitrogenous compounds such as ammonium may be assimilated into amino acid *families*, the carbon skeletons of which originate from common precursors of intermediary carbon metabolism.

The two main fungal storage carbohydrates are glycogen and trehalose. Glycogen is similar to starch with β-1,4-glucan linear components and β-1,6-branches. Trehalose (also known as mycose) is a disaccharide of glucose comprising an α,α-1,1-glucoside bond between two α-glucose units. Both trehalose and glycogen are synthesized following the formation of UDP-glucose, catalysed by UDP-glucose pyrophosphorylase:

$$\text{UTP} + \text{Glucose 1-phosphate} \rightarrow \text{UDP-glucose} + \text{Pyrophosphate}$$

Glycogen is synthesized by glycogen synthase. Glycogen may be metabolized by glycogen phosphorylase when nutrients become limited under starvation conditions, and this contributes to the maintenance metabolism of cells by furnishing energy in the form of ATP. In yeast cells, glycogen breakdown is accompanied by membrane sterol biosynthesis, and this is important for brewing yeast vitality and successful beer fermentations. The other major storage carbohydrate, trehalose, is synthesized from glucose 6-phosphate and UDP-glucose by trehalose 6-phosphate synthase and converted to trehalose by a phosphatase. In addition to a storage role, trehalose is an important translocation material in filamentous forms and is also involved in stress protection in yeasts and fungi, accumulating when cells are subject to environmental insults such as heat shock or osmotic stress, or during plant host–fungal parasite interactions. Polyols, such as mannitol derived from fructose phosphate, are also translocated by fungi.

1.4.5 Fungal Cell Wall Growth

The structural polysaccharides in fungal cell walls include mannans, glucans and chitin and are synthesized from sugar nucleotides substrates formed by pyrophosphorylase enzymes. For example:

$$\text{Glucose 1-phosphate} + \text{UTP} \rightarrow \text{UDP-glucose} + \text{PPi}$$
$$\text{Mannose 1-phosphate} + \text{GTP} \rightarrow \text{GDP-mannose} + \text{PPi}$$

Glucan synthesis involves plasma membrane-associated glucan synthetases for assembly of β-1,3 linkages and β-1,6 branches of cell-wall glucan. Chitin (a polymer of N-acetylglucosamine) is an important fungal cell-wall structural component and is involved in the yeast budding process and in dimorphic transitions from yeast to filamentous forms. Chitin synthetases catalyse the transfer of N-acetylglucosamine from UDP-N-acetylglucosamine to a growing chitin polymer within the fungal cell wall. The mannoproteins predominantly of unicellular forms are pre-assembled within the Golgi and are delivered to the cell wall via vesicles from the vesicle supply centre. Various vesicles containing cell-wall-synthetic enzymes, wall-lytic enzymes, enzyme activators and certain preformed wall components are transported to the tip where they fuse with the plasma membrane and release their contents, which, together with substrates delivered from the cytosol, facilitate synthesis of the growing cell wall.

1.5 Fungal Metabolism

1.5.1 Carbon Catabolism

Being chemoorganotrophs, fungi derive their energy from the breakdown of organic compounds. Generally speaking, fungi, but few yeast species, extracellularly break down polymeric compounds by secreted enzymes prior to utilization of monomers as carbon and energy sources. Owing to their relatively large size (20–60 kDa), enzymes assembled by the Golgi are transported in vesicles to be secreted from sites of cell growth, essentially from extending hyphal tips. Enzymes may either become linked to the cell wall as wall-bound enzymes or may diffuse externally to decay substrates within the local environment.

Some examples follow of hydrolytic, oxidative, peroxidative and free-radical-generating enzyme systems produced by fungi for the degradation of polymeric compounds:

Pectin $\xrightarrow{\text{Pectin lyase, polygalactorunase}}$ Galacturonic acid

Starch $\xrightarrow{\text{Amylases, glucoamylase}}$ Glucose

Inulin $\xrightarrow{\text{Inulinase}}$ Fructose

Cellulose $\xrightarrow{\text{Cellulases}}$ Glucose

Hemicellulose $\xrightarrow{\text{Hemicellulases, xylanase}}$ Xylose, Glucose

Lipids $\xrightarrow{\text{Lipases}}$ Fatty acids

Proteins $\xrightarrow{\text{Proteinases}}$ Amino acids

Chitin $\xrightarrow{\text{Chitinase}}$ N-acetylglucosamine

Lignin $\xrightarrow{\text{Ligninase; manganese peroxidase; laccase; glucose oxidase}}$ Variety of largely phenolic products

Several lipolytic yeasts are known (e.g. *Candida rugosa, Yarrowia lipolytica*) which secrete lipases to degrade triacylgycerol substrates to fatty acids and glycerol.

In wood, the cellulose and hemicellulose components are embedded within a heteropolymeric three-dimensional lignin matrix, thus forming a complex lignocellulose material. Only certain filamentous basidiomycete or ascomycete fungi are able to degrade the recalcitrant lignin component to make available the cellulose or hemicellulose components. These are known as white-rot fungi due to resultant coloration of the delignified wood. Such fungi employ a cocktail of oxidative (including laccases) and peroxidative enzymes, together with hydrogen-peroxide-generating enzyme systems, to attack at least 15 different inter-unit bond types extant within the lignin polymer. The manganese and lignin peroxidase enzyme systems operate by releasing highly reactive but transient oxygen free-radicals, which bombard and react with parts of the lignin molecule, generating a chain of chemical oxidations and producing a range of mainly phenolic end products. White-rot fungi have applications in, for example, upgrading lignocellulose waste for animal feed, paper production and bleaching, the bioremediation of contaminated land and water and (potentially) for biofuel production. Brown-rot and soft-rot (in wet wood) fungi are only able to degrade the cellulose and hemicellulose components of wood. Cellulose decomposition involves the synergistic activity of endoglucanases (that hydrolyse the internal bonds of cellulose), exoglucanases (that cleave cellobiose units from the end of the cellulose chain) and glucosidases (that hydrolyse cellobiose to glucose). Initial attack of cellulose microfibrills within the cell wall may involve the generation of hydrogen peroxide.

Catabolic pathways are oxidative processes which remove electrons from intermediate carbon compounds and use these to generate energy in the form of ATP. The catabolic sequence of enzyme-catalysed reactions that convert glucose to pyruvic acid is known as glycolysis, and this pathway provides fungal cells with energy, together with precursor molecules and reducing power (in the form of NADH) for biosynthetic pathways. Therefore, in serving both catabolic and anabolic functions, glycolysis is sometimes referred to as an amphibolic pathway. Glycolysis may be summarized as follows:

$$\text{Glucose} + 2ADP + 2Pi + 2NAD^+ \rightarrow 2\text{Pyruvate} + 2ATP + 2NADH^+ + 2H^+$$

During glycolysis, glucose is phosphorylated using ATP to produce fructose 1,6-biphosphate, which is then split by aldolase to form two triose phosphate compounds. Further phosphorylation occurs, forming two triose diphosphates from which four H atoms are accepted by two molecules of NAD^+. In the latter stages of glycolysis, four molecules of ATP are formed (by transfer of phosphate from the triose diphosphates to ADP), and this results in the formation of two molecules of pyruvic acid. ATP production (two molecules net) during glycolysis is referred to as substrate-level phosphorylation.

In yeast cells undergoing alcoholic fermentation of sugars under anaerobic conditions, NAD^+ is regenerated in terminal step reactions from pyruvate. In the first of these, pyruvate is decarboxylated (by pyruvate decarboxylase) before a final reduction, catalysed by alcohol dehydrogenase (ADH) to ethanol. Such regeneration of NAD^+ prevents glycolysis from stalling and maintains the cell's oxidation–reduction balance. Additional minor fermentation metabolites are produced by fermenting yeast cells, including glycerol, fusel alcohols (e.g. isoamyl alcohol), esters (e.g. ethyl acetate), organic acids (e.g. citrate, succinate, acetate) and aldehydes (e.g. acetaldehyde). Such compounds are important in flavour development in alcoholic beverages, such as beer, wine and whisky.

Aerobic dissimilation of glucose by fungi leads to respiration, which is the major energy-yielding metabolic route and involves glycolysis, the citric acid cycle, the electron-transport chain and oxidative phosphorylation. In addition to glucose, many carbon substrates can be respired by fungi, including: pentose sugars (e.g. xylose), sugar alcohols (e.g. glycerol), organic acids (e.g. acetic acid), aliphatic alcohols (e.g. methanol, ethanol), hydrocarbons (e.g. *n*-alkanes) and aromatic compounds (e.g. phenol). Fatty acids are made available for fungal catabolism following extracellular lipolysis of fats and are metabolized by β-oxidation in mitochondria.

During glucose respiration under aerobic conditions, pyruvate enters the mitochondria where it is oxidatively decarboxylated to acetyl CoA by pyruvate dehydrogenase, which acts as the link between glycolysis and the cyclic series of enzyme-catalysed reactions known as the citric acid cycle (or Krebs cycle). This cycle represents the common pathway for the oxidation of sugars and other carbon sources in yeasts and filamentous fungi and results in the complete oxidation of one pyruvate molecule to $2CO_2$, 3NADH, $1FADH_2$, $4H^+$ and 1GTP. Like glycolysis, the citric acid cycle is amphibolic, since it performs both catabolic and anabolic functions, the latter providing intermediate precursors (e.g. oxaloacetate and α-ketoglutarate) for the biosynthesis of amino acids and nucleotides. The removal of intermediates necessitates their replenishment to ensure continued operation of the citric acid cycle. The glyoxylate cycle is an example of such an *anaplerotic* reaction and involves the actions of the enzymes pyruvate carboxylase and phosophoenolpyruvate carboxykinase:

$$\text{Pyruvate} + CO_2 + \text{ATP} + H_2O \rightarrow \text{Oxaloacetate} + \text{ADP} + \text{Pi}$$

$$\text{Phosphoenolpyruvate} + CO_2 + H_2O \rightarrow \text{Oxaloacetate} + H_3PO_4$$

During the citric acid cycle, dehydrogenase enzymes transfer hydrogen atoms to the redox carriers NAD^+ and FAD, which become reduced. On the inner membrane of mitochondria, these reduced coenzymes are then reoxidized and oxygen is reduced to water via the electron-transport chain. Energy released by electron transfer is used to synthesize ATP by a process called oxidative phosphorylation. The chemiosmotic theory describes proton pumping across the inner mitochondrial membrane to create a transmembrane proton gradient (ΔpH) and a membrane potential difference. Together, these comprise the proton motive force that is the driving force for ATP synthesis. Each pair of electrons in NADH yields about 2.5 ATP, while residual energy is largely dissipated as metabolic heat. Since mitochondria are impermeable to NADH, this reduced coenzyme generated in the cytoplasm during glycolysis is 'shuttled' across the mitochondrial membrane using either the *glycerophosphate shuttle* (which uses NADH to reduce dihydroxyacetone phosphate to glycerol 3-phosphate) or the *malate shuttle* (which uses NADH to reduce oxaloacetate to malate). These processes enable molecules to be oxidized within mitochondria to yield reduced cofactors which, in turn, are oxidized by the electron-transport chain.

Fungi use molecular oxygen as a terminal electron acceptor in aerobic respiration in different ways (Table 1.10). Some yeasts, including *S. cerevisiae*, exhibit *alternative respiration* characterized by insensitivity to cyanide but sensitivity to azide.

1.5.2 Nitrogen Metabolism

Fungi assimilate simple nitrogenous sources for the biosynthesis of amino acids and proteins. For example, ammonium ions are readily utilized and can be directly assimilated into the amino acids glutamate and glutamine that serve as precursors for the biosynthesis of other amino acids. Proteins can also be utilized following release of extracellular protease enzymes. Glutamate is a key compound in both nitrogen and carbon metabolism, and glutamine synthetase is

Table 1.10 Respiratory chain characteristics of yeasts and fungi (adapted from Walker (1998)).

Type	Typical species	Sensitive to	Insensitive to
Normal respiration	All aerobic fungi	Cyanide and low azide[a]	SHAM[b]
Classic alternative	*Yarrowia lipolytica* (and in stationary-phase cultures of several yeast species)	SHAM	Cyanide, high azide
New alternative	*Schizosaccharomyces pombe*, *Saccharomyces cerevisiae* *Kluyveromyces lactis* *Williopsis saturnus*	High azide	Cyanide, low azide, SHAM

[a]The azide-sensitive pathway lacks proton transport capability and accepts electrons from NADH but not from succinate.
[b]SHAM: salycil hydroxamate. The SHAM-sensitive pathway transports electrons to oxygen also without proton transport and, therefore, does not phosphorylate ADP.

important as it catalyses the first step in pathways leading to the synthesis of many important cellular macromolecules. Other important enzymes of fungal nitrogen metabolism include glutamate dehydrogenase and glutamate synthase (glutamine amide: 2-oxoglutarate-aminotransferase, or GOGAT), the latter requiring ATP. When glutamine synthetase is coupled with glutamate synthase, this represents a highly efficient 'nitrogen-scavenging' process for fungi to assimilate ammonia into amino acids and citric acid cycle intermediates. The particular route(s) of ammonium assimilation adopted by fungi depends on the concentration of available ammonium ions and the intracellular amino acid pools.

Some yeasts and fungi can use *nitrate* as a sole source of nitrogen through the activities of nitrate reductase

$$NO_3^- \rightarrow NO_2^-$$

and nitrite reductase

$$NO_2^- \rightarrow NO_4^+$$

The resulting ammonium ions can then be assimilated into glutamate and glutamine, which represent end products of nitrate assimilation by yeasts.

Urea can also be utilized following its conversion to ammonium by urea aminohydrolase (urea carboxylase plus allophanate hydrolase):

$$NH_2CONH_2 + ATP + HCO_3 \rightarrow NH_2CONHCOO^- \rightarrow 2NH_4^+ + 2HCO_3^-$$

Amino acids can either be assimilated into proteins or dissimilated by decarboxylation, deamination, transamination and fermentation. Amino acid degradation by yeasts and fungi yields both ammonium and glutamate. During fermentation, yeasts may produce higher alcohols or *fusel oils*, such as isobutanol and isopentanol, following amino acid deamination and decarboxylation. These represent important yeast-derived flavour constituents in fermented beverages.

1.6 Fungal Growth and Reproduction

1.6.1 Physical Requirements for Growth

Most yeast and fungal species thrive in warm, sugary, acidic and aerobic conditions. The temperature range for fungal growth is quite wide; but, generally speaking, most species grow very well around 25 °C. Low-temperature psychrophilic fungi and high-temperature thermophilic fungi do, however, exist in nature. Fungal growth at various temperatures depends not only on the genetic background of the species, but also on other prevailing physical growth parameters and nutrient availability. With regard to high-temperature stress (or heat shock) on fungal cells, thermal damage can disrupt hydrogen bonding and hydrophobic interactions, leading to general denaturation of proteins and nucleic acids. Fungi, of course, have no means of regulating their internal temperature, and so the higher the temperature, the greater the cellular damage, with cell viability declining when temperatures increase beyond growth optimal levels. Temperature optima vary greatly in fungi, with those termed 'thermotolerant' growing well above 40 °C. Thermotolerance relates to the transient ability of cells subjected to high temperatures to survive subsequent lethal exposures to elevated temperatures, such that *intrinsic* thermotolerance is observed following a sudden heat shock (e.g. to 50 °C), whereas *induced* thermotolerance occurs when cells are preconditioned by exposure to a mild heat shock (e.g. 30 min at 37 °C) prior to a more severe heat shock. Heat-shock responses in fungi occur when cells are rapidly shifted to elevated temperatures; and if this is sub-lethal, then induced synthesis of a specific set of proteins, the highly conserved 'heat-shock proteins' (Hsps) occurs. Hsps play numerous physiological roles, including thermo-protection.

High water activity a_w is required for growth of most fungi with a minimum a_w of around 0.65. Water is absolutely essential for fungal metabolism, and any external conditions which result in reduced water availability to cells (i.e. 'water stress') will adversely affect cell physiology. The term 'water potential' refers to the potential energy of water and closely relates to the osmotic pressure of fungal growth media. Certain fungal species, for example the yeast *Zygosaccharomyces rouxii*, and some *Aspergillus* species are able to grow in low water potential conditions (i.e. high sugar or salt concentrations) and are referred to as osmotolerant or zerotolerant. By comparison, *S. cerevisiae* is generally regarded as a non-osmotolerant yeast. Mild water stress, or *hypersomotic shock*, occurs in fungi when cells are placed in a medium with low water potential brought about by increasing the solute (e.g. salt, sugar) concentration. Conversely, cells experience a *hypo-osmotic shock* when introduced to a medium of higher osmotic potential (due to reducing the solute concentration). Fungi are generally able to survive such short-term shocks by altering their internal osmotic potential (e.g. by reducing intracellular levels of K^+ or glycerol). Glycerol is an example of a *compatible solute* that is synthesized in order to maintain low cytosolic water activity when the external solute concentration is high. Glycerol can effectively replace cellular water, restore cell volume and enable fungal metabolism to continue. Trehalose, arabitol and mannitol can similarly protect against osmotic stress. Evidence suggests that the accumulation of compatible solutes is attributed not only to their synthesis, but also to control of membrane fluidity, thus preventing their leakage to the external environment.

As for pH, most fungi are acidiophilic and grow well between pH 4 and 6, but many species are able to grow, albeit to a lesser extent, in more acidic or alkaline conditions (around pH 3 or pH 8 respectively). Fungal cultivation media acidified with organic acids (e.g. acetic and lactic acids) are more inhibitory to yeast growth than those acidified with mineral acids (e.g. hydrochloric and phosphoric acids) because organic acids can lower intracellular pH (following their translocation across fungal plasma membranes). This forms the basis of action of weak acid preservatives in inhibiting growth of food spoilage fungi. Many filamentous fungi can alter their local external pH by selective uptake and exchange of ions (NO_3^- or NH_4^+/H^+), or by excretion of organic acids such as oxalic acid.

Other physical parameters influencing fungal physiology include radiation (light or ultraviolet may elicit mycelial differentiation and sporulation in some fungi that produce airborne spores), aeration, pressure, centrifugal force and mechanical shear stress.

1.6.2 Cellular Reproduction

Fungal growth involves transport and assimilation of nutrients followed by their integration into cellular components followed by biomass increase and eventual cell division (as in yeasts) or septation (as in higher fungi). The physiology of vegetative reproduction and its control in fungi has been most widely studied in two model eukaryotes: the budding yeast, *S. cerevisiae*, and the fission yeast, *Schizosaccharomyces pombe*.

Budding is the most common mode of vegetative reproduction in yeasts and multilateral budding is typical in ascomycetous yeasts (Table 1.11). In *S. cerevisiae*, buds are initiated when mother cells attain a critical cell size and this coincides with the onset of DNA synthesis. The budding processes results from localized weakening of the cell wall, and this, together with tension

Table 1.11 Modes of vegetative reproduction in yeasts (adapted from Walker (1998)).

Mode	Description	Representative yeast genera
Multilateral budding	Buds may arise at any point on the mother cell surface, but never again at the same site. Branched chaining may occasionally follow multilateral budding when buds fail to separate	*Saccharomyces, Zygosaccharomyces, Torulaspora, Pichia, Pachysolen, Kluyveromyces, Williopsis, Debaryomyces, Yarrowia, Saccharomycopsis, Lipomyces*
Bi-polar budding	Budding restricted to poles of elongated cells (apiculate or lemon shaped) along their longitudinal axis	*Nadsonia, Saccharomycodes, Haneniaspora, Wickerhamia, Kloeckera*
Unipolar budding	Budding repeated at same site on mother cell surface	*Pityrosporum, Trigonopsis*
Monopolar budding	Buds originate at only one pole of the mother cell	*Malassezia*
Binary fission	A cell septum (cell plate or cross-wall) is laid down within cells after lengthwise growth and which cleaves cells into two	*Schizosaccharomyces*
Bud fission	Broad cross-wall at base of bud forms which separates bud from mother	Occasionally found in *Saccharomycodes, Nadsonia* and *Pityrosporum*
Budding from stalks	Buds formed on short denticles or long stalks	*Sterigmatomyces*
Ballistoconidiogenesis	Ballistoconidia are actively discharged from tapering outgrowths on the cell	*Bullera, Sporobolomyces*
Pseudomycelia	Cells fail to separate after budding or fission to produce a single filament. Pseudomycelial morphology is quite diverse and the extent of differentiation variable depending on yeast species and growth conditions	Several yeast species may exhibit 'dimorphism', e.g. *Candida albicans, Saccharomycopsis figuligera.* Even *S. cerevisiae* exhibits pseudohyphal growth depending on conditions

exerted by turgor pressure, allows extrusion of cytoplasm in an area bounded by a new cell wall. Cell-wall polysaccharides are mainly synthesized by glucan and chitin synthetases. Chitin is a polymer of N-acetylglucosamine, and this material forms a ring between the mother cell and the bud that will eventually form the characteristic *bud scar* after cell division.

Fission yeasts, typified by *Schizosaccharomyces* spp., divide exclusively by forming a cell septum, which constricts the cell into two equal-sized daughters. In *Schiz. pombe*, newly divided daughter cells grow in length until mitosis is initiated when cells reach a constant cell length (about 14 μm). The cell septum in *Schiz. pombe* forms by lateral growth of the inner cell wall (the primary septum) and proceeds inwardly followed by deposition of secondary septa. Cellular fission, or transverse cleavage, is completed in a manner resembling the closure of an iris diaphragm.

In certain yeast species, the presence or absence of pseudohyphae and true hyphae can be used as taxonomic criteria (e.g. the ultrastructure of hyphal septa may discriminate between certain ascomycetous yeasts). Some yeasts grow with true hyphae initiated from *germ tubes* (e.g. *Candida albicans*), but others (including *S. cerevisiae*) may grow in a pseudohyphal fashion when starved of nutrients or when subjected to environmental stress. Filamentous growth of yeasts by hyphal or pseudohyphal extension represents a different developmental pathway that is generally reversible. In other words, cells can revert to yeast unicellular growth in more conducive growth conditions, indicating that a filamentous mode of growth represents an adaptation by yeast to foraging when nutrients are scarce.

What constitutes a cell in filamentous fungi is ambiguous. The apical compartments of higher filamentous fungi are often multinucleate, and so the process of nuclear replication and segregation into a newly extended septated hyphal compartment is known as the duplication cycle. Thus, *Aspergillus nidulans* apical compartments contain approximately 50 nuclei per compartment produced during a 2 h duplication cycle period. Continued septation results in the formation of sub-apical compartments containing fewer nuclei. Hyphae also commonly branch, usually at some distance behind the leading growing hyphal tip and often just behind a septum in higher fungi. The processes that control branching are not fully elucidated, but branch initiation is associated with the appearance of a Spitzenkörper at the site of tip emergence and extension. Branching allows filamentous fungi to fill space in an efficient and appropriate way, according to local environmental circumstances. Therefore, fungi colonizing nutrient-rich substrata branch frequently, producing dense mycelia for resource exploitation, whereas hyphae colonizing nutrient-poor substrata branch less frequently, producing effuse mycelia appropriate for resource exploration.

Rates of branching and tip growth are related to the cytoplasmic volume. Thus, the hyphal growth unit is a measure of the average length of hypha required to support hyphal tip growth. It can be calculated from microscopic preparations growing on agar media as the ratio between the total length of mycelium and the total number of tips. The ratio becomes constant after the initial stages of growth, and is characteristic of each fungal species or strain.

1.6.3 Population Growth

When yeast or fungal cells are inoculated into a nutrient medium and incubated under optimal physical growth conditions, a typical batch growth curve will result comprising lag, exponential and stationary phases. The *lag phase* represents a period of zero population growth and reflects the time required for inoculated cells to adapt to their new physical and chemical growth environment (by synthesizing ribosomes and enzymes). The *exponential phase* is a period of logarithmic cell (or mycelial biomass in the case of filamentous growth) doublings and constant, maximum specific growth rate (μ_{max}, in dimensions of reciprocal time, h^{-1}), the precise value of which depends on the prevailing growth conditions. If growth is optimal and cells double logarithmically, then

$$\frac{dx}{dt} = \mu_{max}^{x}$$

When integrated, this yields

$$\ln x - \ln x_0 = \mu_{max}^{t}$$

(where x_0 is the initial cell mass) or

$$x = x_0 \exp(\mu_{max}^{t})$$

which is the fundamental equation for exponential batch growth. According to these kinetic expressions, a plot of $\ln x$ versus time is linear with the slope being μ_{max}. Calculation of the doubling time t_d of a yeast or fungal culture can be achieved from knowledge of μ_{max} as follows:

$$t_d = \frac{\ln 2}{\mu_{max}} = \frac{0.693}{\mu_{max}}$$

During the exponential phase of balanced growth, cells are undergoing primary metabolism, explicitly those metabolic pathways that are essential for growth of the cell. Industrial fermentations requiring maximum cell biomass production or the extraction of primary metabolites or their products, therefore, aim to extend this phase of growth, often via fed-batch culture (incremental nutrient feeding) or continuous culture techniques (continuous nutrient input with concomitant withdrawal of the biomass suspension).

Following the exponential phase, cells enter a period of zero population growth rate, the stationary phase, in which the accumulated fungal or yeast biomass remains relatively constant and the specific growth rate returns to zero. After prolonged periods in stationary phase, individual cells may die and autolyse (see below). The stationary phase may be defined as cellular survival for prolonged periods (i.e. months) without added nutrients. In addition to nutrient deprivation, other physiological causes may promote entry of fungal cells into stationary phase, including: toxic metabolites (e.g. ethanol in the case of yeasts), low pH, high CO_2, variable O_2 and high temperature. During the stationary phase of unbalanced growth, fungi may undergo secondary metabolism, specifically initiating metabolic pathways that are not essential for growth of cells but are involved in the survival of the organism. The industrial production of fungal secondary metabolic compounds, such as penicillin and the ergot alkaloids, therefore, involves the controlled maintenance of cell populations within a stationary phase of growth.

Filamentous fungi tend to grow as floating surface pellicles when cultivated in static liquid culture. In agitated liquid culture, fungi grow either as dispersed filamentous forms or as pellets of aggregated mycelia subject to species, inoculum size, agitation rate and nutrient availability. Different growth forms will locally experience different micro-environmental conditions that will affect fungal physiology and, hence, fermentation processes. In fungal biotechnology, cell morphology may directly influence fermentation progress. For example, the rheological properties of the growth medium, oxygen transfer and nutrient uptake may adversely affect bioproduct formation.

Yeast or fungal cell immobilization onto inert carriers has many advantages over free cell suspension culture in industrial processes. Cells may be successfully immobilized by entrapment, aggregation, containment, attachment or deposition. Fungal biofilms represent a natural form of cell immobilization resulting from cellular attachment to solid support materials. Yeast biofilms have several practical applications in fermentation biotechnology and are also medically important with regard to colonization of human tissue. Regarding the former case, with dimorphic yeasts such as *Kluyveromyces marxianus*, filamentous cells with a large surface area may be better suited to immobilization than ellipsoidal unicellular yeast forms with a low surface area are. In this latter case of pathogenic yeast biofilms, *C. albicans* has been shown to adhere to surgical devices such as heart pacemakers and catheters, human epithelial cells and dental acrylic.

1.6.4 Fungal Cell Death

An understanding of the death of fungal cells is important from a fundamental viewpoint because fungi, especially yeasts, represent valuable model systems for the study of cellular ageing and apoptosis (programmed cell death). From a practical perspective, cell death in fungi is pertinent in relation to the following situations: industrial fermentation biotechnology (where high culture viabilities are desired), food preservation (regarding inhibition of spoilage fungal growth), food production (promotion of cellular autolysis for yeast extracts) and clinical mycology (where fungal death is the goal in treatment of human mycoses).

Numerous physical, chemical and biological factors influence fungal cell death, which may be defined as complete and irreversible failure of cells to reproduce. Fungi will die if confronted with excessive heat, extreme cold, high-voltage electricity, ionizing radiation, high hydrostatic and osmotic pressures and if exposed to chemical or biological fungicidal agents. When the cells' physiological protection responses are insufficient to counteract the cellular damage caused by physical stress, cells will die. In industrial situations, physical treatments can be used to eradicate contaminant fungi. For example, yeasts exposed to elevated temperatures may lead to their thermal death, and this is exploited in the pasteurization of foods and beverages to kill spoilage yeasts.

There are numerous chemical factors influencing survival of fungi. Several external chemical agents act as fungicides, including toxic organic compounds, oxygen free-radicals and heavy metals. Chemical preservatives are commonly employed as antifungal agents in foodstuffs, including weak acids such as sorbic, benzoic and acetic acids. These agents, which are generally fungistatic rather than fungicidal, act by dissipating plasma membrane proton gradients and depressing cell pH when they dissociate into ions in the yeast cytoplasm. Similarly, sulfur dioxide, which has long been used to eliminate undesirable yeasts (and bacteria) from wine, dissociates within the yeast cell to SO_3^{2-} and HSO_3^{-}, resulting in a decline in intracellular pH, and this forms the basis of its antizymotic action. Fungicidal acids include medium-chain fatty acids (e.g. decanoic acid), which may cause rapid cell death of yeasts and fungi by disruption of cell membrane integrity. Endogenous chemical factors, such as ethanol and other toxic metabolites (e.g. acetaldehyde) produced by fermentative activity, excessive intracellular acidity or alkalinity, inability to protect against oxidative damage or sequester toxic metals, may also prove lethal to fungi. If fungal cells are unable to detoxify or counteract detrimental effects of chemicals, they may die.

Examples of lethal biotic interactions with fungi include direct ingestion (by insects, protozoa), engulfment and lysis (by mycoparasitizing fungi), direct predation (by haustoria-mediated processes) and intoxication (by killer toxin-producing yeasts). *Killer yeasts* secrete proteinaceous toxins that are lethal to other yeasts but to which the killers themselves are immune. Several yeast species have now been identified as possessing killer character, but the best known is the K1 system in *S. cerevisiae*. The K1 toxin from this species acts by binding to cell-wall receptors in sensitive yeast cells, followed by plasma membrane channel formation. This latter event causes disruption of membrane permeability, which leads to the death of sensitive cells. Killer cells synthesize a membrane-bound immunity protein that prevents cellular suicide. In recent years, it has been established that some killer yeasts may also possess antimycotic activity against filamentous fungi. This has led to the potential use of killer yeasts and their toxins as novel antifungal biocontrol agents for combating important fungal pathogens in agriculture. For example, the killer yeast *Pichia anomala* (*Wickerhamomyces anomalus*) has been shown to inhibit the growth of grain-storage fungi (*Penicillium* spp.) and fungal spoilage of fruits (caused by *Botrytis cinerea*).

With regard to endogenous biotic factors influencing fungal cell survival, several physiological, morphological, genetic and biochemical events may take place leading to 'self-inflicted' death. For example, fungal autolysis may be described as cellular self-digestion and occurs when endogenous (vacuolar) hydrolytic enzymes, notably proteases and carbohydrases, cause dissolution of cytoplasmic proteins and cell wall polysaccharides respectively. Autolytic enzymatic activity is encouraged during the production of yeast extracts in the food industry by using high temperatures (e.g. 45 °C), salt (to encourage plasmolysis) and solvents (to promote lipid dissolution). Exogenous hydrolytic enzymes, such as papain, can also be used to accelerate cell-wall breakdown.

Genetic factors also influence fungal cell death. For example, cells may commit suicide following DNA damage, presumably to avoid the risk of producing genetically altered progeny. Cellular ageing and apoptotic cell death have been widely studied in yeasts, especially in *S. cerevisiae*, which is a valuable model organism for understanding molecular genetic basis of the ageing process in eukaryotic cells. Beyond a certain finite limit (termed the Hayflick limit) of cell division cycles (generally around 20 in *S. cerevisiae*), this yeast can generate no further progeny and cells enter a senescent physiological state leading to death. Aged and senescent populations of this yeast can be isolated, together with mutants displaying age-related phenotypes. In *S. cerevisiae*, *UTH* (youth) genes have now been identified that appear to influence both stress resistance and longevity.

1.7 Conclusions

This chapter has highlighted the physiological diversity of yeasts and fungi in terms of morphology, growth, metabolism and cell death. Understanding the ways in which fungi interact with their growth environment is crucial in medical mycology to control fungal pathogens and also in industry to exploit yeasts and fungi for production of biotechnological commodities.

Revision Questions

Q 1.1 Describe the fine structure of the fungal cell envelope and explain its main physiological roles.

Q 1.2 Outline the main nutrients required for the growth of yeasts and fungi and indicate how such nutrients may be accumulated by fungal cells from their growth environment.

Q 1.3 Explain the metabolic fate of glucose (a) by fungi under aerobic growth conditions and (b) by yeasts under anaerobic growth conditions.

Q 1.4 Describe the major physical factors influencing the growth of fungal cells.

Q 1.5 Compare and contrast the modes of cellular reproduction in yeasts and fungi.

Q 1.6 For yeast cells, explain what is meant by exponential growth and describe how the doubling time of a yeast culture population may be determined.

References

Carlile, M.J., Watkinson, S.C. and Gooday, G.W. (2001) *The Fungi*, 2nd edn, Academic Press, London.

Deacon, J. (2005) *Fungal Biology*, 4th edn, Wiley-Blackwell Publishing.

Jørgensen, H., Nielsen, J., Villadsen, J. and Møllgaard, H. (1995) *Biotechnology and Bioengineering*, **46**, 117–131.

Walker, G.M. (1998) *Yeast Physiology and Biotechnology*, John Wiley & Sons, Ltd, Chichester.

Walker, G.M. (2004) Metals in yeast fermentation processes. *Advances in Applied Microbiology*, **54**, 197–229.

Further Reading

Arora, D. (ed.) (2003) *Handbook of Fungal Biotechnology*, 2nd edn, Marcel Dekker, Inc.

Arora, D. (ed.) (2004) *Fungal Biotechnology in Agricultural, Food and Environmental Applications*, Marcel Dekker, Inc.

Daniel, H.-M., Passoth, V. and Walker, G.M. (2011) (eds). *Antonie Van Leeuwenhoek International Journal of Microbiology*, **99**(1) (*Pichia anomala* special issue).

Deak, T. (2008) *Handbook of Food Spoilage Yeasts*, 2nd edn, CRC Press/Taylor & Francis, Boca Raton, FL.

Jennings, D.H. and Lysek, G. (1999) *Fungal Biology; Understanding the Fungal Lifestyle*, Bios.

Moore, D. (1998) *Fungal Morphogenesis*, Cambridge University Press.

Querol, A. and Fleet, G.H. (eds) (2006) *Biodiversity and Ecophysiology of Yeasts*, The Yeast Handbook, vol. **1**, Springer-Verlag, Berlin.

Rosa, C.A. and Peter, G. (eds) (2006) *Yeasts in Food and Beverages*, The Yeast Handbook, vol. **2**, Springer-Verlag, Berlin.

Walker, G.M. (1998) *Yeast Physiology and Biotechnology*, John Wiley & Sons, Ltd, Chichester.

Walker, G.M. (2004) Metals in yeast fermentation processes. *Advances in Applied Microbiology*, 54, 197–229.

Walker, G.M. (2009) Yeasts, in *The Desk Encyclopedia of Microbiology* (ed. M. Schaechter), Elsevier, Oxford, pp. 1174–1187.

Walker, G.M. (2010) *Bioethanol: Science and Technology of Fuel Alcohol*, Ventus Publishing ApS, Copenhagen, ISBN 978-87-7681-681-0 [http://bookboon.com/int/student/chemical/bioethanol-science-and-technology-of-fuel-alcohol].

2

Fungal Genetics

Malcolm Whiteway and Catherine Bachewich

2.1 Introduction

2.1.1 Fungi as Pioneer Organisms for Genetic Analysis

Genetic manipulation of organisms implies the ability to direct the formation of new combinations of traits within an individual. This process has historically been an important human endeavour, providing us with our breeds of domestic animals, economically important plants and industrially important fungi. Genetic manipulation can be as simple as identifying and selecting, within a population, rare individuals that contain interesting traits, but is made more powerful by the ability to enhance the rate of individual variation, and more powerful still when traits identified in different individuals can be combined. Thus, mutagenesis and genetic recombination underpin the process of genetic manipulation. Such genetic manipulation can be used directly for practical ends (new varieties of tomatoes for example), or for more academic aims directed at an understanding of life.

Fungi were amongst the first organisms to be studied scientifically through genetics. Although peas and fruit flies provided the initial evidence for genes and for genetic linkage, some of the earliest fundamental insights into the genetic structure of organisms came from pioneering studies in fungal systems. One exceptional insight developed from analysis of fungal systems, in this case *Neurospora crassa*, was the recognition that individual enzymatic functions were encoded by the information from individual genes, a result rewarded by the 1958 Nobel Prize in Physiology or Medicine to G. Beadle and E. Tatum. Other fundamental advances based on the genetic analysis of fungi included the dissection

Fungi: Biology and Applications, Second Edition. Edited by Kevin Kavanagh.
© 2011 John Wiley & Sons, Ltd. Published 2011 by John Wiley & Sons, Ltd.

of the cell cycle of the yeasts *Saccharomyces cerevisiae* and *Schizosaccharomyces pombe*, and the analysis of *S. cerevisiae* telomeres, work that led to the Nobel Prize in Medicine in 2001 to L. Hartwell and P. Nurse and in 2009 to J. Szostak. These awards, more than 40 years after that to Beadle and Tatum, show that fungal systems have maintained their utility in uncovering important biological truths.

More recently, large-scale, semi-industrialized international efforts have had an enormous impact on the field of fungal genetics. These efforts have been at the forefront of the development of the science of genomics. Many of the primary successes of genomics have come through studies on fungi, in particular the baker's yeast *S. cerevisiae*. Efforts directed at the analysis of the yeast genome have provided the first sequence of a eukaryotic chromosome, the first sequence of an entire eukaryotic genome and the first development of a systematically created collection of null mutants of all the genes of an organism. Building on this pioneering work, the sequences of many fungal genomes are now fully completed or available as unannotated draft sequences.

2.1.2 Significance/Advantages of Fungi as Model Organisms

Fungi have become amongst the pre-eminent models for the genetic investigation of basic cellular processes. There are many intrinsic characteristics of fungal systems that make them ideal model organisms for genetic studies. As unicellular organisms, they can grow in simple defined media and they are easy to culture. Because they are often contain a stable, propagatable haploid phase they are easy to mutate, and in those organisms with a well-characterized sexual cycle the mutants can be readily combined. Finally, because they are eukaryotic cells they exhibit many of the properties and functions characteristic of human cells, and thus served as a better model for many cellular processes than the bacterial systems that had been investigated in depth previously.

The advent of the molecular biological revolution in the 1970s strengthened the importance of fungal systems as models for genetic studies of eukaryotic cell function. Because each individual fungal cell was autonomous, transformation with external DNA allowed the efficient genetic engineering of an entire organism. The rapid development of fungal transformation systems after the initial successes with *S. cerevisiae*, followed by the construction of efficient vectors for the transfer of genes and the cloning of the various components of chromosomes, allowed for effective manipulation of fungal cells. Ultimately, this technology permits the construction of strains of fungi genetically designed to differ from a standard strain by as little as a single selected nucleotide.

This chapter provides an overview of various aspects of fungi and their use as model organisms for genetic analysis. Representative species from the higher fungi, including ascomycetes and basidiomycetes, comprise the focus for discussion, and comparisons between the yeasts (*S. cerevisiae*) and the filamentous organisms (*N. crassa, Aspergillus nidulans, Coprinus cinereus*) are emphasized to introduce both the advantages of particular systems and the diversity within the kingdom. Examples of dimorphic fungi, which live in both a yeast and filamentous form, are also discussed to highlight unique features and variations on themes within the fungi.

2.2 Fungal Life Cycles

2.2.1 Ascomycete Yeast (*Saccharomyces cerevisiae*)

S. cerevisiae is an extremely well studied organism, with a clearly defined and experimentally manipulable life cycle. The life cycle of yeast involves mitotically propagating haploid forms of two distinct mating types, and a diploid form that can either grow vegetatively or can be induced into a meiotic developmental pathway through manipulation of the nutrient conditions of the growth media. The cellular pathways regulating processes such as mitotic proliferation, cell recognition and mating, meiosis and sporulation have been extensively studied on a molecular level, and are generally well understood.

Mitotic growth of yeast cells involves budding (Figures 2.1 and 2.6). During this process, growth of the cell is directed to a specific location on the surface of the mother cell, and a new cell is formed somewhat like blowing up a balloon through a hole in the mother cell. This involves highly polarized growth of the developing daughter cell, implicating both the actin and microtubule-based cytoskeletal networks, and is tightly coordinated with the cell cycle. This coordination ensures that the daughter cell receives a complete copy of the genetic material. Both haploid and diploid cells divide by the budding process, although there are subtle differences in the choice of the sites of bud emergence between haploids and diploids. In addition, some diploid cells can also modify the coordination of the cell cycle and polarized growth to switch to a pseudohyphal growth mode. In this growth pattern, individual cells are more elongate, and the budding pattern leads to the formation of chains of cells rather than compact colonies characteristic of the true budding mode.

Genetic analysis is highly developed in *S. cerevisiae*. When vegetatively growing haploid cells of opposite mating types are brought into proximity, they communicate with each other by diffusible pheromones, synchronize their cell cycles, conjugate and then fuse their nuclei to create nonmating, meiosis-proficient diploids. These diploids can be identified visually in their initial zygote form, and separated from the haploids by micromanipulation, or identified selectively because they contain a pattern of genetic traits not possessed by either haploid parent.

Under rich growth conditions, such diploid cells themselves propagate vegetatively; but under conditions of nitrogen and fermentable carbon limitation,

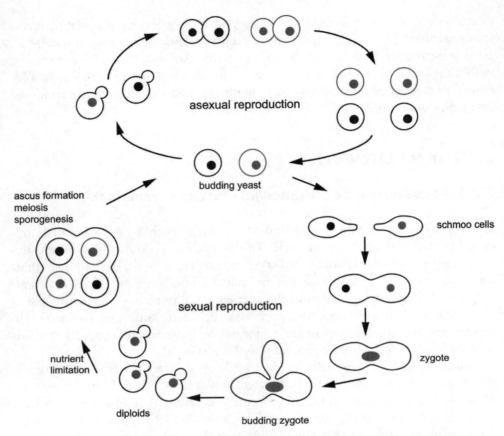

Figure 2.1 Life cycles of *Saccharomyces cerevisiae*.

the diploid cells are induced to initiate meiosis and sporulation. The ability to propagate the diploid allows the amplification of the initial mating product, and provides an essentially unlimited source of potential meiotic events from a single mating.

2.2.2 Ascomycete Filamentous Fungi (*Neurospora crassa* and *Aspergillus nidulans*)

The filamentous fungi differ from the yeasts in that they grow vegetatively as hyphae, which are highly polarized filaments that extend indefinitely at their tips. The hypha initiates new tips in the form of branches from sub-apical regions, and together the growing mass constitutes the mycelium. Hyphae are predominantly multinucleated, with cross-walls called septa dividing the hypha into compartments. The compartments are connected through pores in the septa and, therefore, display cytoplasmic continuity. The hypha functions primarily in acquisition of nutrients and exploration of the environment. Enzyme secretion at the tip assists both processes. In pathogenic fungi, the hyphal growth form can also be important for virulence. In filamentous fungi, vegetative hyphal growth initiates from a spore. Spores are products of either sexual (ascospores, basidiospores) or asexual (conidia) reproduction. Conidia are typically produced from a differentiated structure called a conidiophore, whereas ascospores and basidiospores are produced within an ascus or basidium respectively contained within the fruiting body called an ascocarp or basidiocarp.

During asexual reproduction in the ascomycetes, such as *A. nidulans* (Figures 2.2 and 2.6), a spore containing a single nucleus (monokaryotic) germinates into a mutinucleate, homokaryotic hypha. The hypha grows and develops branches

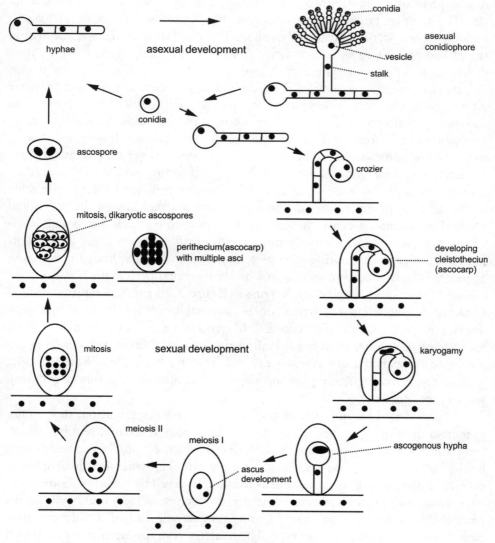

Figure 2.2 Life cycles of *Aspergillus nidulans*.

for a period of time and then initiates a specialized branch called the coni-diophore. Development of the conidiophore involves numerous different cell types, and is investigated as a model developmental process. The nucleus divides mitotically within the conidiophore, allowing the ultimate production of asex-ual, haploid conidia. Upon release, conidia germinate into vegetatively growing hyphae, and the cycle continues. The factors that trigger initial conidiophore development in *Aspergillus* are not clear, but involve the supply of carbon and nitrogen. The process can normally only occur in cultures grown on solid media with an air interphase; conidiation does not occur in liquid.

Sexual reproduction in *A. nidulans* begins when vegetatively growing hyphae fuse to create a heterokaryon, or dikaryotic hypha (Figure 2.2). The dikary-otic hyphae differentiate into a developing fruiting body called a cleistothe-cia. The fruiting body is a complex structure composed of many cell types, including both sterile and fertile hyphae. The dikaryotic fertile hyphae within the cleistothecium develop into hooked structures called croziers, which then differentiate into developing asci. Karyogamy or nuclear fusion occurs within the crozier, creating a diploid. The diploid undergoes meiosis and the four meiotic products then undergo mitosis, creating eight haploid ascospores. The ascospores undergo another round of mitosis and are thus binucleate. Thou-sands of asci are contained within a cleistothecium and are fragile, hampering their individual isolation. Upon release, the ascospores germinate into hyphae as described. *Aspergillus* is homothallic, or self-fertile, and sexual reproduc-tion can be initiated within one colony containing genetically identical nuclei. In the absence of heterokaryon formation with another strain, the individual strain differentiates a cleistothecium as described, into which the hypha devel-ops into a crozier and an ascogenous hypha. Unlike *S. cerevisiae*, *A. nidulans* does not undergo any mating-type switching. *A. nidulans* hyphae can also grow as heterokaryons and diploids as part of a parasexual cycle, which will be dis-cussed later in this chapter. In *N. crassa* (Figure 2.3), asexual reproduction is triggered by circadian rhythyms, or an internal clock mechanism, and pro-duces both macro and microconidia. Macroconidia are produced first from aerial hyphae and are used for subculturing strains, whereas microconidia are produced later in the growth process and have poor viability. Macroconidia germinate into vegetatively growing hyphae, but also serve a function during sexual reproduction.

The sexual cycle is initiated in response to nitrogen strarvation, or changes in temperature or light. *N. crassa* is heterothallic and, therefore, requires genet-ically different mating partners. Macroconidia or microconidia produced from hyphae serve as the 'male' and produce a pheromone, which is a hydrophobic peptide. The opposite strain serving as the female develops a fruiting body inter-mediate called a protoperithecia. A polarized structure called a trichogyne grows from the protoperithecium of one mating-type female and fuses with the male conidia of the opposite mating type. The nucleus from the latter moves through the trichogyne into the ascogonium within the protoperithecium, which is then

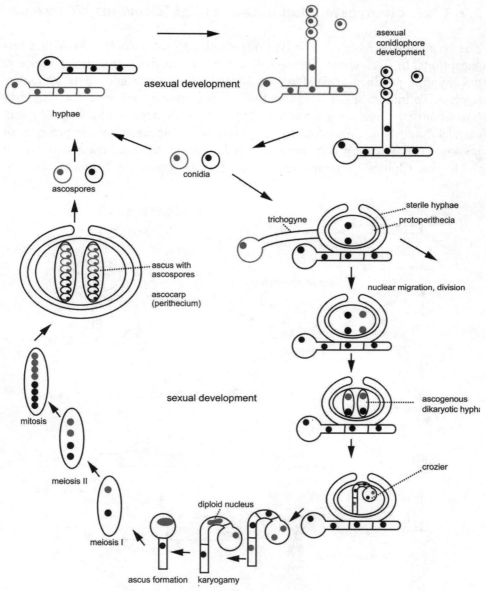

Figure 2.3 Life cycles of *Neurospora crassa*.

referred to as the perithecium. Nuclei from both mating partners divide within a developing dikaryotic ascogenous hyphal structure. The ascogenous hypha develops a crozier, where nuclear fusion or karyogamy takes place, followed quickly by meiosis within the developing ascus. Mitosis and subsequent ascosporogenesis results in eight spores within an ascus within the perithecium. Acsi are long and slender in *Neurospora*, allowing for individual dissection and separation of ordered ascospores.

2.2.3 Basidiomycete Filamentous Fungi (*Coprinus cinereus*)

The basidiomycete life cycle is typically similar to the ascomycetes with a few exceptions. In *C. cinereus* (Figure 2.4), a typical mushroom fungus, monokaryotic hyphae produce asexual spores called oidia, which germinate and form hyphae. To initiate sexual reproduction, monokaryotic hyphae fuse at their tips (anastomosis) to create a dikaryotic hypha. The dikaryotic hypha grows vegetatively, and is distinguished from hyphae of ascomycetes by the presence of hooked cells or clamp connections, which connect septated compartments of the hypha. Changes in temperature and light can trigger the hypha to undergo

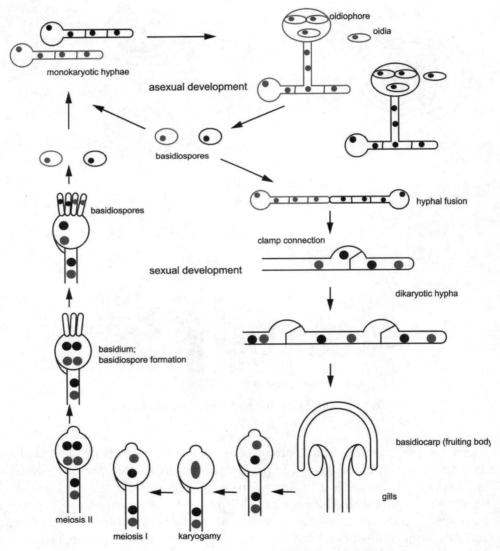

Figure 2.4 Life cycles of *Coprinus cinerus*.

differentiation into the fruiting structure, or basidiocarp. Within the basidiocarp, basidium fomation (the equivalent of an ascus) and karyogamy takes
place. Meiosis produces the basidiospores, which hang off the basidium contained within the gills of the mushroom cap, as opposed to being encased as in
the asci of ascomycetes. Haploid basidiospores are then released and germinate
into monokaryotic hyphae and continue the cycle.

In dimorphic basidiomycetes that exist in both yeast and hyphal forms, such as
the human pathogen *Cryptococcus neoformans* (Figure 2.5), vegetative growth
occurs via budding yeast. Sexual reproduction involves the differentiation of
yeast cells into hyphae upon exposure to pheromone, resulting in dikaryotic
hyphae. Basidia differentiate from the ends of the hyphae, in which karyogamy
followed by meiosis and mitosis occurs, producing haploid basidiospores. Upon

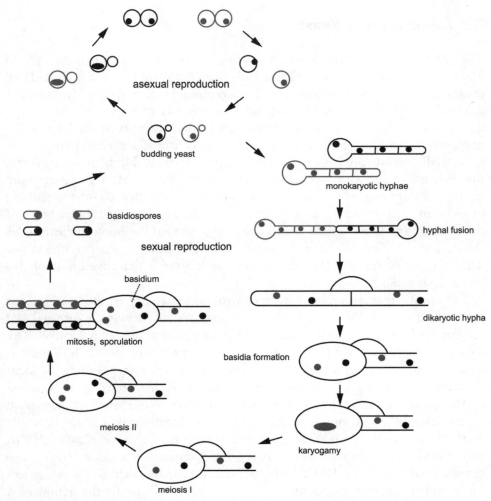

Figure 2.5 Life cycles of *Cryptococcus neoformans*.

release, the spores grow as budding yeast. The yeast cells can also undergo asexual sporulation in response to nitrogen limitation or desiccation. Under these conditions, yeast cells differentiate into hyphae, which differentiate monokaryotic basidia at their tips. Mitosis occurs within the basidia, producing haploid spores, which then germinate into yeast cells.

Filamentous fungi exhibit many variations in their lifecycles, and some do not exhibit a known sexual phase (the Deuteromycetes). Others exploit parts of the life cycle for pathogenesis, particularly in the pathogenic basidiomycetes. For example, the dimorphic corn smut fungus *Ustilago maydis* exists as a non-pathogenic yeast form, but upon mating the yeast cells differentiate into dikaryotic hyphae, which are associated with virulence.

2.3 Sexual Analysis: Regulation of Mating

2.3.1 Ascomycete Yeast

The yeast *S. cerevisiae* has a simple mating system, with cells of two haploid mating types termed **a** and α. These cells can conjugate to form a diploid cell containing both **a** and α information. The **a**/α diploid is not capable of mating but can initiate meiosis to form four haploid products, two of which are mating-type **a** and two of which are mating-type α. Laboratory strains typically have stable mating types and are termed heterothallic. In contrast, most wild strains are homothallic and do not have stable mating types. Instead, during mitotic growth, the cells are capable of switching their mating types; thus, a growing spore colony develops cells of both **a** and α. These cells mate with each other, resulting in a colony that grows up as an **a**/α diploid. The regulation of the stability of the mating type is controlled by a single locus, termed the homothallism locus (*HO*). The *HO* locus contains a functional endonuclease, whereas the stable mating types of the heterothallic strains result from a defective allele of this locus, designated *ho*.

The reason that an endonuclease controls the stability of the mating type is the result of a sophisticated genetic exchange system involving expressed and unexpressed cassettes of information. The mating type of the cell is controlled by a single locus termed *MAT* close to the centromere of chromosome 3. When this locus contains the *MAT***a** allele the cells are of the **a** mating type; when the locus contains the *MAT*α allele the cells are mating-type α. If there is no information at *MAT* then the cells select **a** as the default mating type. However, the typical *S. cerevisiae* cell also contains an extra copy of both mating-type genes, typically **a** information near the right telomere of chromosome 3 at a locus called *HMR*, and α information at the left telomere of chromosome 3 at locus *HML*. The sequences at *HML* and *HMR*, although structurally identical to the sequences for **a** and α information at *MAT*, are not expressed due to the action of a series of proteins designated silent information regulators. In the presence of a functional *HO* endonuclease the information is exchanged between the *HMR* or *HML* loci and the *MAT* locus as often as once per cell division. This exchange is recombinational; the *HO* endonucleases make a double-strand cut at the *MAT* locus and gene conversion then transfers the information from one of the silent loci to the *MAT* locus. The same machinery that keeps the information at *HML* and *HMR* unexpressed blocks the cutting of these DNA sequences by *HO*, so the informational exchange is typically unidirectional.

The *MAT* locus defines the mating type of the cell through direct transcriptional control. Each allele, *MAT*a and *MAT*α, expresses transcription factors, and these transcription factors control the expression of blocks of genes defining the two cell types. *MAT*α encodes two transcription factors, α1 and α2. The role of α1 is to stimulate the expression of α-specific genes such as *STE3*, encoding the receptor for the a-factor mating pheromone, and *Mfα1* and *Mfα2*, the genes specifying the α-factor mating pheromone. The α2 transcription factor is a repressor, serving to repress a-specific gene expression in the *MAT*α cells and, together with the a1 factor, to shut off haploid gene expression in *MAT*a/*MAT*α diploid cells. Amongst the key genes shut off in the diploid cell is *RME1*, which encodes a repressor of meiosis; this regulatory circuit ensures that it is the a/α diploid cell that is uniquely capable of meiosis and sporulation.

2.3.2 Filamentous Ascomycetes

N. crassa is heterothallic and requires two mating types, A and a, for sexual reproduction. The mating-type loci, *mat A* and *mat a*, control gene expression required for the mating process. The DNA sequences at the mating loci of opposite mating types are very different and, therefore, are regarded as 'idiomorphs' as opposed to alleles. The *mat a* idiomorph is 3.2 kb in length, and encodes one gene called *mat a-1* while the *mat A* idiomorph is 5.3 kb long and encodes three genes, including *mat A-1, A-2* and *A-3*. The *mat a-1* and *mat A-3* genes encode for HMG box-containing DNA binding proteins, and are the major regulators of mating in both strains. Homologues of such factors are required for mating in other organisms, including some filamentous fungi and the fission yeast *S. pombe*. The *mat A-1* product is also a DNA binding protein and similar to α1p from *S. cerevisiae*. *mat A-3* encodes a protein with little homology in other organisms and of unknown function but which, together with *mat A-2*, is required for ascosporogenesis. *mat A-2* and *A-3* are expressed constitutively during both vegetative growth and sexual development. In contrast to the situation in *S. cerevisiae*, the downstream targets of the mating regulatory genes are unknown. They presumably control expression of the mating pheromone, but other processes must also be regulated, including nuclear migration, nuclear compatibility and fruiting body development. Since filamentous fungi utilize multiple different cell types for sexual reproduction and contain complex fruiting body structures, including different thalli for male and female mating partners, the regulation and function of the mating type loci must be more complex than in yeast. In contrast to *S cerevisiae*, mating type-switching does not occur in *N. crassa*.

The mating loci in *Neurospora* regulate additional processes, including vegetative, heterokaryon compatibility. Vegetative hyphae can fuse to form a heterokaryon, but only if they arise from opposite mating-type strains. If hyphal fusion occurs between incompatible cells, the fused hyphal compartment seals off from the rest of the hyphae through deposition of cross-walls, and the compartment undergoes a type of programmed cell death characterized by DNA fragmentation, organelle and cytoplasmic breakdown and vacuole production. Mutants of the *mat a* strain, which lost the incompatibility response, were affected in the *mat a-1* orf, and analysis of the *mat a-1* protein identified domains important for mating versus vegetative incompatibility.

Several filamentous fungi, including *A. nidulans*, are homothallic or self-fertile, where a colony derived from a single spore is able to undergo sexual reproduction. Although the genetic basis of homothallism is not fully understood, recent completion and analysis of the genome from *A. nidulans* has uncovered the presence of many conserved elements of mating from heterothallic species, suggesting that sexual reproduction may be regulated by similar genes in 'selfing' fungi. For example, MAT-2 and MAT-1 genes encoding an HMG box-containing DNA binding protein and an α1p domain homologue respectively have now been identified. In addition, genes encoding homologues to the hydrophilic pheromone alpha factor in *Saccharomyces*, the mating pheromone protease *KEX2*, and pheromone receptors *STE2* and *STE3* are present in the genome. However, no homologue to a factor mating pheromone was detected, despite the fact that its receptor was present. The MAT genes and factors associated with a pheromone response MAPK pathway were shown to be important for sexual development, suggesting that similar pathways underlie self and non-self mating.

2.3.3 Filamentous Ascomycete Dimorphic Fungi

Candida albicans (Figure 2.6) is an important human pathogen and has been extensively studied for this reason. *C. albicans* had been classified as an asexual Deuteromycete, but recent genomic studies have provided convincing evidence for the potential for a sexual cycle. However, although a well-defined mating system has been identified that allows the conjugation of mating-type locus homozygous diploid cells, there is currently no evidence for a functional meiotic pathway that allows reductional division and a return to the diploid state from the tetraploid.

The detection of the potential mating ability of *C. albicans* arose through analysis of the genome sequence. A region of the genome was detected that encoded genes similar to those found at the mating type locus of *S. cerevisiae*. Further analysis of this region uncovered a more complex locus than that found

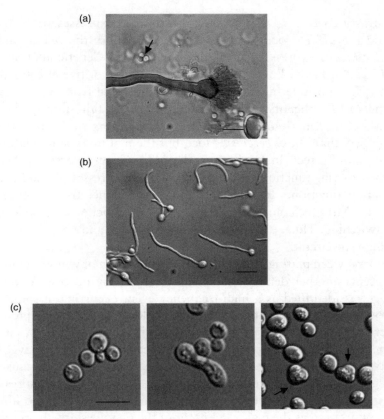

Figure 2.6 (a) Developing conidiophore composed of a vesicle giving rise to phialides and conidia in the filamentous fungus *A. nidulans*. Released conidia are indicated by an arrow. (b) Hyphae growing from yeast cells of the dimorphic fungus *C. albicans*. (c) Various stages in the life cycle of the yeast *S. cerevisiae*. The first panel demonstrates vegetatively growing yeast, the second panel shows a zygote and the third panel shows demonstrates asci (arrows). Bars: 10 μm.

in *S. cerevisiae*; in addition to the candidate cell-type regulating transcription factors, there were other genes whose function implied no obvious link to the mating process. Such complex loci are found in other fungi, such as *C. neoformans*, and in algae such as *Chlamydomonas reinhardtii*; analysis across the spectrum of mating proficient lower eukaryotes suggests that many organisms contain mating-type loci that mix genes with roles in controlling the mating type with genes involved in other cellular processes. In *C. albicans* these genes include oxysterol-binding proteins (*OBPa* and *OBPα*), poly-A polymerases (*PAPa* and *PAPα*) and phosphatidyl inositol kinases (*PIKa* and *PIKα*). The a and α versions of each protein are somewhat divergent, but the ability to homozygose each *MTL* allele shows that both versions of the proteins are capable of supporting cell viability.

In contrast to *S. cerevisiae*, where the *MATa* locus expresses one transcription factor and the *MATα* locus encodes two factors, both the *MTLa* and *MTLα* loci of *C. albicans* express two transcription factors. Genetic analysis suggests that *MTLa2*, an HMG box-containing protein, is a positive effector of *MTLa* functions. This positive function (and gene) is missing from *S. cerevisiae* cells, where the *MATa* phenotype is expressed as the default state in the absence of any *MAT* information. This makes the *C. albicans* *MTL* loci structurally more complex than the *S. cerevisiae* loci, but the mating-type regulatory circuit is actually more direct. In *C. albicans*, the *MTLa* locus expresses *MTLa2* that directs the a mating functions and the *MTLα* locus expresses *MTLα1* controlling the α mating functions. In the diploid state, the other transcription factors, Mtla1p and Mtlα2p, combine to repress mating functions, as well as white-opaque switching. Thus, each locus has both positive and negative roles within the mating-type circuit.

The link between mating ability and the phenomenon of white-opaque switching represents another detail of the *C. albicans* mating process. White-opaque switching was identified as a high-frequency event occurring in some *Candida* cells through which cells changed their morphology, some of their physiological characteristics and their colony morphology. It is now appreciated that the ability to undergo this switching was occurring in cells that had homozygosed their mating type, thus relieving the repression of the switching process caused by the a1/α2 regulatory molecule. Efficient mating in *Candida* cells requires that the cells be in the opaque state, rather than the white state. Recent work has clarified this connection, as the major regulator of the opaque state, the transcription factor Wor1p, is repressed in the a1/α2 expressing cells, but has the potential to be transcribed in *MTL* homozygotes. Because WOR1 is auto-regulated, it can set up a positive loop of high WOR1 expression generating the opaque state, which is epigentically stable. However, fluctuations in Wor1p levels or function can break this loop, leading to a stable low-WOR1-expression state that generates the white form of cell.

2.3.4 Filamentous Basidiomycetes

The basidiomycetes are novel in that they have multi-allelic mating type genes. In *C. cinereus*, for example, more than 12 000 mating types exist, demonstrating the diversity in mating systems and complex levels of control within the filamentous fungi. *Coprinus* has two unlinked mating type loci, A and B, which are polymorphic and contain sub-loci called α and β. The mating system, therefore, is described as being tetrapolar. The α and β loci are redundant, but recombination occurs between them. Together, these loci contribute hundreds of specificities to A and B loci, creating thousands of different mating types.

The A locus encodes genes for homeodomain proteins, and regulates nuclear pairing, clamp connection formation and septation. At Aα, two classes of homeodomain proteins are produced. The HD1 and HD2 classes contain homologues to α2p and a1p respectively from *S. cerevisiae*. HD1 and HD2 proteins from different specificity loci dimerize and subsequently regulate the dikaryon. The B loci encode six pheromone genes and three pheromone receptor genes. In contrast to ascomycetes, where pheromone signalling stimulates the formation of mating-specific morphological shapes, pheromone signalling in *Coprinus* stimulates fusion of the monokaryotic hyphae to initiate the sexual cycle. The B locus also regulates nuclear migration and attachment between the clamp connection and the corresponding sub-apical hyphal compartment. As in the filamentous ascomycete *N. crassa*, the targets of the mating loci genes are currently not well characterized.

2.4 Unique Characteristics of Filamentous Fungi that are Advantageous for Genetic Analysis

2.4.1 Parasexual Analysis

Parasexual genetics involves examination of recombination in the absence of sexual reproduction, and has been helpful in mapping genes to chromosomes. The unique feature of heterokaryosis, or maintaining two genetically distinct nuclei within one thallus in filamentous fungi, allows for this type of analysis. The parasexual cycle has been extensively utilized in *A. nidulans*, and involves heterokaryon formation, followed by karyogamy to produce a diploid that then undergoes spontaneous mitotic recombination. Thus, genetic recombination can occur within the vegetatively growing hypha. Diploids can be differentiated from heterokaryons in parasexual analysis in *Aspergillus* based on spore colour. Heterokaryons formed from fusion of strains containing white or yellow spores will produce conidia of either colour. Sections of the colony that undergo karyogamy and form a heterozygous diploid, however, can be recognized by the resulting spore colour green, since the recessive mutations in spore colour leading to white and yellow spores will complement each other. The diploid is then isolated and forced to undergo haploidization through treatment of drugs, such as benomyl, to induce chromosome loss, and the resulting haploid products are analysed for evidence of mitotic crossing over. In *A. nidulans*, master strains containing unique markers per chromosome are used as a reference during 'crossing' with the test strain.

Parasexual analysis has recently been utilized to great advantage in *C. albicans*, a diploid pathogen that does not have a known sexual phase involving meiosis and, therefore, could not be analysed through traditional genetic analysis involving crossing of strains and sexual reproduction. *C. albicans* can mate and contains the necessary mating genes, as described earlier, but the resulting tetraploid products of mating break down to diploids through spontaneous chromosome loss, not meiosis. Therefore, *C. albicans* may naturally use a parasexual cycle to produce recombinant individuals. Parasexual analysis has been used for genetic linkage and construction of new strains in this organism, and holds promise for future mutagenic analysis.

2.4.2 Gene Silencing

A unique feature that has greatly facilitated genetic/molecular analysis in the filamentous fungus N. crassa is the process of repeat induced point (RIP) mutation. If more than one copy of a gene is introduced in tandem into the haploid prior to sexual reproduction, the tandem copies are inactivated through GC to AT mutations when passed through the sexual cycle. Therefore, gene inactivation can be achieved by simply introducing additional copies, and allowing the strain to undergo sexual reproduction. N. crassa was one of the first eukaryotic systems in which a form of RNA interference (RNAi) was investigated. The fungus demonstrates the process called quelling, where genes that are introduced at heterologous locations are silenced. This silencing involves degradation of mRNA through several factors, including homologues of Argonaute and Dicer which are involved in RNAi in other systems. RIPing and quelling, therefore, are very useful for investigating gene function in N. crassa, and for understanding the related mechanisms of gene silencing in other eukaryotes, including plants and worms. Because of the utility of RNAi, recent efforts have been made to transfer the molecular machinery to organisms like S. cerevisiae and C. albicans that do not naturally posses the capacity.

2.5 Genetics as a Tool

2.5.1 Tetrad Analysis

2.5.1.1 Saccharomyces cerevisiae

The classic strategy for genetic analysis of S. cerevisiae meiosis involves tetrad analysis. Each meiotic event from a typical diploid cell generates four haploid spores. These spores are arranged in an ascus sac that is degraded by enzymatic treatment with an endoglucanase. This liberates the spores, which are then placed in separate locations on rich media plates by micromanipulation and allowed to divide to form a spore colony. The spore colonies are then analysed to determine the distribution of the markers introduced into the initial cross. Recombination occurs when the diploid undergoes meiosis. During meiosis I, chromosomes from each parent pair up, then duplicate, creating two chromatids that remain attached. Chiasma formation between paired homologous chromosomes at this stage results in recombination of genetic material. Homologous chromosomes then separate to opposite poles of the meiotic spindle, and nuclear division results in two diploid nuclei. Independent assortment of chromosomes occurs at this stage. Meiosis II follows, which involves the splitting of the attached chromatids to opposite poles, and another round of nuclear division, resulting in four haploid nuclei. A critical advantage of the yeast meiotic process for genetic analysis is that all four products of the meioses are detected and available for analysis. This avoids questions of statistics in the analysis of segregation patterns, and allowed for the identification of 'non-Mendelian' segregation patterns characterized by a $3:1$ rather than $2:2$ distribution of heterozygous markers. The recombinational replacement of information from one allele to the other that generates this pattern is termed gene conversion.

Because there is no pattern to the position of the spores within the ascal sac, *S. cerevisiae* does not provide ordered tetrads as are found in some of the filamentous fungi. However, the identification of markers tightly linked to centromeres permits the indirect ordering of the spores relative to the actual meiotic event, because the centromere, and therefore all markers tightly linked to the centromere, segregate in the first meiotic division.

A modification of the classic tetrad analysis involves *selected tetrads*. This approach has been used during the analysis of meiotic recombination within a single locus. The frequency of intra-allelic recombinants is low, so selection is used to identify those infrequent meiotic events where a recombination has taken place. Typically, such intragenic recombination studies involve heteroalleles of an auxotrophic marker, so all the nonrecombinant products are auxotrophs. In contrast to standard tetrad analysis, the glucanase-treated ascal sacs are not spread on plates containing a rich growth medium, but rather are spread on plates that lack the nutrient required by the auxotrophic cells. The separated ascal groups are monitored microscopically to detect sets in which at least one member of the tetrad begins to germinate. These tetrads are then micromanipulated to permit subsequent analysis of the genetic structure of all four spores – this approach enriches for tetrads in which a recombination event is known to occur within the gene under study. This allows for a fine structure analysis of the recombination process, and the sophisticated ability to monitor all the consequences of the meiotic events has been critical to the development of models of the process of meiotic recombination.

In situations where the pattern of marker segregation in individual asci is not important, *random spore analysis* can be applied. In this approach, populations of asci are digested with glucanase en mass, and the mixture of haploid meiotic products and unsporulated diploid cells are spread on plates. These plates are designed to select against the initial diploid cells, typically by containing a recessive drug resistance marker that is only uncovered in the haploid cells. These haploid segregants can be rapidly screened to identify a cell containing a desired combination of markers, or the population can be scored to determine the overall patterns of segregation. An interesting recent development of the random spore strategy is found in the synthetic genetic array (SGA) approach pioneered by C. Boone and collaborators. In this approach, whole-genome wide screens for

synthetic lethal interactions are created by robotic replica plating, and sophisticated use of mating-type-specific gene expression is used to permit the growth of
only a desired haploid cell type.

2.5.1.2 Neurospora crassa

N. crassa is the pioneering organism for genetic analysis in microorganisms, pre-
dating work with bacteria and with the yeast *S. cerevisiae*. It is an attractive
model genetic system, since it is haploid and asci are large enough to remove
ascospores, allowing the recovery of all products of meiosis and determining recombination of the parental genes within the progeny. *Neurospora* is particularly
attractive for tetrad analysis, since the ascospores are ordered within the linear
ascus, allowing genes to be mapped in relation to the centromere. In *N. crassa*,
meiosis is followed by another mitotic division, producing eight haploid nuclei,
or four pairs of sister nuclei. Sporogenesis then produces eight ascospores within
the ascus (Figure 2.7). As in *Saccharomyces*, the individual spores are isolated

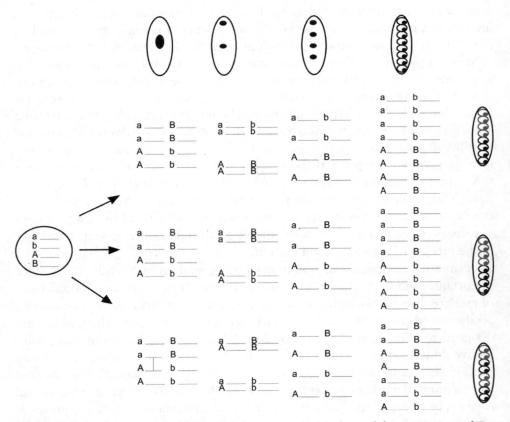

Figure 2.7 Chromosome assortment in *N. crassa* (adapted from Davis and De
Serres (1970)).

and tested for traits such as nutrient requirements. Alternatively, spores from many asci are collected and analysed as a random sample.

Owing to independent assortment of chromosomes at meiosis I, if two strains that differ at loci A and B mate (A/b × a/B) and A and B are located very close together, the chances of recombination between these loci is small, and the resulting meiotic products will be 50 % parental type (A/b and a/B). Such close genes are described as being linked. If A and B are far apart, crossing over or chiasma formation can occur during meiosis I, resulting in a proportion of recombinant progeny. The proportion increases the farther apart the genes, to a maximum of 50 % (25 % Ab, 25 % aB, 25 % AB, 25 % ab). In addition, if A and B are on different chromosomes, 50 % recombinant progeny are expected. The tetrad of *Neurospora* is ordered, where the products of the first division are located within one half of the ascus. This feature contributes to the efficiency and convenience of tetrad analysis in this organism. To set up strains for sexual reproduction and subsequent genetic analysis, the strain designated as the female is grown first, and conidia from the male are spread over the female culture. Perithecia develop within a few days, and acsi are mature and start to discharge ascospores by 10 days. Spores collect on the side of the incubation tube and can be taken for random spore analysis. Alternatively, a perithecium is dissected out and crushed in a water droplet to release asci. An ascus is pressed with a needle to discharge a spore, break the ascus and subsequently release the remaining spores. Spores are teased apart in order and separately analysed.

Prior to sequencing of the genomes, tetrad analysis was used for linkage analysis and mapping of genes. Reference strains with known markers were crossed with a test strain containing the gene of interest to detect any linkage, and to map the gene relative to the known position of the marker gene in the reference strain.

2.5.1.3 *Aspergillus nidulans*

Analysis of ascospores is different in *A. nidulans*, since traditional tetrad analysis is not as feasible as in *N. crassa*. However, Mendelian segregation for all unlinked nuclear markers is demonstrated in random spore samples. Sexual reproduction for genetic analysis involves growing two strains of different auxotrophy and other genetically distinct markers, such as spore colour, on media that selects against the individual strains and, thus, forces heterokaryon formation. The plate containing the resulting heterokaryons is sealed to prevent any aeration, and after 2 to 3 weeks the fruiting bodies or cleistothecia form. Since the asci are extremely fragile, they are not dissected out from the cleistothecium, so traditional tetrad analysis is not usually performed. Instead, a cleistothecium is rolled onto a semi-solid water–agar surface to remove other cellular tissue, then crushed and spread across a plate to observe segregation of markers such as ascospore colour. Some cleistothecia are derived from self-fertilization, which is evident upon analysis of ascospore colours of the products. A stock of the spores from the crushed cleitothecium is maintained and tested on different selective media to determine the genotype of the individual products and extent of recombination.

2.5.2 Molecular Methods for Genetic Screens

2.5.2.1 Transformation

S. cerevisiae was the first fungus to be transformed with exogenous DNA, and the development and refinement of this technology has led to the ability to effectively manipulate the genome of this yeast. Initial proof of the transformation potential of S. cerevisiae involved the construction of a yeast strain that contained two separate point mutants within a single locus, preventing any reversion of the gene to functionality, followed by introduction of a bacterial plasmid containing a cloned copy of the inactivated gene. A technique initially developed for protoplast (cells digested of their cell walls) fusion was modified to allow uptake of the DNA into the yeast cells, and proof that the exogenously added DNA was the source of the restored function of the missing gene was provided by the detection of the sequence of the bacterial plasmid in the clones that contained the restored function. There have been many modifications of the initial transformation protocol; currently, the use of electroporation or treatment of cells with lithium salts to trigger DNA uptake has essentially replaced the initial protocol of generating spheroplasts which had to regenerate their walls in an osmotically stabilized medium.

Transformation of filamentous fungi is performed either in protoplasts or in asexual conidia. In the former, protoplasts are typically transformed using polyethylene glycol and calcium to facilitate entry of DNA. Conidia in some fungi, including *Aspergillus fumigatus*, can be transformed without degrading the cell wall, through electroporation. Lithium-acetate-based transformation has also been used. Transformation efficiencies and optimum methods vary between fungi.

2.5.2.2 Plasmids, Transforming DNA

Transformation requires that the DNA contains a selectable marker, which is normally a nutrition or drug-resistance gene. For example, if *pryG* encoding for orotidine-5′-phosphate decarboxylase, which is part of the uridine biosynthesis pathway in *A. nidulans*, is used as a marker on transforming DNA, the resulting positive transformants will grow on media lacking uridine and uracil, whereas untransformed cells will not. The use of such selectable markers is common to all fungal transformation systems.

The initial transformation of yeast cells was not highly efficient, and involved sequences that integrated into the genome. The great utility of episomal DNA sequences for the transformation of bacterial cells led to the search for equivalent tools for the manipulation of *S. cerevisiae*. These tools took two forms; episomes based on the backbone of the endogenous yeast 2μ plasmid, and episomes that contained origins of replication from the chromosomes. The 2μ plasmids contained sequences for efficient segregation, and thus were more stable than those based solely on the autonomous replication sequences (ARS) elements derived from chromosomes. However, a further introduction of centromeric sequences, which provide efficient segregation and maintain a low plasmid copy number, allow ARS-based plasmids to be very stable. Such plasmids can be maintained for many generations in the absence of selection and now provide the workhorse tools for the molecular manipulation of yeast cells. The selection of the plasmids typically involves nutritional markers – the standard markers (*URA3*, *HIS3*, *TRP1* and *LEU2*) represent genes that were initially cloned by complementation of *Escherichia coli* mutations, and were available in cloned form prior to the development of yeast transformation. Dominant drug-resistant markers are also available for selection of plasmids; resistance to G418 has been a useful marker in *S. cerevisiae*. Essentially all commonly used yeast transformation plasmids include selection markers and replication origins for propagation of the plasmid in *E. coli*, so the plasmids can be shuttled between the prokaryotic and eukaryotic hosts.

Most filamentous fungi lack the ability to maintain extrachromosomal plasmids, in contrast to the yeast *S. cerevisiae*, so transformed DNA typically integrates homologously or heterologously in the genome. The frequency of homologous recombination increases with increasing length of homologous DNA. Transforming DNA, therefore, typically involves a vector background containing an *E. coli* origin of replication and ampicillin resistance marker to allow replication and selection in bacteria for plasmid propagation, as well as a fungal-specific marker and the desired gene sequence.

Transformation is performed to accomplish things such as knocking out a gene, replacing a gene with a mutated version or modifying gene expression. A gene knockout construct typically contains a marker gene surrounded by 5′- and 3′-flanking DNA of the gene of interest, so that the linear ends can recombine with the endogenous 3′- and 5′-flanking ends of the endogenous gene, allowing its replacement with the marker gene. Gene expression can be controlled by recombining a regulatable promotor in front of the endogenous ORF. In *S. cerevisiae*, the promotor for a galactose-regulated gene such as GAL1 is often used, where galactose or glucose in the medium regulates overexpression or repression respectively. In *A. nidulans*, the *alcA* promotor is commonly used; this becomes overexpressed if the cells are grown on media containing ethanol as a carbon source or repressed when cells are grown on media containing glucose.

2.5.2.3 *Genetic Screens*

In a post-genomic fungal world where genomes are sequenced and annotated, and genes no longer need to be mapped to chromosomes through traditional genetic techniques, genetic analysis is still a very powerful tool, particularly in construction of mutant screens. Genetic screens in filamentous fungi have uncovered an enormous amount of information about diverse cellular processes, and in many cases identified the first examples of conserved genes and their functions. Pioneering work in *N. crassa* by Beadle and Tatum demonstrated that individual genes encoded for individual enzymes, bringing together genetic and biochemical analyses for the first time. Subsequent screens in this organism have provided novel information on gene conversion, recombination, circadian rhythms, gene silencing and DNA methylation. Genetic screens in *A. nidulans* and *S. cerevisiae* have and continue to uncover novel information in diverse areas, including cell cycle regulation, cellular motors and cytoskeletal dynamics, signalling and development.

Classical screening involves mutagenizing cells, typically with ultraviolet light, radiation or chemicals, then allowing growth of survivors. This approach was used by Beadle and Tatum in 1941 to uncover the metabolic mutants in *N. crassa* (Figure 2.8). Single conidia colonies were exposed to X-irradiation on complex medium and transferred to minimal medium. Growth on complex versus minimal medium was screened, and any colony that could not grow on minimal medium was considered to contain a nutritional mutation. The strain was maintained on complex media, but subsequently tested for the restoration of growth on minimal media containing defined additives, such as tyrosine, leucine or alanine. If growth was restored only when tyrosine was added, for example, the specific mutation could then be identified. From this approach, strains containing

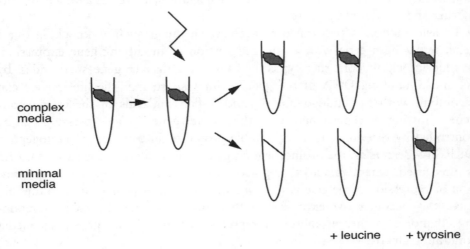

Figure 2.8 Genetic screen identifying metabolic mutants in *N. crassa*.

mutations in vitamin B_6, vitamin B_{12} and *para*-aminobenzoic acid were uncovered. The mutant strains were crossed with wild-type parental strains to ensure that only a single gene was mutated.

Another classical mutagenesis screen identified many genes that control a circadian rhythm in *N. crassa*. Circadian rhythms are present in fungi to humans, and are biological processes that are sensitive to light and temperature and, therefore, oscillate every 24 h in the absence of environmental signals. The timing of conidia formation in *N. crassa* is regulated by such an internal clock, and genes involved in responding to the clock were uncovered by using mutagenesis and race tube assays. When inoculated at one end of a long glass horizontal tube, called a race tube, *Neurospora* hyphae grow to the other end, creating a periodic banding pattern along the tube from the coloured conidia that form every 24 h. Mutagenized conidia were placed at one end of the tube and changes in banding pattern, reflecting changes in day length, relative to control strains were screened. The *frequency*, *period* and *chrono* genes were uncovered, most of which have homologues involved in clock functions in *Drosophila melanogaster* and humans.

In *S. cerevisiae*, classical screens were applied to identify the cell division cycle (*cdc*) genes that uncovered many of the key controlling elements underlying what is now considered the universal eukaryotic cell cycle. Because defects in the cell cycle blocked cellular proliferation, it was necessary to identify conditional mutations, in this case temperature-sensitive mutants that arrested with a uniform terminal phenotype. In general, the uniform terminal phenotype arose because a particular cellular function necessary for completion of the cell cycle was missing at the restrictive temperature. Key proteins identified through this screening included Cdc28p, the cyclin-dependent protein kinase controlling both mitosis and the G1 to S transition, Cdc35p, which is adenylyl cyclase, Cdc12p, which is a septin, and Cdc9p, which is DNA ligase. Novel cellular processes were uncovered through the analyses of these mutations, and genes identified initially as *cdc* mutants form the underpinning of much of the current cell biology of *S. cerevisiae*.

Other strategies for gene identification in *S. cerevisiae* involved modifications of the classical screen approach that added enrichment protocols to improve the frequency of mutant identification. For example, screening for mutations that blocked the process of secretion was made more efficient by treating the mutagenized cell population to a density enrichment prior to looking for temperature-sensitive mutants.

The traditional screen for mutants has become more powerful and specific through introduction of various types of selection. A classic screen designed to identify mutants of the cell cycle in *A. nidulans* was designed by Ron Morris in 1975 (Figure 2.9a). Conidia were mutagenized with ultraviolet light and then allowed to grow at 32 °C. Cells were replica spotted onto media at 42 °C to screen for temperature sensitivity. Cells that could not grow at the restrictive temperature were analysed for phenotype and stained to visualize nuclei. Several classes

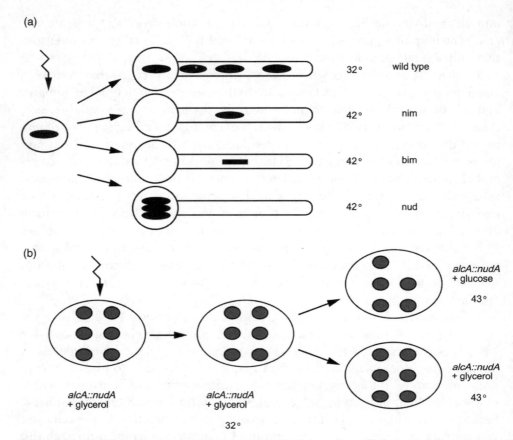

Figure 2.9 Selected genetic screens identifying cell cycle mutants in *A. nidulans* (adapted from Casselton and Zolan (2002)).

of mutants were uncovered and characterized as the NIM (never-in-mitosis), BIM (blocked-in-mitosis) and NUD (nuclear distribution) mutants. These acronyms have subsequently been used for mutants in other systems, including *S. pombe* and *S. cerevisiae*. The genes responsible for the mutant phenotypes were identified through complementation analysis. Several of the genes, including NIMA kinase, which is essential for the G2/M transition, and BIMC, which is a kinesin motor, were the founding members of families of related proteins subsequently found in other organisms, from yeast to man. Another gene identified in the screen, NUDF, has a homologue in humans that, when mutated, is the cause of the human neuronal disease lissencephaly.

In other landmark genetic screens performed in *A. nidulans*, resistance to the anti-microtubule drug benomyl was incorporated to select for mutations in tubulin. The subsequent *benA* mutants were confirmed to be β1- and β2-tubulin. Suppression analysis of the *benA* mutants was then used to identify mutations that affected proteins that interacted with β-tubulin. Based on work done in bacteriophage, if mutations of a protein prevented its interaction with another protein and thereby inactivated its function, a compensatory mutation in the second factor which restores the ability of the two proteins to interact would also restore function. The *benA* mutant, therefore, was subjected to another round of mutagenesis, and temperature-insensitive revertants were isolated. Through subsequent analysis, α-tubulin was identified, which forms a dimer with β-tubulin. Suppression screening of the *benA* mutant also identified the *mipA* mutants, which, upon mapping and cloning, provided the first example of γ-tubulin in any organism.

A classic selection approach was applied to identify mutants defective in the mating process in *S. cerevisiae*. This selection was based on the observation that a functioning mating response pathway resulted in cells arresting the cell cycle in the presence of the mating pheromone produced by cells of the opposite cell type. Thus, mutagenized *MATa* cells were treated with the mating pheromone α-factor, and the only cells that could grow to form colonies would be those that were insensitive to the pheromone. This selection allowed the identification of many of the key kinases and regulators of this pathway.

The mutagenesis screen has more recently been adapted to identify synthetic lethals, which can also uncover potential functionally interacting proteins. In one example, a strain of *A. nidulans* in which nudA (dynein) was placed under control of the *alcA* regulatable promotor was utilized (Figure 2.9b). On glucose, dynein expression was off, but dynein was expressed on glycerol. After ultraviolet mutagenesis, the strain was grown on glycerol at 32 °C and then replica spotted onto glycerol or glucose at the restrictive temperature of 42 °C. The ability of temperature-sensitive mutations to grow on glycerol (+dynein) but not on glucose (−dynein) was screened. The resulting strains were back-crossed to wild-type and *nudA* strains to isolate single and double mutants. The mutants, called *sld* for synthetic lethal with dynein, were cloned and determined to be homologues of Bub1p and Bub3p spindle checkpoint factors.

Insertional mutagenesis is highly advanced in *S. cerevisiae*. Initially, circular plasmids containing selectable markers were transformed into cells and were found to integrate at homologous sites quite efficiently. Markers inserted into repetitive sequences such as the ribosomal DNA allowed the analysis of both mitotic and meiotic recombination. The recognition that double-stranded breaks both dramatically enhanced the frequency of insertion and provided for efficient targeting to the homologous site in the genome greatly improved the technology of directed gene replacement. Several strategies are currently available for disrupting gene function by replacing part or all of a gene with foreign DNA, thus inactivating the function of the gene of interest. The use of such homology-driven gene inactivation has been extensively applied to investigations of individual genes. A modification of this approach has been used to provide efficient insertional mutations for random yeast sequences. In this approach, a library of random yeast DNA inserts in the 10 kB range is constructed in an *E. coli* plasmid vector. This library is transformed into *E. coli* and the *E. coli* strain is then subjected to a transposon mutagenesis with a modified bacterial transposon containing at least a selectable marker for *S. cerevisiae*. Plasmids are selected which have picked up a transposon insertion, and many of these insertions have occurred in the yeast DNA portion of the plasmid. The yeast inserts are liberated from the vector by restriction digest and the digestion products used to transform yeast cells. Selection for the yeast marker on the transposon allows detection of integration events, and these integration events represent transposon insertions into essentially random regions of the yeast genome – the overall distribution dependent on the randomness of the initial library of yeast sequences and the randomness of the transposon insertions into this library. Strategies that made use of the endogenous yeast transposable element to do the transposon hops directly in yeast have also been developed, but were limited by the nonrandom characteristics of the insertion locations.

Insertional mutagenesis has also been applied to genetic screens in filamentous fungi. In REMI (restriction enzyme mediated insertions), a plasmid that has been linearized is transformed into protoplasts in the presence of the restriction enzyme that was used for cutting. The enzyme is thought to cleave the genomic DNA at restriction sites that are compatible with the ends of the cut plasmid, allowing integration at multiple sites throughout the genome and creation of multiple, potentially mutagenized transformants. To determine where the plasmid integrated, and the identity of the mutagenized gene, genomic DNA from the strain is isolated and cut with enzymes that would not cut within the plasmid itself. Fragments of DNA that contain the plasmid plus flanking genomic DNA are allowed to ligate and then transformed into *E. coli* to rescue the plasmid using a plasmid-specific drug resistance marker.

A more efficient system for random mutagenesis incorporates a tag on the mutagenic DNA, as is the case with the transposon mutagenesis library used in *Saccharomyces*. In the system termed TAGKO, a cosmid library containing the genome of *Magneporthe griseae* was mutagenized with a transposon. The individual mutated cosmids were sequenced off of the transposon and annotated. The cosmids with known mutated genes were then transformed into the fungus to determine the effect of replacing the endogenous gene with the mutated version. Transposon mutagenesis is also routinely used with other filamentous fungi, including *A. nidulans*.

2.6 Conclusion

Fungi have important interactions with humans. Many of these organisms are economically significant, such as the baker's or brewer's yeast *S. cerevisiae* generating our bread and alcohol, or medically important, such as the human pathogens *C. albicans* and *A. fumigatus*. They represent, as well, experimental systems that have been extensively exploited to investigate the molecular details of eukaryotic cell function. In particular, the fungi include some of the most highly developed eukaryotic genetic systems. From early work in *Neurospora* leading to recognition of the relationship of genes and enzymatic functions, to more recent work in the yeasts *S. cerevisiae* and *S. pombe* that has uncovered the molecular basis of control of cell proliferation, we can see that the isolation and characterization of genetic variants of fungal cells has revolutionized our understanding of cellular function.

A key characteristic of genetic systems is that they can be exploited to create new combinations of variations. This ability is highly developed in fungal systems. The capacity to identify all the products of the meiotic event has allowed for such a sophisticated analysis of meiotic recombination that fungal systems have played leading roles in the development of molecular models of DNA recombination. Only through the analysis of complete individual meioses would we be able to detect the gene conversions and post-meiotic segregations that established the existence of hetroduplex DNA during the recombination process.

We have only touched the surface of the uses of fungi in uncovering details of the function of eukaryotic cells. Our current work focuses on a few fungi, but the ability to sequence whole genomes promises to allow many more organisms to be investigated through the tools of genetics and genomics. In the future we should be able to study the characteristics of more and more of the fungal diversity, and investigate the current model organisms with even greater sophistication. Based on the success of the investigations of the prior decades, we can expect the fungi to continue to be in the forefront of research in providing new information on a diversity of cellular processes in the future.

Acknowledgement

We acknowledge the assistance of Andre Migneault (BRI/NRC).

Revision Questions

Q 2.1 What are some of the advantages of using fungi as model systems to investigate biological questions?

Q 2.2 Describe some key structural differences between yeast and filamentous fungi.

Q 2.3 What are some differences in ascus structure between *S. cerevisiae* and *N. crassa*, and what are the advantages of some of these features for genetic analysis?

Q 2.4 Describe the first genetic screen used in fungi and the main finding of the study.

Q 2.5 Describe different selection criteria used in devising genetic screens in fungi.

Q 2.6 Genetic screens have not been extensively used in the analysis of *C. albicans*. Why?

Q 2.7   In *S. cerevisiae*, the genes *NUP60* and *TRP1* are located close to centromeres of chromosomes 1 and 4 respectively. A strain with the mutant *TRP1* locus (*trp1*) and a functional NUP60 locus (*NUP60*) has the genotype *trp1 NUP60*. This strain was crossed to a strain with a functional *TRP1* gene and a mutant *NUP60* gene (*TRP1 nup60*). The segregation of these two markers was examined in 20 dissected tetrads. The patterns of spores could be: parental ditypes (PD), where the four spores consist of two *trp1 NUP60* and two *TRP1 nup60* spores; nonparental ditypes (NPD), where the four spores consist of two *trp1 nup 60* and two *TRP1 NUP60* spores; or tetratypes (TT), where all four spore combinations are present. In the 20 tetrads, would tetratype tetrads be rare or common? Would this change if the starting parents were the double mutant *trp1 nup60* and the wild-type *TRP1 NUP60*?

Q 2.8   What are some key differences in the mating loci of *N. crassa* and *C. cinereus* compared with that of *S. cerevisiae*?

Q 2.9   What is parasexual genetic analysis, where has it been used and what information can it provide?

Q 2.10   What is a major difference between transformation plasmids used in *S. cerevisiae* and those used in most filamentous fungi, and how is this difference advantageous for genetic analysis?

References

Casselton, L. and Zolan, M. (2002) The art and design of genetic screens: filamentous fungi. *Nature Reviews Genetics*, 3(9), 683–697.

Davis, R.H. and De Serres, F.J. (1970) Genetic and microbiological research techniques for *Neurospora crassa*. *Methods in Enzymology A*, 17, 79–143.

Further Reading

Bennett, R.J. and Johnson, A.D. (2003) Completion of a parasexual cycle in *Candida albicans* by induced chromosome loss in tetraploid strains. *Embo Journal*, 22(10), 2505–2515.

Casselton, L. and Zolan, M. (2002) The art and design of genetic screens: filamentous fungi. *Nature Reviews Genetics*, 3(9), 683–697.

Davis, R. H. and De Serres, F.J. (1970) Genetic and microbiological research techniques for *Neurospora crassa*. *Methods in Enzymology A*, 17, 79–143.

Davis, R.H. and Perkins, D.D. (2002) Timeline: *Neurospora*: a model of model microbes. *Nature Reviews Genetics*, 3(5), 397–403.

Deacon, J.W. (1997). *Modern Mycology*, Blackwell Science Ltd, Oxford.

Herskowitz, I. (1989) A regulatory hierarchy for cell specialization in yeast. *Nature*, 342(6251), 749–757.

Kues, U. (2000) Life history and developmental processes in the basidiomycete *Coprinus cinereus*. *Microbiology and Molecular Biology Reviews*, 64(2), 316–353.

Morris, N.R. and Enos, A.P. (1992) Mitotic gold in a mold: *Aspergillus* genetics and the biology of mitosis. *Trends in Genetics*, **8**(1), 32–37.

Timberlake, W.E. (1990) Molecular genetics of *Aspergillus* development. *Annual Review of Genetics*, **24**, 5–36.

Timberlake, W.E. (1991) Temporal and spatial controls of *Aspergillus* development. *Current Opinion in Genetics & Development*, **1**(3), 351–357.

Timberlake, W.E. and Marshall, M.A. (1988) Genetic regulation of development in *Aspergillus nidulans*. *Trends in Genetics*, **4**(6), 162–169.

Tong, A.H., Evangelista, M. *et al.* (2001) Systematic genetic analysis with ordered arrays of yeast deletion mutants. *Science*, **294**(5550), 2364–2368.

3
Fungal Genomics

David Fitzpatrick and Edgar Mauricio Medina Tovar

3.1 Introduction

Genomics is defined as the study of an organism's complete genome sequence. The first complete (nonviral) genome to be sequenced was the bacterium *Haemophilus influenzae* in 1995. Today, more than 1300 bacterial genomes have been sequenced. Baker's yeast (*Saccharomyces cerevisiae*) was the first eukaryote to have its genome completely sequenced (released in 1996). Since then, over 250 eukaryote genomes have been completed, including our own (in 2001). Because of their relatively small genome size, roles as human/crop pathogens and importance in the field of biotechnology, 102 fungal genomes (Table 3.1) have been sequenced to date, accounting for approximately 40 % of all available eukaryotic genomic data. Some species, such as *S. cerevisiae* have actually had multiple strains sequenced. Presently, the majority (78 %) of fungal species that have been sequenced belong to the Ascomycota phylum; furthermore, there is a significant bias towards species that are pathogens of humans. Reduced costs and recent improvements associated with new sequencing technologies (Section 3.2) should mean that a wider range of evolutionarily, environmentally and biotechnologically interesting organisms will become available in the coming years.

This abundance of genomic data has moved the fungal kingdom to the forefront of eukaryotic genomics. While some of the species sequenced are closely related, others have diverged 1 billion years ago. This enables us to use fungi to study evolutionary mechanisms associated with eukaryotic genome structure, organization and content. Furthermore, doing a direct comparison between two or more closely related pathogenic and nonpathogenic species, a process called comparative genomics (Section 3.4), permits us to locate metabolic pathways or genes associated with virulence.

Fungi: Biology and Applications, Second Edition. Edited by Kevin Kavanagh.
© 2011 John Wiley & Sons, Ltd. Published 2011 by John Wiley & Sons, Ltd.

Table 3.1 List of fungal genomes currently sequenced.

Species	Lineage	No. genes
Allomyces macrogynus	Chytridiomycetes	17 600
Batrachochytrium dendrobatidis	Chytridiomycetes	8 732
Spizellomyces punctatus	Chytridiomycetes	8 804
Mucor circinelloides	Zygomycota	10 930
Phycomyces blakesleeanus	Zygomycota	16 528
Rhizopus oryzae	Zygomycota	17 459
Alternaria brassicicola	Ascomycota	10 688
Ashbya gossypii	Ascomycota	4 717
Aspergillus carbonarius	Ascomycota	11 624
Aspergillus clavatus	Ascomycota	9 120
Aspergillus flavus	Ascomycota	12 587
Aspergillus fumigatus	Ascomycota	9 887
Aspergillus nidulans	Ascomycota	10 560
Aspergillus niger	Ascomycota	8 592
Aspergillus oryzae	Ascomycota	12 063
Aspergillus terreus	Ascomycota	10 406
Blastomyces dermatitidis	Ascomycota	9 522
Botryotinia cinerea	Ascomycota	16 448
Candida albicans	Ascomycota	6 205
Candida dubliniensis	Ascomycota	5 928
Candida glabrata	Ascomycota	5 202
Candida guilliermondii	Ascomycota	5 920
Candida lusitaniae	Ascomycota	5 941
Candida parapsilosis	Ascomycota	5 823
Candida tropicalis	Ascomycota	6 258
Chaetomium globosum	Ascomycota	11 124
Coccidioides immitis	Ascomycota	10 654
Coccidioides posadasii	Ascomycota	10 124

Table 3.1 (*Continued*)

Species	Lineage	No. genes
Cochliobolus heterostrophus	Ascomycota	9 633
Cryphonectria parasitica	Ascomycota	11 184
Debaryomyces hansenii	Ascomycota	6 272
Fusarium graminearum	Ascomycota	13 321
Fusarium oxysporum	Ascomycota	17 608
Fusarium verticillioides	Ascomycota	14 195
Histoplasma capsulatum	Ascomycota	9 251
Kluyveromyces lactis	Ascomycota	5 076
Kluyveromyces waltii	Ascomycota	5 350
Lachancea thermotolerans	Ascomycota	5 091
Lodderomyces elongisporus	Ascomycota	5 799
Magnaporthe grisea	Ascomycota	11 109
Microsporum canis	Ascomycota	8 765
Microsporum gypseum	Ascomycota	8 876
Mycosphaerella fijiensis	Ascomycota	10 313
Mycosphaerella graminicola	Ascomycota	10 933
Nectria haematococca	Ascomycota	15 707
Neosartorya fischeri	Ascomycota	10 403
Neurospora crassa	Ascomycota	9 908
Neurospora discreta	Ascomycota	9 948
Neurospora tetrasperma	Ascomycota	10 640
Paracoccidioides brasiliensis	Ascomycota	9 136
Penicillium chrysogenum	Ascomycota	12 773
Penicillium marneffei	Ascomycota	10 638
Pichia pastoris	Ascomycota	5 040
Pichia stipitis	Ascomycota	5 807
Podospora anserina	Ascomycota	10 601
Pyrenophora triticirepentis	Ascomycota	12 169

(*continued*)

Table 3.1 (*Continued*)

Species	Lineage	No. genes
Saccharomyces bayanus	Ascomycota	9 417
Saccharomyces castelli	Ascomycota	4 677
Saccharomyces cerevisiae	Ascomycota	5 885
Saccharomyces kluyveri	Ascomycota	5 321
Saccharomyces kudriavzevii	Ascomycota	3 768
Saccharomyces mikatae	Ascomycota	9 057
Saccharomyces paradoxus	Ascomycota	8 955
Schizosaccharomyces cryophilus	Ascomycota	5 057
Schizosaccharomyces japonicus	Ascomycota	4 814
Schizosaccharomyces octosporus	Ascomycota	4 925
Schizosaccharomyces pombe	Ascomycota	5 010
Sclerotinia sclerotiorum	Ascomycota	14 522
Sporotrichum thermophile	Ascomycota	8 806
Stagonospora nodorum	Ascomycota	16 597
Talaromyces stipitatus	Ascomycota	13 252
Thielavia terrestris	Ascomycota	9 815
Trichoderma atroviride	Ascomycota	11 100
Trichoderma reesei	Ascomycota	9 143
Trichoderma virens	Ascomycota	11 643
Trichophyton equinum	Ascomycota	8 560
Trichophyton rubrum	Ascomycota	8 625
Trichophyton tonsurans	Ascomycota	8 230
Uncinocarpus reesii	Ascomycota	7 798
Vanderwaltozyma polyspora	Ascomycota	5 367
Verticillium alboatrum	Ascomycota	10 220
Verticillium dahliae	Ascomycota	10 535
Yarrowia lipolytica	Ascomycota	6 448
Zygosaccharomyces rouxii	Ascomycota	4 991

Table 3.1 (*Continued*)

Species	Lineage	No. genes
Agaricus bisporus	Basidiomycota	10 438
Coprinopsis cinerea	Basidiomycota	13 394
Cryptococcus gattii	Basidiomycota	6 210
Cryptococcus neoformans	Basidiomycota	6 967
Heterobasidion annosum	Basidiomycota	12 299
Laccaria bicolour	Basidiomycota	19 036
Moniliophthora perniciosa	Basidiomycota	13 560
Phanerochaete chrysosporium	Basidiomycota	10 048
Pleurotus ostreatus	Basidiomycota	11 603
Postia placenta	Basidiomycota	9 113
Schizophyllum commune	Basidiomycota	13 181
Serpula lacrymans	Basidiomycota	14 495
Tremella mesenterica	Basidiomycota	8 313
Melampsora laricis-populina	Basidiomycota	16 831
Puccinia graminis	Basidiomycota	20 566
Sporobolomyces roseus	Basidiomycota	5 536
Malassezia globosa	Basidiomycota	4 286
Ustilago maydis	Basidiomycota	6 522

3.1.1 The Fungal Kingdom

Fungi are eukaryotic organisms (contain a nucleus and membrane-bound organelles) and form one of the kingdoms of life. They lack chlorophyll and are saprobic (live on dead organic matter). Traditionally, fungi were thought to be closely related to plants; however, recent phylogenetic studies have shown that fungi are more closely related to animals than plants. Fungi are ubiquitous and can be beneficial (useful in biotechnology), harmful (cause disease) or mutualistic (symbionts with plants). The exact number of fungal species is unknown, but it is estimated to be 1.5 million. Phylogenetic analysis (Figure 3.1) has revealed that there are four distinct phyla within the fungal kingdom; they are the Chytridiomycota, Zygomycota, Ascomycota and Basidiomycota.

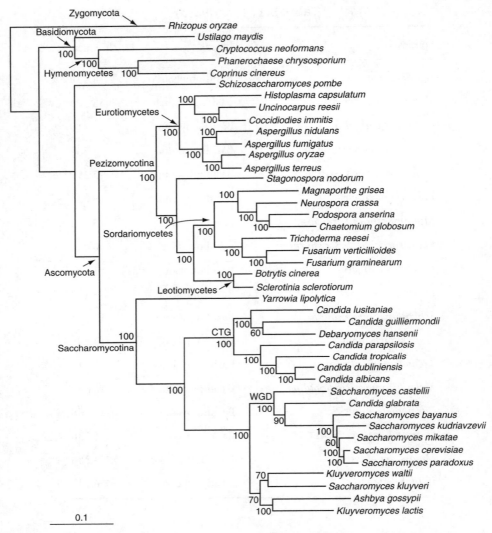

Figure 3.1 Fungal phylogeny based on 42 complete genomes. The Ascomy-cota, Basidiomycota and Zygomycota make up three of the four fungal phyla and are present as monophyletic clades.

The Chytridmycota (chytrids) is the only fungal phylum to produce zoospores and requires water for their dispersal. They are an ancient group of organisms and are thought to have changed little since fungi first diverged from the last common ancestor of all eukaryotes. Most chytrids live in soil or freshwater, although some are found in marine environments, where they have important roles in the decomposition of organic matter. The chytrid *Batrachochytrium dendrobatidis* has been shown to be responsible for a disease in amphibians called chytridiomycosis, which is responsible for declining frog populations in tropic regions.

The Zygomycota reproduce sexually and form thick-walled sexual spores called zygospores. Zygomycetes are morphologically diverse and account for 1 % of all described fungal species. They are also the most ecologically diverse phyla of fungi, living as saprophytes on dung, fruit and soil. They can also found in the gut of arthropods and some are pathogens of plants, animals and other fungi. Some well-known members include *Mucor* and *Rhizopus* species, which cause bread mould and fruit rots respectively.

The Ascomycota is the largest fungal phylum, accounting for approximately 65 % of all know fungal species. The distinguishing feature of this phylum is an ascus. The ascus is the sexual spore-bearing cell where meiosis followed by one round of mitosis occurs to generate eight (or a multiple of eight) ascospores. Ascospores have thick walls and, therefore, are resistant to adverse conditions; but under favourable conditions they will germinate to form a haploid fungus. Three subphyla have been described in the Ascomycota; they are the subphyla Saccharomycotina, Pezizomycotina and Taphrinomycotina. The Saccharomycotina lack an ascoma, resulting in naked asci, and include important species such as *S. cerevisiae* (brewer's yeast) and *Candida albicans* (human pathogen). Members of the Pezizomycotina include all filamentous fungi (moulds) and include species such as *Aspergillus fumigatus* (human pathogen) and *Penicillium chrysogenum* (produces penicillin antibiotic). The Taphrinomycotina phylum includes many diverse morphologies, including the fission yeast form of *Schizosaccharomyces pombe*.

The Basidiomycota accounts for approximately 35 % of the known fungal species. A number of Basidiomycetes are instantly recognizable, as they produce elaborate fruit bodies, including puffballs and mushrooms. Well-known edible Basidiomycota mushrooms include *Agaricus bisporus* (common mushroom) and *Pleurotus ostreatus* (oyster mushroom). The ability to degrade lignin (found in plant cell wall) by certain members (e.g. *Armillaria mellea*) of the Basidiomycota is significant, as few microbes have this ability. Fungi that can degrade lignin are interesting in a biotechnological sense, as they have the potential to detoxify and delignify waste products.

3.2 Genome Sequencing

3.2.1 Sanger Sequencing

Fredrick Sanger won the Nobel Prize in 1958 for determining the amino acid sequence of the protein insulin. After this, he turned his attentions to developing sequencing methods for RNA and DNA. In 1977 Sanger published a method for DNA sequencing commonly referred to as dideoxy sequencing (or chain termination) and won his second Nobel Prize for this work in 1980.

Sanger sequencing relies on DNA polymerase (a replication enzyme) to synthesize a new strand of DNA, which in turn can reveal the sequence of the target DNA strand. DNA polymerase replicates a new DNA strand complementary to a piece of single-stranded DNA, by linking the 5'-hydroxyl end of a free nucleotide to the 3'-OH group of the nucleotide at the end of a primer. A primer is a small piece of single-stranded DNA that can hybridize to one strand of the template DNA and be extended by successive additions of nucleotides. As well as a supply of nucleotide triphosphates (dNTPs), the Sanger method also requires 2',3'-dideoxynucleotide triphosphates (ddNTPs) in small quantities relative to dNTPs. ddNTPs contain no reactive 3'-OH and, therefore, terminate DNA synthesis once they are incorporated into the primer extension.

A typical reaction mixture contains dATP, dTTP, dCTP, dGTP and one ddNTP (ddGTP, for example). Primer extension continues until an unmatched nucleotide is paired with a complementary ddNTP. Many fragments, each ending with a ddNTP of varying length, are produced from such a reaction (Figure 3.2).

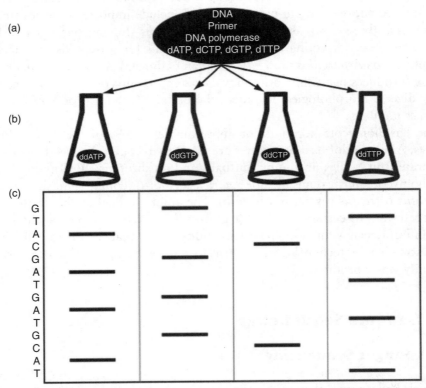

Figure 3.2 Schematic of the Sanger sequencing method. (a) Four separate DNA extension reactions are carried out. Materials required include single-stranded DNA, DNA polymerase, DNA primers and all four dNTPs. One of the dNTPs is radioactively labelled to enable visualization in (c). (b) Each of the four reactions contains a different dideoxynucleoside triphosphate (ddNTP). Synthesis continues until a ddNTP is incorporated, terminating extension reaction. (c) Products are separated based on size on a polyacrylamide gel and the sequence can be determined.

Radioactive sulfur or phosphorus isotopes are incorporated into the newly synthesized DNA template via labelled dNTPs, therefore making all fragments detectable by radiography. Fragments can then be separated based on length with polyacrylamide gel electrophoresis and the sequence can be determined (Figure 3.2). To determine the relative position of all four nucleotides it is necessary to run four reactions (each with a different ddNTP) in parallel (Figure 3.2).

3.2.2 Next-Generation Sequencing

When Fred Sanger and co-workers first developed their enzyme-based chain termination method for DNA sequencing they could not have predicted the massive advances in sequencing technology that have taken place in recent years. Next-generation sequencing (NGS) refers to novel commercial technologies which make it possible to generate millions of sequence reads (hundreds of base pairs in length) in a single sequencing reaction. With respect to fungi, NGS has been used to resequence targeted strains (such as *S. cerevisiae*), sequence *de novo* genomes (such as the xylose fermenter *Pichia stipitis*), analyse transcriptomes and characterize fungi in environmental samples. In the following two sections we will examine two of the most popular NGS platforms: Roche/454 GS FLX pyrosequencer and the Illumina genome analyser.

3.2.2.1 Roche/454 GS FLX Pyrosequencer

The first commercial NGS was introduced in 2004 by 454 Life Sciences (now Roche Diagnostics). It utilizes pyrosequencing, a technique that ultimately emits light (using the firefly enzyme luciferase) after each incorporation of a nucleotide by DNA polymerase. With the latest instrument and sequencing kits and reagents it is possible to generate more than 1 million reads (average length 400 bases) in a single 10 h run.

The Roche/454 sequencer has three basic steps: single-stranded template DNA library preparation, emulsion-based clonal amplification of the library and sequencing by synthesis.

The DNA library preparation stage fragments sample DNA into small single-stranded DNA fragments (300–800 base pairs). Universal adapters specific for 3' and 5' ends are added to each fragment. Each universal adapter is 44 bases in length and consists of a 20-base PCR primer, a 20-base sequencing primer and an initiating 4-base (TCGA) sequence.

For the clonal amplification stage the single-stranded DNA library is mixed with small DNA capture beads (~35 µm in size). The beads contain one of the adapter primers and ligate a single-stranded DNA library fragment. The ratio of capture beads to library DNA is chosen to ensure that each bead binds a single DNA fragment. The bead-bound library complexes are emulsified with amplification reagents, resulting in microreactors containing just one bead with one unique sample–library fragment. In parallel, each library fragment is amplified using thermal cycling within its own microreactor. The end result is several million copies of unique library DNA per bead. At the end of this phase the emulsion is broken down. DNA-positive beads are enriched and deposited onto a PicoTiterPlate (PTP) (a solid surface containing wells (~44 µm)). The DNA-positive beads are overlaid with packing and enzyme (luciferase and sulfurylase) beads.

The final step is sequencing by synthesis. Nucleotides are flown across the PTP sequentially in a fixed order. Nucleotides that are complementary to the template strand are incorporated by the DNA polymerase, extending the DNA strand. If the template DNA contains three adjacent guanines (G), three cytosines (C) will be incorporated into the sequencing strand. As incorporation of nucleotides occurs at different rates, strands extend at different rates. Nucleotide incorporation generates free pyrophosphate, which is converted to ATP by the sulfurylase enzyme. ATP results in the oxidation of luciferin by the enzyme luciferase and light whose intensity is proportional to the number of bases incorporated is emitted. Light photons are captured by a charge-coupled-device camera and signal intensity per nucleotide is used to determine the sequence of template DNA.

3.2.2.2 Illumina Genome Analyser

The Illumina genome analyser (IGA) was released in 2006 and currently can output 300 gigabases in a single run. Currently, read lengths (up to 150 bp) are shorter than those of 454 sequencing. The major advantage of the IGA over other sequencing platforms is the quantity of data produced at low cost. The sequencing process used by the IGA has three main steps: DNA library preparation, generation of clonal clusters and sequencing by synthesis.

The DNA library is prepared by fragmenting sample DNA by sonication (ultrasound) or nebulization (vaporizing) and sequencing adapters are ligated to the fragments (Figure 3.3a).

Clonal clusters are then generated by immobilizing sequencing templates on a flow cell. The flow cell is composed of silica and has eight lanes running lengthways. Separate DNA libraries can be loaded into different lanes, enabling eight individual sequencing runs per slide. Adapter-ligated template is pumped into the flow cell and template DNA is captured by forward/reverse 'lawn' primers that are covalently linked to the flow cell. Free ends of DNA template attach to lawn primers forming U-shaped bridges (Figure 3.3b). Unlabelled nucleotides are added and solid-phase bridge amplification occurs, resulting in double-stranded clonal clusters. Reverse strands are removed from double-stranded DNA and sequencing primers are hybridized to free 3′ ends; this step ensures all clusters are sequenced in the same direction from the same end. The flow cell is then transferred to the IGA for sequencing.

IGA sequencing by synthesis involves the incorporation of fluorescent terminator deoxynucleoside triphosphate (dNTP) (Figure 3.3c). During each sequencing

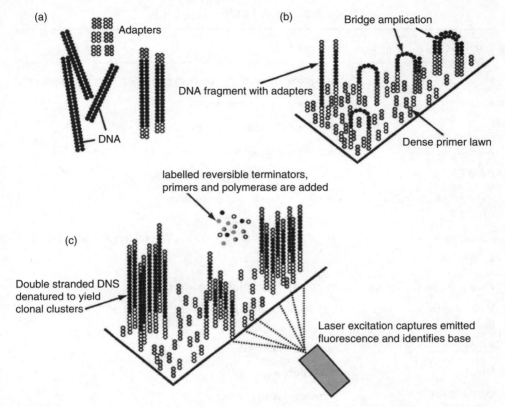

Figure 3.3 Schematic of the Illumina genome analyser sequencing technology. (a) DNA is fragmented and adapters are ligated to both ends of the fragments. (b) Single-stranded fragments bind randomly to the surface of the flow cell; see main text. (c) Sequencing by synthesis; see main text.

cycle, DNA polymerase incorporates a single dNTP to each of the growing nucleic acid chains (Figure 3.3). After each cycle, the IGA images the relevant fluorescent dye identifying the base and then cleaves the terminator dye so that addition of the next nucleotide can proceed. The sequence lengths of all clusters are identical as they are governed by the number of cycles (nucleotide incorporation, imaging and cleavage) undertaken.

3.3 Bioinformatics Tools

3.3.1 Locating Homologues

Sequence similarity searches are an essential component of genomic studies. They allow researchers to identify homologues, conserved structural motifs and help assign putative functions to unannotated genes in *de novo* genomes. The

Table 3.2   Useful online resources.

Database	URL address
FGI	http://www.broadinstitute.org/annotation/fungi/fgi/index.html
SGD	http://www.yeastgenome.org/
CGD	http://www.candidagenome.org/
AspGD	http://www.aspgd.org/
CADRE	http://www.cadre-genomes.org.uk/
Aspergillus fumigatus database	http://old.genedb.org/genedb/asp/
CandidaDB	http://genodb.pasteur.fr/cgi-bin/WebObjects/CandidaDB
NCBI	http://www.ncbi.nlm.nih.gov/
Fungal Genomes Central	http://www.ncbi.nlm.nih.gov/projects/genome/guide/fungi/
Wellcome trust Sanger Institute	http://www.sanger.ac.uk
EMBL	http://www.embl.de/
DDBJ	www.ddbj.nig.ac.jp/
SWISSPROT	http://expasy.org/sprot/
Fungal Tree of Life	http://aftol.org/
CGOB	http://cgob.ucd.ie/

last 15 years has seen an exponential increase in the quantity of genetic data available in public databases such as NCBI (Table 3.2). To utilize this deluge of genetic data, it is imperative we have efficient similarity search techniques.

3.3.1.1 *Global and Local Alignments*

The methods used to infer homology can be categorized into two main types. A global alignment attempts to align two sequences over their entire length. Global sequence alignment works best when the sequences being compared are approximately the same length and highly similar. A local alignment, on the other hand, attempts to align two sequences at regions where high similarity is observed instead of trying to align the entire length of the two sequences being compared. Local alignments are usually more meaningful than their global counterparts, as they align conserved domains that may be important functionally even though the matched region is only a small proportion of the entire sequence length.

Needleman and Wunsch first implemented dynamic programming in 1970 to align sequences globally. Dynamic programming is a computational technique that determines the highest scoring alignment between two sequences. Smith and Waterman later adapted the Neeleman and Wunsch method to align sequences locally. Both methods utilize a scoring matrix where rows and columns correspond to the bases/amino acids being aligned. The first row and column of the matrix are filled with zeroes; the remaining cells are filled iteratively with values dependent on neighbouring cell values. If a matrix cell corresponds to an identical base/residue, a match score is added to the score from the neighbouring diagonal square. Alternatively, the maximum score is determined from cells above by adding a gap penalty. Gap penalties are generally negative numbers. Local alignments are produced by starting at the highest scoring cell in the matrix and following a trace path to a cell that scores a zero.

3.3.1.2 BLAST

The basic local alignment search tool (BLAST) is the most commonly used method for locating homologues in a sequence database. BLAST is both sensitive and efficient at locating regions of sequence similarity between nucleotide or protein sequences.

The BLAST algorithm begins by 'seeding' the search with a small subset of letters (query word) from the query sequence (Figure 3.4). The query word and related words (where conservative substitutions have been introduced) are located. All words are scored by a scoring matrix and this yields the 'neighbourhood' (Figure 3.4). BLAST uses a neighbourhood threshold T to determine which words are closely related to the original query word. Increasing the value of T implies that only closely related sequences are considered, while decreasing it allows for distantly related sequences to be considered.

The original query word is aligned to a word above the neighbourhood threshold (Figure 3.4). The BLAST algorithm then proceeds to extend the alignment in both directions, tracking the alignment score by addition of matches, mismatches and gaps. The maximal length of the alignment is determined by the number of positions aligned versus the cumulative score of the alignment. The alignment extension continues until the number of mismatches starts to decrease the cumulative score of the alignment; if this decrease is large enough (above a predefined value X, Figure 3.4), the alignment procedure ceases and the resultant alignment is called the high-scoring segment pair (HSP). A score threshold is defined by the algorithm, and if the HSP clears this score then the alignment is reported in the BLAST result file.

Finally, the biological significance of an HSP is determined. BLAST uses the E-value to calculate the number of HSPs that that would have a score greater than S by chance alone. Lower values of E imply greater biological significance; in essence, E can infer whether the HSP is a false positive.

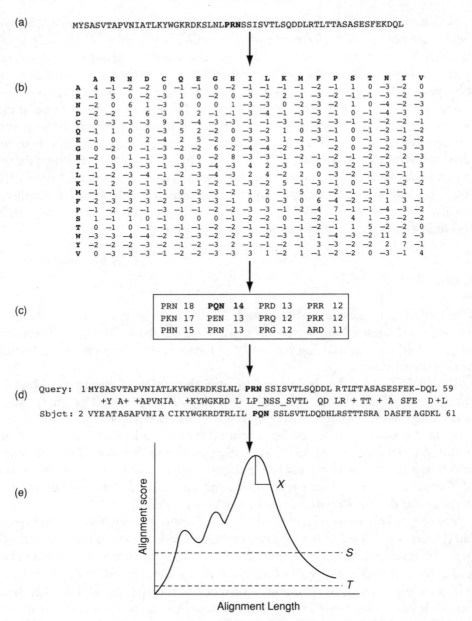

Figure 3.4 Steps taken by the BLAST algorithm when searching a database. The query sequence (a) is compared with a scoring matrix (b) and scores for query words of a given length (three in this case) are calculated (c), query words greater than a certain threshold *T* are used to search the database. (d) The algorithm attempts to extend the alignment either side of the query word that has a hit in the target database. (e) Extension continues until the alignment score falls off more than the allowable significance decay *X*.

3.3.1.3 FASTA

Like BLAST, FASTA is a program for rapid alignment of pairs of protein or DNA sequences and was the first widely used algorithm utilized for database similarity searching.

FASTA begins by locating subsequences above a particular word length from the database sequence to subsequences of the query sequence. In FASTA, the word length parameter is termed ktup and it is equivalent to W in BLAST searches. FASTA generates diagonal lines on a dotplot where residues match up (Figure 3.5). The FASTA algorithm next locates diagonal regions in the alignment matrix that contain as many ktup matches as possible with short distances separating them (Figure 3.5). The top 10 highest scoring diagonal regions are retained and correspond to high scoring local alignments that do not contain gaps.

FASTA then determines which of the adjacent diagonals can be joined together thereby increasing the overall length of the alignment. For each diagonal that are connected a joining penalty is invoked and the overall score is determined by addition of the net scores of individual diagonals minus the joining penalties. The score of the enlarged diagonals is referred to as init$n$. All enlarged diagonals are ranked based on score and the highest scoring ones are aligned optimally using a local alignment strategy. Finally, FASTA assesses the significance of the alignment by randomly generating sequences of similar length and composition as the query sequences and calculates the probability that an alignment would be seen my random chance.

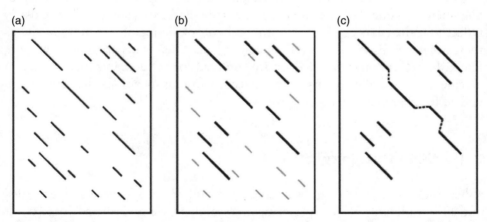

Figure 3.5 Steps taken by the FASTA algorithm when searching a database. (a) Words common to the query and target sequence are located. FASTA connects words close to one another and these are represented by diagonal lines. (b) The top 10 diagonals are selected for further analysis. (c) Diagonals are aligned optimally using a local alignment strategy.

3.3.2 Multiple Sequence Alignment

Multiple sequence alignment (MSA) is a method that allows us to infer the interrelationships between DNA or protein families. While pairwise alignments are useful for locating homologues in databases and illustrating conservation between two sequences, they are not as informative as MSA. MSA has the ability to locate conserved residues/domains amongst thousands of sequences that can provide insights into important evolutionary and physiochemical processes. MSA is the first step in phylogenetic analysis and is commonly used when designing primers for DNA amplification.

Multiple sequence alignment is much more computationally intensive and difficult than the pair-wise strategy employed by BLAST and FASTA. One of the most commonly used MSA algorithms is CLUSTAL, and it utilizes progressive alignment to efficiently align all sequences of interest. CLUSTAL follows three steps.

1 An initial assessment of how closely related different sequences are to one another by performing pair-wise alignments.

2 A guide tree is generated based on the pairwise alignment scores.

3 Sequences are aligned progressively guided by the phylogenetic tree. Closely related sequences are aligned first, and then additional sequences and groups are aligned.

CLUSTAL refines its progressive alignments by implementing a number of alignment penalties. For example, gap insertion and extension penalties exist to reflect that the chances of a gap within a hydrophilic region are more likely, as these are generally loops or random coil regions where gaps are more common. Similarly, residue-specific penalties are enforced so that domains that are rich in glycine are more likely to have an adjacent gap than positions that are rich in valine, for example.

3.3.3 Gene Ontology

When the first comparison between two complete eukaryotic genomes (yeast and nematode worm) was performed, researchers were surprised to discover a high proportion of genes displayed orthology between these two distantly related organisms (diverged ~1.6 billion years ago). Orthologues are genes that are derived from a common ancestor and commonly have the same function. Following from this, knowledge of the biological role of an orthologue in one species can be used to illuminate the putative function of the other orthologue. However, organizing biological data from multiple species databases is a major challenge and is made harder when different databases use different terminologies to describe the same process.

To overcome these difficulties, the Gene Ontology (GO) Consortium was set up in 2000 with the goal of producing a structured, precisely defined, common, controlled vocabulary for describing the roles of genes and gene products in any organism. Ontology terms provide a framework for storing and querying different databases using the same search terms. The GO Consortium provides detailed annotations for 12 important model organisms (*Arabidopsis thaliana*, *Caenorhabditis elegans*, *Danio rerio*, *Dictyostelium discoideum*, *Drosophila melanogaster*, *Escherichia coli*, *Gallus gallus*, *Mus musculus*, *Rattus norvegicus*, *Saccharomyces cerevisiae* and *Schizosaccharomyces pombe*). Collectively, those 12 species are referred to as the GO reference genomes. The GO consists of over 26 000 terms arranged in three branches.

1 **Cellular component:** an individual component of a cell, but part of some larger object, such as an anatomical structure (nuclear membrane, for example).

2 **Biological process:** describes broad biological goals, such as mitosis or purine metabolism.

3 **Molecular function:** describes the roles carried out by individual gene products, examples include transcription factors and DNA binding.

The annotation of newly sequenced fungi can be greatly accelerated by comparisons to the GO reference genomes. *De novo* genes can be assigned putative functions based on sequence similarity to existing genes in one of the model organisms. The fact that two of the model organisms are fungi (*S. cerevisiae* and *S. pombe*) makes the GO resource highly applicable to genome annotation in newly sequenced fungal genomes.

3.4 Comparative Genomics

3.4.1 Gene Families Associated with Disease

Comparative genomic analyses have shown that certain gene families are important for virulence in some fungal species. For example, a comparative analysis of 34 fungal genomes identified gene families that are specific to fungal plant pathogens (*Botryotinia cinerea*, *Ashbya gossypii*, *Magnaporthe grisea*, *Sclerotinia sclerotiorum*, *Stagonospora nodorum*, *Ustilago maydis* and *Fusarium graminearum*). These families have expanded in terms of number (through duplication) during the evolution of phytopathogens. The same study also predicted the set of secreted proteins encoded by each phytopathogen and located gene families that were significantly enriched in the secretome (proteins secreted

from a cell) of these species. Not surprisingly, many of the protein families identified are associated with pathogenic processes such as plant cell-wall degradation
and biosynthesis of toxins. Similarly, the complete genome sequence of the corn
smut fungus *U. maydis* uncovered a large set of secreted proteins, many of which
are arranged in clusters. These genes account for ~20 % of all proteins secreted
from *U. maydis*. Furthermore, the deletion of individual clusters seriously affects
virulence, implicating the importance of these extracellular proteins.

In *Candida* species there are a number of gene families that are particularly enriched in highly pathogenic species (*C. albicans*, *C. parapsilosis* and *C. tropicalis*)
compared with nonpathogenic species. For example, comparative analysis has
shown that the Hyr/Iff proteins are present in all *Candida* species, but they are
present in large numbers in the *Candida* pathogens (11, 17 and 18 copies respectively). Members of this family are components of the cell wall and, based
on sequence similarity, are known to be evolving rapidly. They most likely play
a role in host/fungal recognition, as rapid evolution of cell-wall proteins is a
common escape mechanism employed by microbial pathogens. Another family enriched in pathogenic *Candida* species is the agglutinin-like sequence (Als)
family. ALS genes are well characterized in *C. albicans* and are important for
adhesion to host cells, plastic surfaces and biofilm formation. The ability to bind
to plastic surfaces is a major problem in a hospital setting, as it allows *Candida*
species to enter the blood stream via medical devices, such as intravenous drips;
similarly reduced susceptibility to antifungal drugs is observed when *Candida*
species grow as biofilms.

3.4.2 Synteny

The term synteny was traditionally used by geneticists to indicate the presence of
two or more loci on the same chromosome. In the postgenomic era the concept of
synteny has been expanded to address the relative order of genes on chromosomes
that share a common evolutionary history. Two regions are considered syntenic
if multiple consecutive genes are found in a conserved order between the two
genomes under consideration (Figure 3.6). Synteny between two species may
break down due to genome rearrangements in one or both species since they last
shared a common ancestor.

Comparative fungal genomic analyses have shown that syntenic structure is
generally conserved between very closely related fungal species, but it reduces
as species become more distantly related. For example, the main subdivisions of
Saccharomycotina yeasts share minimal synteny conservation between one another. However, large syntenic blocks are observed when members of the same
subdivision are compared. For example, an analysis of nine *Candida* genomes
(using the *Candida* genome order database (CGOB, Table 3.2)) showed they
shared a high proportion of syntenic blocks. Conservation of gene order between

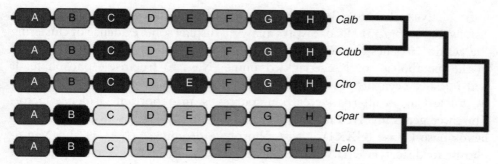

Figure 3.6 Synteny of eight orthologues in five *Candida* species. Genes A, D and F are syntenic in all species; this is represented by conservation of colour between them. Genes B, C, G and H are syntenic in *C. albicans*, *C. dubliniensis* and *C. tropicalis*. Orthologues of B, C, G and H are present in *C. parapsilosis* and *L. elongisporus* and are syntenic with one another, although they are located in different genomic locations relative to the first three species. The degree of synteny between the five species closely matches the phylogeny of these species.

closely related species correlates with phylogenetic analyses (Figure 3.6). Other studies have shown that homologous chromosomes between the *Saccharomyces* '*sensu stricto*' species are almost collinear, differing only by a small number of translocations and large inversions (segment of DNA is reversed). A comparison of two Basidiomycete genomes (*Coprinopsis cinerea* and *Laccaria bicolor*) showed they share extensive regions of synteny. The largest syntenic blocks occur in regions with low meiotic recombination rates and contain no transposable elements (cause translocations).

3.4.3 *In Silico* Metabolic Analysis

The availability of fungal genomes allows us to compare and contrast the metabolic repertoire of different species *in silico*. Detailed information from a metabolic pathway in one organism allows us to locate similarities or differences in another. Comparative metabolic analysis provides us with insights into potential disease mechanisms between pathogenic and nonpathogenic fungal species. Similarly, they also allow us to investigate the metabolic differences that allow one species to break down a particular substrate while another cannot.

Comparative studies of fungal species have shown that the genomic locations of certain genes are not random. For example, an analysis of the *S. cerevisiae* genome revealed that there is a significant tendency for genes from the same metabolic pathway to cluster in its genome. Similarly, genes involved in secondary metabolism are often clustered in the genomes of filamentous fungi (such as *Aspergilli* species).

An analysis of synteny between nine *Candida* genomes showed that approximately 20 % of metabolic pathways analysed display some evidence of clustering (lie within 10 genes of one another). One of the clustered pathways is involved in the metabolism of *N*-acetylglucosamine (Nag) to fructose-6-phosphate. It had initially been proposed that the ability of pathogenic strains of *Candida* to utilize Nag as alternative carbon sources is an important virulence factor. The three genes involved in the conversion of Nag to fructose-6-phosphate are hexokinase kinase (HXK1), Nag-6-phosphate deaminase (NAG1) and Nag-6-phosphate deacetylase (DAC1). These act sequentially on Nag and are present in *C. albicans* in a cluster termed the Nag regulon. Synteny analysis showed that the Nag regulon is conserved in nearly all *Candida* species. The conservation of the Nag regulon in pathogens like *C. albicans*, *C. tropicalis* and *C. parapsilosis* and nonpathogens such as *C. dubliniensis*, *Lodderomyces elongispors* and *Debaryomyces hansenii* suggests that the ability to utilize Nag is not a virulence factor. *S. cerevisiae* is missing the Nag regulon and cannot utilize Nag; however, it has been shown that expression of *C. albicans NAG* genes in *S. cerevisiae* enables it to utilize Nag.

3.4.4 Horizontal Gene Transfer

Horizontal gene transfer (HGT) is the exchange of genes between different strains or species. HGT introduces new genes into a recipient genome that are either homologous to existing genes or belong to entirely new sequence families. Bacterial genomic sequencing has revealed that HGT is prominent in bacterial evolution and has been linked to the acquisition of drug resistance and the ability to catabolize certain amino acids that are important virulence factors. There are numerous methods to detect genes that have been transferred horizontally into a genome, including locating genes with an atypical base or codon usage pattern. Another approach is to perform a similarity search of candidate genes against a database and locate unexpected top database matches. These approaches have the advantage of speed and automation, but do not have a high degree of accuracy. Some notable flaws with the similarity-based approach of detecting HGT were brought to attention when the initial publication of the human genome reported that there were 223 genes that have been transferred from bacterial pathogens to humans. These findings were based on top hits from a BLAST search, but subsequent phylogenetic analyses showed these genes were not recently transferred from bacterial species through HGT. Indeed, the most convincing method to detect HGT is by phylogenetic inference. Topological disagreement (incongruence) between trees inferred for one gene family and that inferred for another can often be parsimoniously explained only by invoking HGT.

The process of gene transfer has been assumed to be of limited significance to fungi. However, the availability of fungal genome data (Table 3.1) and subsequent comparative genomic analyses are showing the importance of HGT in the genome evolution of fungi. For example, *S. cerevisiae* has acquired 13 genes (from bacteria) via HGT since it diverged from its close relative *A. gossypii* (Figure 3.1). This number corresponds to a small minority of the *S. cerevisiae* genome (less than 1 %). However, these 13 genes have contributed to important functional innovations, including the ability to synthesize biotin, to grow under anaerobic conditions and to utilise sulfate from several organic sources. Other documented examples of HGT in fungi include the acquisition of bacterial metabolic genes by *C. parapsilosis* and the acquisition of a toxin gene (ToxA) by *Pyrenophora tritici-repentis* from *Stagonospora nodorum* resulting in *Pyrenophora* infestations of wheat.

Unlike prokaryotes, the mechanisms of gene transfer into fungi are poorly understood. To date, no DNA uptake mechanism has been identified. Interkingdom conjugation between bacteria and yeast has been observed, however, and *S. cerevisiae* is transformant competent under certain conditions. However, HGT is probably facilitated by the fact that fungi are saprobes that live in close proximity with other organisms.

3.5 Genomics and the Fungal Tree of Life

3.5.1 Phylogenetics

The goal of phylogenetics is to arrange a set of populations, species, individuals or genes into a logical arrangement that infers the evolutionary relationships amongst them. Evolutionary relationships infer the historical development of species and are usually presented as an evolutionary tree (Figure 3.1). Traditional methods of fungal systematics, such as vegetative cell morphology, sexual states, physiological responses to fermentation and growth tests, can assign fungal species to particular genera and families. The fungal fossil record is poor, however, and fungi exhibit few morphological characters; therefore, an alternative approach is desirable. Fungal sequence data (RNA, DNA and protein) have been used successfully to infer evolutionary relationships amongst species. In many cases, aligned sequences (Section 3.3.2) are processed as a distance matrix. Species that are most closely related will have a small distance, while distantly related species will have a larger distance measure. Phylogenetic algorithms such as UPGMA, minimum evolution and neighbour joining are used to represent distance matrices as phylogenetic trees.

The choice of phylogenetic markers for inferring the fungal tree of life is a contentious issue. Ideally, a phylogenetic marker should be ubiquitous throughout the species under consideration, present in single copy, have slowly evolving sites and be unlikely to undergo HGT. For this reason a significant majority of accepted relationships between fungal organisms are determined using 18S ribosomal DNA. However, single-gene analyses are dependent on the phylogenetic markers having an evolutionary history that reflects that of the entire organism, an assumption which is frequently violated. Also, individual genes contain a limited number of sites and, in turn, limited resolution. An alternative approach to single-gene phylogenies are multigene phylogenies. These attempt to combine all available phylogenetic markers. There are two commonly used methods to do this: concatenated multigene phylogeny reconstruction and supertree analysis.

3.5.1.1 Concatenated Multigene Phylogenies

Multigene concatenation essentially appends many aligned genes together to give a large super alignment. Combining the data increases its informativeness, helps resolve nodes, basal branching and improve phylogenetic accuracy. Numerous species phylogenies have been derived by concatenation of universally distributed genes. Recently, the Fungal Tree of Life consortium (Table 3.2) used six housekeeping genes (18S rRNA, 28S rRNA, 5.8S rRNA, elongation factor 1-alpha and two RNA polymerase II subunits (RPB1 and RPB2)) from 199 fungal species to reconstruct the evolutionary history of the fungal kingdom. As well as showing the evolutionary history of all fungal phyla, this analysis showed that the loss of spore flagella from early diverging fungi (similar to extant chytrids) coincided with the development of novel spore dispersal mechanisms leading to the diversification of terrestrial fungi.

3.5.1.2 Supertrees

Supertree methods take all input trees and generate a single representative species phylogeny (Figure 3.7). Individual input trees are derived from single genes. Comparative fungal genomic analyses have shown that less than 1 % of all fungal genes are universally distributed. This situation implies that when we reconstruct multigene phylogenies we are ignoring 99 % of the genes found in fungi. Ideally we would use 100 % of the gene data. Supertree methods enable us to do this.

Supertree methods generate a phylogeny from a set of input trees that possess fully or partially overlapping sets of taxa (Figure 3.7). Therefore, supertree methods take as input a set of phylogenetic trees and return a phylogenetic tree that represents the input trees. This type of analysis yields a phylogeny that maximizes the number of genes used and, therefore, is truly representative of the entire genome. A supertree analysis of 42 complete fungal genomes identified 4805 individual gene families (Figure 3.1). Individual phylogenies for each gene family were reconstructed and the complete set was summarized by supertree techniques. This analysis showed that within the Saccharomycotina a monophyletic (single) clade containing C. albicans and close relatives is evident (Figure 3.1). Species within this clade translate the codon CTG as serine rather than leucine. A second monophyletic clade (Figure 3.1) containing genomes that have undergone a whole-genome duplication (S. cerevisiae and close relatives) is

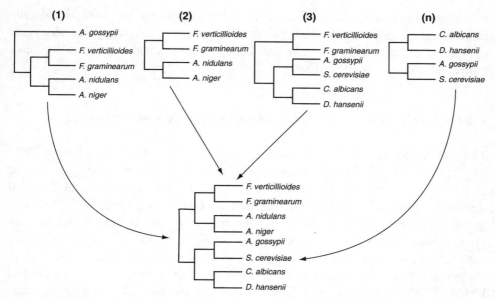

Figure 3.7 Representation of supertree reconstruction. Supertree methods take individual gene trees and express them as a single representative phylogeny. Thousands of trees (expressed as *n*) can be used as input for supertree techniques.

also evident. Supertree techniques are becoming more popular in phylogenetic analysis and will be useful in reconstructing the fungal tree of life as additional fungal genomes become available.

3.6 Online Fungal Genomic Resources

3.6.1 NCBI: Fungal Genomes Central

The National Centre for Biotechnology Information (NCBI; Table 3.2) is a resource that houses public sequence databases (such as GenBank), supporting bibliographic/biological annotation, bioinformatics tools and associated applications.

The NCBI has a dedicated fungal genome section (Fungal Genomes Central (FGC)). The FGC has direct links to the 309 fungal genome sequencing projects currently underway or completed, as well as links to raw sequence reads generated by fungal sequencing centres such as the BROAD and Wellcome Trust Sanger Institute (Table 3.2). The complete sequences of 80 fungal mitochondria are also available for download. A fungal BLAST search utility is present enabling users to search genes of interest against all available fungal genomic data housed by the NCBI. A system for automated detection of homologues amongst completely sequenced fungal genomes is available; more than 9000 fungal gene families are currently catalogued. Finally, the FGC has a taxonomy search engine containing the names and phylogenetic lineages of all fungal organisms that have molecular data in NCBI databases.

3.6.1.1 Saccharomyces, *Candida* and *Aspergillus* Genome Databases

The *Saccharomyces* Genome Database (SGD) (Table 3.2) went online in 1997 and is a specialized database dedicated specifically to *S. cerevisiae*. It is housed at the Stanford Human Genome Center and currently receives over 200 000 database hits a week. SGD provides users with access to the complete *S. cerevisiae* genome, its genes and their products, mutant phenotypes and the literature supporting these data. It should be noted that SGD is not a primary sequence database (contains information of the sequence alone) but instead collects DNA and protein sequence information from primary providers (such as GenBank, EMBL, DDBJ and SwissProt (Table 3.2)) and assembles all available information into datasets that are useful for molecular biologists. Therefore, SGD is considered a composite database, as it amalgamates a variety of different primary database sources and cuts out the need to search multiple resources.

SGD is highly annotated, and supporting literature linked to each gene is curated by dedicated SGD curators. Weekly automated searches of PubMed locate literature associated with *S. cerevisiae* genes or products and these are refined by curators who assign a given publication with appropriate genes. SGD provides an excellent text-based search interface that allows users to search by gene name, gene information, protein information, author name or full text. Amongst other things, SGD also allows users to perform BLAST database searches, view yeast metabolic pathways, search yeast-specific literature, view gene expression data from multiple microarray studies and view genes relative positions on chromosomes.

SGD organizes gene information around locus pages. The gene name and associated systematic name are shown at the top of each locus page. Information about the feature of the gene is also given; genes can be 'verified', meaning there is experimental evidence to show they are expressed or 'uncharacterized' implying a lack of experimental evidence. The 'description' section details important information known about the gene and associated products. Each gene product is assigned gene ontology terms that describe its molecular functions, location within the cell and putative biological processes in which it participates. A mutant phenotype section is also visible on the locus page. This section lists the type of mutation and any corresponding observable phenotype. Links to sequence information and literature describing the gene of interest are also available from the locus page.

The *Candida* Genome Database (CGD) went online in 2005 and is the central resource for researchers studying *Candida* pathogenesis and genetics. Before the launch of CGD, three independent web sites contained information about the

Candida genome sequence and associated gene products. The Stanford Genome Technology Center (Table 3.2) sequenced and distributed the genome; CandidaDB (Table 3.2) contained annotated genes for early assemblies of *C. albicans*, as did the *Candida* Working Annotation group (Table 3.2). The information available in these three web sites was initially pooled together and has subsequently been expanded on. CGD is based on the SGD framework; therefore, the software, user interfaces and data structure in CGD are identical to those described for the SGD above.

The *Aspergillus* Genome Database (AspGD, Table 3.2) went online in 2009 and is an online genomic resource for scientists studying the genetics and molecular biology of *Aspergilli* species. Currently, there are a number of databases containing information for multiple *Aspergillus* genomes. For example the Central *Aspergillus* Data Repository (CADRE) database (Table 3.2) contains clinical and patient-oriented information, the *Aspergillus* genome site at the Broad Institute (Table 3.2), the *Aspergillus fumigatus* database (Table 3.2) and also other web sites that focus on sequencing projects of one or several *Aspergillus* species. AspGD aims to link the resources of these individual databases and complement them by implementing in-depth manual curation of the primary scientific literature associated with the data. As with the CGD, AspGD is based on the SGD framework described above. AspGD is currently focusing on high-quality cuaration of *A. nidulans*, the best-characterized species of the *Aspergilli*, but will add information for other *Aspergilli* species (*A. fumigatus*, *A. flavus*, *A. oryzae*, *A. niger*, *A. clavatus*, *A. terreus* and *Neosartotya fischeri*) in the near future.

3.6.2 The Fungal Genome Initiative

The Fungal Genome Initiative (FGI, Table 3.2) is a partnership between the Broad Institute of Harvard and MIT and the broad fungal research community. It is directed by a steering committee of fungal geneticists and biologists who back in 2000 realized the slow pace of fungal genome sequencing projects was a major barrier to fungal biomedical and evolutionary research. A strategy where multiple organisms would be sequenced simultaneously as part of a cohesive strategy instead of individual ad hoc projects was conceived. The primary selection criteria for sequencing were (a) the importance of the organism in human health and commercial activities, (b) the value of the organism as a tool for studies of fungal diversity and comparative genomics and (c) presence of genetic resources and an established research community.

Initially, 15 fungi were selected for sequencing covering three broad aspects of fungal diversity:

1 Medical; e.g. *Rhizopus oryzae*, cause of mucormycosis.

2 Commercial; e.g. *Aspergillus flavus*, source of aflatoxin in food.

3 Evolutionary; e.g. *Neurospora discreta*, study of population genetics.

A subsequent 48 genomes have been targeted. In order to utilize the strengths of a comparative approach, strategic clusters were chosen (*Candida, Cryptococcus, Aspergillus, Fusarium, Histoplasma, Coccidioides, Penicillium, Neurospora, Schizosaccharomyces, Puccinia* and *Schizosaccharomycetes*).

To date, over 50 fungal genomes have been sequenced by the FGI and all sequence data is can be accessed publicly via the FGI homepage (Table 3.2). Information pertaining to genome maps and basic statistics about genome size, gene density and GC content are available. As well as providing standard database search tools such as BLAST, the FGI also incorporates text-based searches that allow researchers to locate genes based on a particular function or protein domain. Users can also select closely related organisms (from the same cluster) and view synteny maps.

3.7 Conclusion

The majority of fungi that have been sequenced to date are important biological pathogens (*C. albicans*, *A. fumigatus* and *Cryptococcus neoformans*, for example) or helpful species involved in brewing/fermentation (*S. cerevisiae* and *Aspergillus niger*, for example). Because of this, there is an unintentional bias in terms of the phylogenetic distribution of species sequenced. Owing to falling sequencing costs and their relatively small genome size, a deluge of fungal genomic data from all fungal phyla is expected in the years ahead. This data will allow us to address many new questions about fungal evolution and pathogenicity and will undoubtedly help uncover novel proteins with medical and biotechnological potential.

Revision Questions

Q 3.1   List the four major phyla of the fungal kingdom.

Q 3.2   Outline the steps taken when sequencing DNA using the Sanger method.

Q 3.3   What is next-generation sequencing?

Q 3.4   What information is contained in online databases such as the *Saccharomyces* genome database (SGD), *Candida* genome database (CGD) and the *Aspergillus* database (AspGD)?

Q 3.5   List the main difference between a global and local alignment.

Q 3.6   What is a phylogenetic supertree?

Q 3.7   What is the gene ontology (GO)? List and explain the three main branches of GO.

Q 3.8   Explain how comparative genomics has the potential to uncover the genetic basis of disease.

Q 3.9   What is gene synteny?

Q 3.10   What is HGT?

Further Reading

Journal Articles

Altschul, S.F., Gish, W., Miller, W. *et al.* (1990) Basic local alignment search tool. *Journal of Molecular Biology*, **215**, 403–410.

Fitzpatrick, D.A., Logue, M.E., Stajich, J.E. and Butler, G. (2006) A fungal phylogeny based on 42 complete genomes derived from supertree and combined gene analysis. *BMC Evolutionary Biology*, **6**(1), 99.

Fitzpatrick, D.A., O'Gaora, P., Byrne, K.P. and Butler, G. (2010) Analysis of gene evolution and metabolic pathways using the *Candida* Gene Order Browser. *BMC Genomics*, **10–11**(1), 290.

Hall, C. and Dietrich, F.S. (2007) The reacquisition of biotin prototrophy in *Saccharomyces cerevisiae* involved horizontal gene transfer, gene duplication and gene clustering. *Genetics*, **177**(4), 2293–2307.

The Gene Ontology Consortium (2000) Gene Ontology: tool for the unification of biology. *Nature Genetics*, **25**(1), 25–29.

Thompson, J.D., Higgins, D.G. and Gibson, T.J. (1994) CLUSTAL W: improving the sensitivity of progressive multiple sequence alignment through sequence weighting, position-specific gap penalties and weight matrix choice. *Nucleic Acids Research*, **22**, 4673–4680.

Books

Janitz, M. (2008) *Next-Generation Genome Sequencing*, Wiley-Blackwell.

Baxevanis, A.D. and Ouellette, B.F. (2005) *Bioinformatics: A Practical Guide to the Analysis of Genes and Proteins*, 3rd edn, Wiley-Interscience.

4
Fungal Genetics:
A Post-Genomic Perspective

Brendan Curran and Virginia Bugeja

4.1 Introduction

Ushered in by the exponential accumulation of DNA sequences in databases throughout the world, the post-genome era is characterized by the application of computer technology to a deluge of information arising from the large-scale parallel analysis of biological molecules. With information accumulating from an ever-increasing network of resources, the challenge now is to reintegrate these molecular details to reveal the secrets of the dynamic processes they mediate within the cell, an experimental and theoretical approach referred to as systems biology.

4.1.1 The Yeast *S. cerevisiae*: A Cornerstone of Post-Genomic Research

A forerunner of the much more ambitious project to sequence the human genome, the yeast *Saccharomyces cerevisiae* entered the history books in 1996 as the first eukaryotic organism to have its entire genome sequence deposited in a computer database. With other eukaryotic genomes entering the databases, and the development of computational tools for capturing, storing, displaying, distributing and comparatively analysing the rapidly accumulating information, biologists were able for the first time to analyse and compare entire eukaryotic genomes – the post-genomic era had begun. However, the DNA sequence of *S. cerevisiae* was just the starting point for large-scale molecular analysis of eukaryotic cells.

Fungi: Biology and Applications, Second Edition. Edited by Kevin Kavanagh.
© 2011 John Wiley & Sons, Ltd. Published 2011 by John Wiley & Sons, Ltd.

Within a very few years this extremely tractable model organism rapidly yielded a whole series of molecular secrets on a global scale: each of its genes was systematically deleted in search of phenotypes; technology to allow its global mRNA profiles to be identified was developed; all possible protein–protein interactions were examined; and cellular metabolites exhaustively characterized. In short this simple eukaryote became the key to post-genomic research.

4.1.2 Of -*Omics* and Systems Biology

Whereas pre-genomic research was characterized by hypothesis-driven sequential experiments, post-genomic research is driven by the massively parallel analysis of biological information, followed by pattern recognition within datasets. *Genomics*, the accumulating and analysis of massive amounts of DNA sequence data, was possible because an appropriate technology platform was put in place. The components of this 'platform' included automated preparation of DNA, automated sequencing of multiple DNA fragments and the computational tools to store and then process the data generated. As the science of examining genomes is referred to as genomics, so to the complementary global analysis of other biological molecules is given the -omic suffix: *transcriptomics* concerns the accumulation of information on RNA sequences and their expressed levels; *proteomics* deals with protein sequences, protein structures, protein levels and protein interactions, both with DNA and other proteins. Each one of these has required the development of automated and computationally intensive technology platforms to complement the high-throughput DNA sequence and analysis platform, which characterizes genomics. These include *DNA array technology*, which allows multiple DNA or RNA sequences to be simultaneously identified, and *mass spectrometry* for the identification of multiple protein samples. Once again the yeast *S. cerevisiae* has led the way in the development of these platform technologies. Much more than this, however, *S. cerevisiae* has now become a central player in the development of an entirely new approach to biological research – *systems biology*. This newly emerging field uses a cross-disciplinary approach involving biology, chemistry, physics, mathematics, computer science and engineering to develop working models of how these molecules interact to generate biological phenomena.

4.2 Genomics

4.2.1 Analysing Encoded Information

Genomics, the study of whole genomes, encompasses: (1) searches for patterns of relatedness within and between genome sequences and (2) attempts to ascribe specific biological functions to particular DNA sequences within those

genomes. The successful completion of fungal genomes, in particular the annotated genomes of the three model organisms *S. cerevisiae*, *Schizosaccharomyces
pombe* and *Neurospora crassa*, excited much interest within and beyond the
fungal community.

Three hundred and fifty-two fungal genome sequencing projects can be accessed at http://www.ncbi.nlm.nih.gov/genome of which 19 have already been
completed (Table 4.1), 149 are in the assembly process, and a further 184 are
'in progress'. Fungal sequencing projects to date include model organisms, fungi
with relevance to healthcare and fungi of agricultural and commercial importance. Each genome project is powerful in its own right; however, comparative
analysis of fungal genomes is set to revolutionize our understanding of this
ancient and evolutionarily diverse group of organisms. Moreover, given the underlying unity of biological information, many of these findings will also be
relevant to our understanding of plant and animal biology.

4.2.2 Pattern Recognition within and between Genome Sequences

Without ever revealing anything specific about biological function, computer
analysis of a complete genome sequence provides lots of interesting and revealing facts about its topology and evolutionary history. For instance, computer
programs use pattern recognition to identify tracts of DNA that start with a
methionine codon and run for another 99 codons without hitting a nonsense
one (Figure 4.1). Such sequences, referred to as open reading frames (ORFs), are
then annotated as highly likely to encode proteins. More advanced programs even
allow for the presence of introns by being able to recognize intron–exon bound-
aries. Sequence recognition programs provide a detailed map of the genome and
lots of useful information, including: the number of genes, the size of the gene
sequences, the presence or absence of introns, the spacing of genes within the
chromosomes and lots more besides.

Such programs identified the precise position of approximately 6000 genes
in the *S. cerevisiae* genome – the best labours of yeast geneticists during the
preceding 40 years had revealed the position of less than 1000 genes! Sequence
analysis also revealed that the yeast genome is extremely compact, with 70 %
consisting of protein-encoding sequences – protein encoding sequences that are
almost completely devoid of introns. In addition, the number, and chromosomal
locations, of genes encoding non-coding RNA species, such as ribosomal RNA,
small nuclear RNAs and transfer RNA (tRNA) were also identified. A similar
pattern-recognition analysis revealed that the *S. pombe* genome carried 600 fewer
protein-encoding genes (approximately 4900 in all), whereas their multicellular
filamentous cousin *N. crassa* requires twice that number: 10 000 genes. Unlike
S. cerevisiae, the protein-encoding genes in both *S. pombe* and *N. crassa* carry
introns. The percentage of the genome consisting of protein-encoding sequences

Table 4.1 Details of the 19 fungal genomes completed to date. Almost all of the completed fungal genomes are Ascomycetes, two are basidiomycetes[a] and two are neither of these[b].

Organism	Genome size (Mb)	Release date	Centre/consortium and useful URL
Cryptococcus gattii WM276[a]	18.3	01/12/2011	Canada's Michael Smith Genome Sciences Centre http://www.bcgsc.ca/project/cryptococcus/
Encephalitozoon intestinalis ATCC 50506[b]	2.22	08/16/2010	University of British Columbia, Canada http://www.botany.ubc.ca/keeling/
Saccharomyces cerevisiae S288c	12.08	02/03/2010	Saccharomyces Genome Database http://www.yeastgenome.org/
Aspergillus nidulans FGSC A4	29.8	09/24/2009	Eurofungbase (Eurofung) http://eurofung.net/
Lachancea thermotolerans CBS 6340	10.38	06/05/2009	Genolevures http://www.genolevures.org/#consortium
Zygosaccharomyces rouxii CBS 732	9.76	06/05/2009	Genolevures http://www.genolevures.org/#consortium
Pichia pastoris GS115	9.2	05/25/2009	Unit for Molecular Glycobiology, VIB, UGent, Belgium http://bioinformatics.psb.ugent.be/genomes/view/Pichia-pastoris
Candida dubliniensis CD36	14.61	02/16/2009	Wellcome Trust Sanger Institute http://www.sanger.ac.uk/
Magnaporthe oryzae 70-15	3.99	01/30/2006	North Carolina State University (NCSU) http://www.cifr.ncsu.edu/

Organism			
Aspergillus oryzae RIB40	37.08	12/20/2005	NITE http://www.bio.nite.go.jp/dogan/project/view/AO
Cryptococcus neoformans var. *neoformans* JEC21[a]	19.05	01/07/2005	TIGR http://www.jcvi.org/cms/research/projects/tdb/overview/
Kluyveromyces lactis NRRL Y-1140	10.68	07/02/2004	Genolevures http://www.genolevures.org/#consortium
Yarrowia lipolytica CLIB122	20.50	07/02/2004	Genolevures http://www.genolevures.org/#consortium
Debaryomyces hansenii CBS767	12.22	07/02/2004	Genolevures http://www.genolevures.org/#consortium
Candida glabrata CBS 138 CBS138	12.28	07/02/2004	Genolevures http://www.genolevures.org/#consortium
Ashbya gossypii ATCC 10895	8.76	03/06/2004	Zoologisches Institut der Univ. Basel, Switzerland http://evolution.unibas.ch/
Schizosaccharomyces pombe 972h-	12.57	02/21/2002	*S. pombe* European Sequencing Consortium (EUPOM) http://www.sanger.ac.uk/Projects/S_pombe/EUseqgrp.shtml
Encephalitozoon cuniculi GB-M1[b]	2.49	11/24/2001	Genolevures http://www.genoscope.cns.fr/spip/Encephalitozoon-cuniculi-whole.html
Saccharomyces cerevisiae S288c	12.08	10/25/1996	Wellcome Trust Sanger Institute http://www.sanger.ac.uk/Projects/S_cerevisiae/

```
                   GTC
                   TGT
                   ATG
         5 -ATGTCGAATTCGCCTATAG - 3
         3 -TACAGCTTAAGCGGATATC - 5
                          ATC
                          TAT
                          ATA
```

Figure 4.1 Finding an ORF. Finding the ATG methionine codon requires a computer search of six different reading frames. An ORF continues for a further 99 codons before hitting a nonsense codon.

is also lower in both than in *S. cerevisiae*. Finally, the gene density of the latter two is also significantly lower than that found in their cousin (Table 4.2).

Having identified the majority of the ORFs in a genome, a computer-aided comparison that aligns each ORF against all other ORFs within the same genome can reveal much about the evolutionary history of a genome. One such study revealed that an ancient genome duplication had occurred during the evolutionary history of *S. cerevisiae*. ORF comparisons within the genome sequence identified 55 segments of chromosomes carrying three or more pairs of homologous genes – despite this being a haploid genome This supported a model in which two ancestral diploid yeast cells, each containing about 5000 genes, fused to form a tetraploid. Most of the duplicate copies were then subsequently lost by deletion as this species evolved, leaving it in its haploid phase with approximately 5500 genes. It is estimated that protein pairs derived from this duplication event make up 13 % of all yeast proteins. The same study also revealed that transposable elements play an important role within the genome: there are 59 such elements, constituting 2.4 % of the entire genome. On the other hand, *intragenomic* searches in the *S. pombe* genome sequence failed to find evidence of large-scale *genome duplications*. However, they did find evidence suggesting that *gene duplication* played a key role in the evolution of this yeast. In fact, as many as 10 % of the ORFs can be defined as having paralogous sequences within this genome (paralogous genes being homologous genes that arose by gene duplication within the same species). ORF comparisons also reveal that transposable elements are

Table 4.2 Genome topology of model fungi.

Fungal species	Organismal complexity	Genome size (Mb)	Protein encoding sequences	Gene density (excluding introns)
S. cerevisiae	Budding yeast	12.07	5 500	1 every 2.09 kB
S. pombe	Fission yeast	13.8	4 900	1 every 2.53 kB
N. crassa	Filamentous fungus	41	12 000	1 every 3.1 kB

Table 4.3 Evolutionary history of model fungal genomes.

Fungal species	Evidence of former genome duplication	Evidence of gene duplication	Evidence of transposon activity
S. cerevisiae	yes	yes	abundant
S. pombe	no	yes	abundant
N. crassa	no	no	very little

important to this yeast: there are 11 intact transposable elements, which account for 0.35 % of the genome (Table 4.3).

Within-genome comparison of ORFs in *N. crassa*, the multicellular distant cousin to both of these yeasts, fails to find evidence for large-scale genome duplication, gene duplication or transposon activity. The level of redundancy amongst ORFs of *N. crassa* is very low: less than 4 % can be defined as having a paralogous ORF within the genome. This analysis suggests that the genome did not undergo duplication, that gene duplication is rare and that transposon activity is extremely limited (Table 4.3). The paucity of transposable elements and paralogues in the *Neurospora* genome can perhaps be explained by the process of RIP (repeat-induced point mutation), which involves the hypermutation of duplicated sequences of more than 1 kb in length during sexual development in this fungus. However, these results pose the interesting biological question: Can *Neurospora* currently utilize gene duplication as a means of gene diversification?

4.2.3 Assigning Biological Functions to Fungal Genome Sequences

Although analyses of individual genome sequence databases can be interesting and informative in their own right, biologists are more interested in using the sequence to understand the biology of the cell. *Functional genomics*, the name given to this process of *assigning functions to ORF sequences*, is multifaceted. Potential functions can be attributed to some ORFs by identifying similar sequences of known function in existing databases (*in silico analysis*). Alternatively, ORFs of unknown function can be disrupted in order to produce a phenotype (*reverse genetics*). Failing this, clues to the potential biological significance of unknown ORFs can be gleaned by large-scale comparisons of genomes from a variety of different organisms. This so-called *comparative genomics* can identify potential gene functions by virtue of the fact that certain sequences are present in some genomes but absent from others (*identification by association*).

4.2.4 Identification by *In Silico* Analysis

Even before genome-sequencing projects were initiated, research groups around the world had already set up databases containing large numbers of annotated DNA, RNA and protein sequences. As these grew exponentially, computational techniques were developed to make these resources readily available and searchable. These databases, which frequently annotate DNA sequences with encoded protein functions, provided invaluable information on the potential or actual function of proteins encoded by newly identified ORFs. Basic local alignment sequence tool (BLAST) searches of these resources were used to annotate about 50 % of the ORFs in the *S. cerevisiae* genome. This allowed biologists to discover the percentage of genes that this simple eukaryotic cell dedicated to various aspects of cellular biology: there are in excess of 600 metabolic proteins, in excess of 400 proteins involved in intracellular trafficking or protein targeting and approximately 200 transcription factors. Specific genes whose existence in *S. cerevisiae* had hitherto been in doubt were also identified in this way: histone H1 was found on chromosome XVI (Figure 4.2) and a yeast gamma tubulin gene, which had previously eluded yeast geneticists despite intensive efforts, was identified on chromosome XII.

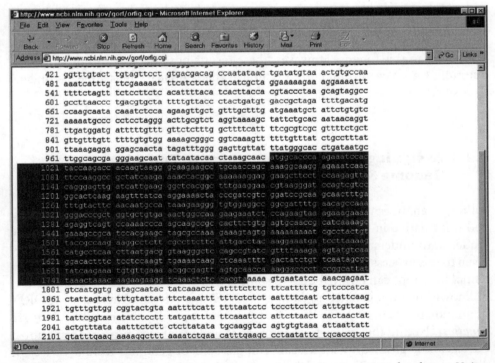

Figure 4.2 The sequence of the H1 gene in the *S. cerevisiae* database. Using an appropriate search, the bases encoding the H1 gene (blocked in above) were found as part of the sequence of chromosome 16.

4.2.5 Identification by Reverse Genetics

Although *in silico* analysis is useful, it can only annotate ORFs from new genome sequences using functions that had previously been identified elsewhere. Indeed, because it frequently identifies protein domains (rather than the entire protein sequence) the annotation is often no more than a general assignment (e.g. ORF 'X' encodes a 'protein kinase' or ORF 'Y' encodes a 'transcription factor'). A specific and unequivocal assignment is possible, however, if a phenotype can be identified when a particular ORF is deleted or disrupted. It was for this reason that an international consortium of scientist undertook the daunting task of using *S. cerevisiae*'s ability to undergo homologous recombination to produce yeast strains in which specific ORFs had been deleted. It was hoped that such specific deletions would provide insight as to the biological function of the affected ORF. The EUROFAN project deleted and grossly characterized the mutant phenotypes from 758 ORFs of unknown function. A number of other European laboratories complemented this by undertaking a more focused range of deletion mutants. However, a transatlantic consortium undertook the most ambitious project: they used PCR to generate hybrid DNA molecules consisting of a selectable marker flanked by the 5' and 3' ends of the ORF of interest. They then transformed this into wild-type cells and selected for recombinants that had integrated the PCR fragment into the target ORF (Figure 4.3). By repeating this with each ORF in turn, they systematically delete all 6000 or so genes. A number of mutant yeast libraries have also been generated using transposons to randomly generate mutations by insertional mutagenesis. Many of these strains have subsequently been characterized for phenotypes. Of the approximately 3000 ORFs of unknown function after the yeast sequence was released in 1996,

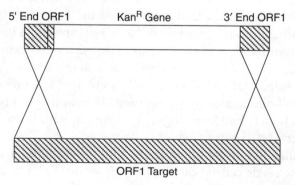

Figure 4.3 ORF knockout in the yeast *S. cerevisiae*. Appropriate PCR primers are used to generate hybrid DNA molecules consisting of the Kan^R-selectable marker flanked by blocks of DNA consisting of the 5' and 3' ends of the ORF of interest. Homologous recombination in transformants knocks out the ORF of interest.

such approaches have to date identified biochemical or physiological functions for more than 1000 of them.

Gene knockout technologies using a variety of molecular approaches are also valuable in the study of other fungi. With appropriate selectable markers and transformation procedures available for many pathogenic fungi, including *Candida albicans*, *Candida dubliniensis*, *Cryptococcus neoformans*, *Aspergillus fumigatus*, *Histoplasma capsulatum* and *Coccidioides immitis*, the application of gene knockout technology to investigate the basis of their pathogenicity is awaited with great interest. Such practical application of the basic approach taken in the investigation of *S. cerevisiae* will empower the development of antimycotic drugs in post-genomic research in these organisms. High-frequency homologous recombination is possible in a number of these, and complete or partial genome sequences for a number of pathogens are already available. However, the relationship between gene knockouts and pathogenicity is complicated by a number of factors, including the fact that high-frequency homologous recombination is not possible in all cases: different isolates, and the same isolate under different nutrient conditions, vary in their virulence; and, in the case of *Candida* (a diploid), knockouts require two independent deletion events. Nevertheless, the application of reverse genetics to some of these organisms is already underway.

However, as it is not feasible to analyse all fungi in this way, alternative strategies are being devised. One such strategy uses a process of 'guilt by association' to link ORFs of unknown function (so-called orphans) with biological function.

4.2.6 Identification by Association

When the fission yeast *S. pombe* entered the history books as the second completed fungal genome it became the sixth eukaryotic genome to enter the fully annotated genome databases. It also presented the first opportunity to undertake a comparative genomic analysis of fungal organisms, namely *S. cerevisiae* and *S. pombe* – both with one another and also with other genomes. Using the genome sequence of *Caenorhabditis elegans* as a simple *multicellular* eukaryote, this analysis revealed that 681 ORFs (14 %) were uniquely present in *S. pombe*, 769 (16 %) were homologous to *S. cerevisiae* ORFs and about two-thirds of the *S. pombe* ORFs (3281) had homologues in common with both *S. cerevisiae* and *C. elegans* (Figure 4.4).

With the availability of the *Neurospora* genome sequence it became possible to compare these single celled yeasts with their multicellular cousins. Given that *Neurospora* produces at least 28 morphologically distinct cell types, and has a preponderance of 11 000 overwhelmingly non-paralogous genes, it is perhaps not surprising that a comparative genomic analysis reveals that a large proportion of *Neurospora* genes do not have homologues in the yeasts *S. cerevisiae* and *S. pombe*. Analysis of other genome databases reveals that, in relation to yeasts,

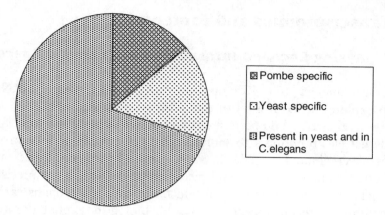

Figure 4.4 Using comparative genomics to assign biological significance to ORFs. BLAST searches of all the *S. pombe* ORFs revealed that 681 ORFs (14%) were uniquely present in *S. pombe*, 769 (16%) were homologous to *S. cerevisiae* ORFs and approximately 3300 *S. pombe* ORFs (70%) had homologues in common with both *S. cerevisiae* and *C. elegans*.

this particular filamentous fungus possesses an expanded group of sugar transporters, transcription factors, environmental sensing pathways and a diversified metabolic machinery: *Neurospora* possess 38 proteins with a cytochrome P450 domain; *S. cerevisiae* and *S. pombe* possess fewer than five such proteins.

As more fungal genomes become available it will become possible to identify subgroups of genes that are associated with specific types of fungal biology. Already, relatively crude comparisons between very distantly related fungi reveal much about the subset of fungal-specific genes, and genes that are yeast or multicellular specific. Evermore accurate predictions become possible as more genomes and, indeed, more closely related genomes become available. But the power of comparative fungal genomics does not stop with revealing the biology of this important and ubiquitous group of organisms. Over 200 predicted *Neurospora* proteins exhibit significant similarity to human gene products associated with disease states in humans. Many of these have counterparts in *S. cerevisiae* and *S. pombe*, including 23 that are cancer-related proteins. Such findings lend support to the notion that fungi could be good model organisms for studying the associated human disease pathways – it has even been used to model neurodegeneration.

Annotated genomes reveal much about the topology, evolutionary history and biological potential of cells, but not much about how this develops into an integrated biological entity. This requires a global view of how two-dimensional (2-D) DNA information manifests in four-dimensional (4-D) time and space. Experimental approaches to characterize global changes in cellular molecules, as they obey the central dogma 'DNA makes RNA makes protein' have, therefore, been devised.

4.3 Transcriptomics and Proteomics

4.3.1 Analysing Encoded Information in Time and Space

In the pre-genome era, biologists used northern blot analysis of mRNA transcripts to explore the expression of the information stored in the DNA in the cell. Likewise, they relied on pattern changes in one-dimensional and 2-D polyacylamide gel electrophoresis to monitor changes in cellular protein profiles. Researchers were limited to analysing, at best, time-lapsed gene expression from a very limited number of genes, and piecemeal revelations as to precise three-dimensional (3-D) cellular context in which the encoded proteins carried out their functions. In the post-genomics world, biologists exploit the availability of genome sequences to examine simultaneous changes in gene expression profiles for every gene in the genome using array technology. Moreover, high-throughput protein analysis provides detailed information on cellular location, and protein–protein interactions, for thousands of characterized and uncharacterized proteins. The post-genomic era is characterized, therefore, by massively parallel molecular analyses of information-rich molecules.

4.3.2 Transcriptomics

If genomics deals with the relationship between DNA sequence and its encoded function, transcriptomics looks at RNA sequences and their expressed levels. If the pre-genomic technique of northern blot analysis of mRNA acted as an index of expression for one or two genes, the post-genomic world demands a more global perspective on gene expression patterns. As was the case with fungal genome analysis, global fungal transcriptome analysis began with *S. cerevisiae*.

4.3.3 Dissecting the Diauxic Shift Using a Yeast Microarray

In a classic paper, DeRisi *et al.* (1997) PCR-amplified each of the 6400 distinct ORFs described in the first edition of the yeast genome and printed these unique DNA molecules onto glass slides using a simple robotic printing device. The resulting DNA microarrays, in which each of the known DNA sequences was attached at a particular grid reference on an 18 mm square area of a glass slide, was then a multi-gene probe. This was then used simultaneously to explore the expression profile of mRNA as yeast cells underwent a diauxic shift from fermentative growth to aerobic respiration. mRNA samples taken over a 21 h period as the cellular metabolism changed from glucose fermentation to ethanol respiration were reverse-transcribed into cDNA molecules labelled with a red fluorescent dye. Green fluorescent cDNA was also prepared from the mRNA at the first time point to serve as reference. The expression level of each gene was measured by hybridizing fluorescently labelled cDNA to the probe, visualizing the fluorescence pattern using a confocal microscope and using a computer to analyse the image of the relative intensity of the spots. At the initial time point, the green and red signals were equal and all spots appear yellow. At later time points, red colour indicates gene expression increased relative to the reference, while green colour indicates gene expression decreased relative to the reference (Plate 4.1).

The analysis revealed that a staggering 28 % of all yeast genes underwent a significant alteration in gene expression level as a result of a diauxic shift: 710 genes were induced and 1030 genes repressed by a factor of 2 or more. Moreover, cluster analysis identified groups of genes whose pattern of expression changed in association with one another. Such co-ordinated gene regulation points towards a common promoter element – a co-ordinated regulation that helps to identify possible cellular roles for ORFs that encode proteins of unknown function. For example, the sequences upstream of the named genes in (Plate 4.2) all contain stress response elements (STREs). When the promoter sequences of the 13 additional (not previously recognized as stress-inducible) genes that shared this expression profile were examined, nine were found to contain one or more recognizable STRE sites. This suggests that many of these 13 additional genes, which were unidentified heretofore, have a role that is linked to stress response.

The same DNA microarrays were also used successfully to identify genes whose expression was affected by deletion of the transcriptional co-repressor *TUP1* or overexpression of the transcriptional activator *YAP1*, thereby demonstrating the feasibility and utility of this approach to the dissection and characterization of regulatory pathways and networks on a genome-wide basis.

4.3.4 The Vocabulary of Transcriptomics

Encouraged by the successful demonstration of this approach to global gene expression studies, other yeast studies rapidly followed. Then, as genomes became available, other fungi were explored in the same way. Fungal researchers found themselves embracing the new vocabulary of post-genomic analyses as they set about exploiting this approach in their organism of choice. Copy DNA (cDNA) libraries, expressed sequence tags (ESTs), serial analysis of gene expression (SAGE), macro- and micro-arrays, BLAST searches and sequence alignments (Table 4.4) all entered the fungal literature within 5 years of the DeRisi *et al.* (1997) paper. *S. cerevisiae* has led the way in the microarray-based analysis of fungal transcriptomes because it was the first available fungal genome and because its genes lack introns, therefore allowing the ORFs to be PCR-amplified directly from genomic DNA. A number of other global approaches to mRNA analyses are also available, including methods based on cDNA, SAGE and EST analysis.

Next-generation DNA sequencing is poised to replace all of these techniques. Next-generation sequencing (NGS) refers to a series of new technologies in which the DNA sequence is recorded as the DNA molecules are synthesized in real time.

Table 4.4 Post-genomic vocabulary.

BLAST	An algorithm that searches a sequence database for sequences that are similar to the query sequence. There are several variations for searching nucleotide or protein databases using nucleotide or protein query sequences
cDNA	A DNA molecule synthesized by reverse transcriptase using an mRNA molecule as template. Hence, cDNA lacks the introns found in genomic DNA
NGS	Refers to a series of technologies in which the DNA sequence is recorded as the DNA molecules are synthesized in real time.
ESTs	Short cDNA sequences that are derived from sequencing of all the mRNAs present in a cell. ESTs represent the expression profile of the cell at the time point of RNA isolation
Microarray	An ordered grid of DNA sequences fixed at known positions on a solid substrate, e.g. glass slide
ORF	The sequence of codons, in DNA or RNA, that extends from a translation start codon to a stop codon
SAGE	Extremely short ESTs that are linked together as DNA chimeras consisting of 15 base pair sequences from 40 different mRNAs. The sequence analysis of thousands of these 40 X15bp chimeras permits a quantitative estimation of the mRNAs in the original sample
Sequence alignment	A linear comparison of nucleotide or protein (amino acid) sequences. Alignments are the basis of sequence analysis methods and are used to identify conserved regions (motifs)

Sequence data is collected across millions of simultaneous reactions and powerful bioinformatics tools used to interpret them. This completely circumvents the traditional dideoxy-nucleotide (Sanger) sequencing, which required labelled DNA fragments to be physically resolved by electrophoresis. The commercial companies involved in this include: 454 Genome Sequencer (Roche), Illumina (formerly Solexa) genome analyser, the SOLiD system (Applied Biosystems/Life Technologies) and the Heliscope (Helicos Corporation), and such technology will revolutionize SAGE and EST analysis. Moreover, because it can use mRNA as a primary template directly, NGS also has significant advantages over microarrays because the genome sequence is no longer needed to set up the DNA probe on the chip.

4.3.5 From Clocks to Candidosis: the 'When' of Inherited Information

DeRisi *et al.*'s seminal paper revealed the power of transcriptomics in dissecting out complex differential gene expression *when* cells are undergoing diauxic growth and *when* regulatory gene expression levels are altered. In short, it reveals the information stored in the genome as a dynamic process. This technology allows scientists simultaneously to analyse the expression levels of any number of individual genes as they change in time, thereby identifying co-ordinated gene expression, and most critically understanding what happens *when* environmental changes and gene perturbations impact on the flow of information through the cell.

Knowing the *when* of a gene's expression often provides a strong clue as to its biological role. SAGE and microarray technologies have already been used to: identify the set of fungal genes most highly expressed during cerebrospinal fluid infection by *C. neoformans*; identify the pattern of differential gene expression that solved the problem of why glucose metabolism by *Trichoderma reesei* is so different to that of *S. cerevisiae*; identify 18 genes of unknown function implicated in the production of carcinogenic aflatoxins by *Aspergillus parasiticus*.

Conversely, the pattern of genes expressed in a cell *when* it has been subjected to various treatments can identify critical points of great biological significance. This approach has been used to identify potential anti-mycotic targets to combat the increasing threat posed by *Candida* infections. Microarray analysis has been used: to identify the two transcription factors *EFG1* and *CPH1* induced when *C. albicans* is shifted to 37 °C in the presence of serum; to establish that the final transcriptional profile found in some clinical isolates resistant to fluconazole could also be reproduced by allowing resistance to evolve in the laboratory; and to demonstrate that the antimycotic itraconazole affected the expression of 296 ORFs in this fungus.

Transcriptomics has been used to accelerate research projects which could be solved without this technology, but in other cases it allows fungal biologists to explore biological phenomena that are quite simply too complicated to approach in any other way. Moreover, this applies to fungal biology at both basic and applied levels. In an example of basic research, *S. pombe* workers used expression data for each *pombe* gene to reveal that many conserved genes are expressed at high levels, whereas a disproportionate percentage of the poorly expressed genes are organism specific. This analysis was only possible because the expression data were available and database searches were possible of all ORFs in *S. pombe*, *S. cerevisiae* and the nematode worm *C. elegans*. In another example of basic research that was impossible by pre-genomic methods, microarrayed cDNA sequences of 1000 different genes revealed that circadian clock control in *N. crassa* cover a range of cellular functions rather than preferentially belonging to specific pathways.

And then sometimes, unexpectedly, what was basic research to begin with turns out to have applied aspects: microarray analysis in *C. albicans* of the genes induced by α-factor in mating competent type-**a** cells revealed 62 genes. Interesting in their own right as an insight into the cryptic mating pathway of this fungus, the realization that seven of these genes encoded cell-surface or secreted proteins that had previously been shown to be required for full virulence of *C. albicans* unexpectedly revealed a new potential target for the development of an antimycotic aimed at the factor(s) regulating the mating pathway.

An impressive example of purely applied research, on the other hand, is a microarray produced by the Agilent company. The first of its kind, this array carries 7137 rice ESTs and all 13 666 predicted ORFs from the major rice fungal pathogen *Magnaporthe grisea*. Therefore, it uniquely allows simultaneous analysis of gene expression profiles in both the host and pathogen. This microarray has been used in a number of studies since it first appeared in 2005. It has been responsible for identifying: a serine vacuolar protease involved in pathogenicity; the suite of genes induced by the Con7 transcription factor (which has a role in pathogenicity); a comprehensive list of genes thought to be involved in the formation of the flattened hypha (appressorium) that this parasitic fungus uses in order to penetrate the host tissues.

4.4 Proteomics

The 2-D and 3-D analyses of biological information are extremely powerful. However, it is the 4-D manifestation of that information in time and space that biologists seek to define and understand. Small wonder that proteomics, which seeks to study biology in that extra dimension, faces challenges that are orders of magnitude beyond those posed by the global analysis of 2-D and 3-D information in DNA and mRNA respectively. One has only to look at Table 4.5 to realize that, unlike genomics and transcriptomics, proteomics is in fact an entire suite of distinct but intricately interrelated technologies. Even though 2-D gel electrophoresis provided a platform for the global analysis of cellular proteins as early as 1978, it revealed nothing of the spatial arrangement of the proteins in the cell, little of post-translational modifications, only

Table 4.5 Proteomics is much more challenging than transcriptomics or genomics.

DNA (information)	RNA (information in time)	Protein (information in time and space)
Nucleotide sequence	Nucleotide sequence	Amino acid sequence
	Level in cell	Level in cell
	Modifications (splicing)	Modifications (glycosylation, phosphate groups, etc.)
		3-D structure
		Function
		Location
		Interactions

visualized moderately abundant proteins and told the researcher nothing of the protein's structure and function. Developments such as micro-sequencing of the gel-separated proteins and the use of sensitive glyco-specific stains addressed some of these limitations in the pre-genomic era, but global analysis of proteins in the post-genomic era seeks to build on genomic and transcriptomic projects to provide databases of information that address all of the parameters listed in Table 4.5 and more.

4.4.1 Protein Sequence and Abundance

At its most basic level proteomics seeks to identify at a given moment in time which ORFs in a genome are expressed as proteins in a cell, and what level of each one is present. Unlike RNA and DNA sequences that can be enzymatically amplified, large-scale parallel analysis of proteins must deal with unavoidable problems of limited sample material and abundance variation over six orders of magnitude. Nevertheless, a combination of tryptic digestion of protein mixtures coupled with chromatographic separation of the resulting peptides allows increasingly sophisticated mass spectrometry (MS) to identify the presence and abundance of ever smaller amounts of protein from increasingly complex protein mixtures. Peptide fingerprints (the spectrum of ionized molecules generated by the peptides in an MS system) can then be used to identify the same pattern already entered into the databases from MS analysis of known proteins. Alternatively, a further round of MS can be performed on each peptide to identify its constituent amino acids and these sequences can be compared with entries in protein and DNA databases using the appropriate database searching tools. In a less high-throughput system, protein mixtures can first be separated by 2-D gel electrophoresis, protein spots excised, digested with trypsin and subjected to MS analysis in the same way.

This type of proteomic analysis, which essentially generates the same type of information on the protein level as transcript analysis does at the RNA level, has already been used to probe fungal biology. 2-D electrophoresis followed by MS analysis was used to assess the effect of concanamycin A, an antibiotic produced by *Streptomyces*, on protein levels in the filamentous fungus *Aspergillus nidulans*. Twenty spots were identified and five excised for tryptic digestion followed by estimation of their complete amino acid sequence by MS. The functions of four of those proteins were identified using the protein sequence to search ORF databases; a fifth was identified as being homologous to a protein in *A. niger*, but of unknown function. In a more extensive analysis, to identify the extracellular proteins secreted by *A. flavus* when provided with the flavonoid rutin as the only source of carbon, approximately 100 spots were identified by 2-D electrophoresis. Trypsin digestion and MS analysis allowed 22 of these to be identified by searching peptide fingerprint databases. Over 90 protein spots remained unidentified, however, indicating that these proteins are either novel proteins or proteins whose peptide fingerprints are not yet available.

A recurring theme in protein sequence and abundance analyses is that proteomic findings differ with respect to transcriptomic findings. In some cases mRNAs are present but no corresponding protein is found, indicating that the protein was in very low abundance, was unstable during extraction or that those transcripts were the subjects of post-transcriptional regulation. On other occasions proteins are present but no corresponding RNA is found – indicating a less than representative RNA sample or a protein with an unusually long half-life. However, despite these differences, large-scale parallel protein analyses are becoming a central theme in the fungal community. Indeed, progress has been so rapid that, in the case of *S. cerevisiae*, global protein analysis projects have managed to address, with some success, the much more formidable question of how the proteins present in the cell interact with one another to form complexes, and how complexes are organized on a cellular scale. Moreover, given the power of comparative genomics, the findings in this model organism are already being extended.

4.4.2 From Locations to Interactomes

Once the presence and abundance of a protein has been determined for a given situation, the next parameter in the 4-D arrangement of biological information is the cellular location of the protein in question. Thereafter, a more complete description requires information on (1) the proteins that it interacts with, (2) the complexes it is involved with and (3) how these complexes interact within the cell. The yeast *S. cerevisiae* is presently in the forefront of turning this dream into reality. This is because it is an extensively characterized model organism with a well-developed gene expression system, and its extremely accurate homologous recombination system allows novel gene constructs to be precisely targeted into the correct chromosomal locus, thereby ensuring that the gene is subjected to appropriate native gene regulation. Moreover, the vast majority of its ORFs are known and many have been characterized.

Using these cellular attributes, Huh *et al.* (2003) used PCR to amplify hybrid DNA molecules consisting of the coding sequence for GFP fused in-frame with the 3′ end of the coding sequence for each one of 4100 yeast ORFs (Figure 4.5). They then used homologous recombination to target each hybrid molecule back into the correct chromosomal locus, thereby creating 4100 yeast strains, each one carrying a different GFP-tagged gene sequence. Then, using fluorescence microscopy to find out where they were in the cell, they managed to classify these proteins into 22 distinct subcellular locations and in doing so provided localization information for 70 % of previously unlocalized proteins – a number that constitutes about 30 % of the yeast proteome.

Hybrid DNA molecules encoding fusion proteins were also the basis of the first systematic search for interacting proteins in a yeast cell. Using a slight

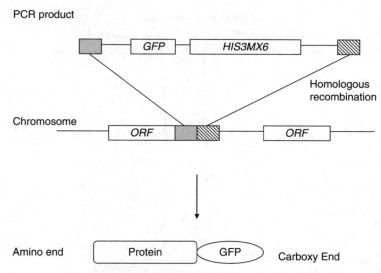

Figure 4.5 Creating a fluorescently labelled protein. PCR was used to create a hybrid DNA consisting of a selectable marker (*His 3*) and the coding sequence for GFP fused in-frame with the two adjacent sequences of DNA from the 3′ end of the coding sequence of the target ORF. Homologous recombination to target each hybrid molecule back into the correct chromosomal locus, thereby creating a GFP tagged gene sequence.

variation of the two-hybrid system, described in detail elsewhere in this book (see Chapter 8 on heterologous protein production), Hua *et al.* (2003) mated 192 cells expressing different 'prey' proteins with each of 6000 strains expressing different 'bait' proteins and used reporter gene expression in the resulting diploid cells as an index of interaction. Although all two-hybrid studies identify false-positive interactions and fail to identify weakly interacting proteins ones, they do provide strong clues as to possible protein interactions and functions. In this study, two proteins of unknown function were found to be intimately linked to arginine metabolism, a hitherto unknown cell cycle control circuit was discovered and small networks of interactions traced a series of protein interactions joining a protein involved in the formation of double-stranded DNA breaks and one involved in the formation of the synaptonemal complex during meiosis. In a subsequent paper using these and related data from other studies, Schwikowski *et al.* (2001) compiled a list of about 2700 protein interactions in *S. cerevisiae* and found that 1548 yeast proteins could be depicted in a single large network (Figure 4.6). Moreover, within the network it was found that proteins which could be allocated to specific cellular functions (e.g. DNA synthesis, amino-acid metabolism, structural proteins, etc.) showed a high level of interactions and, therefore, clustered together into functional groups. This allowed them to ascribe potential functions to almost 40 yeast proteins previously of unknown function. Their global interactome revealed the 'great interconnectedness' of the

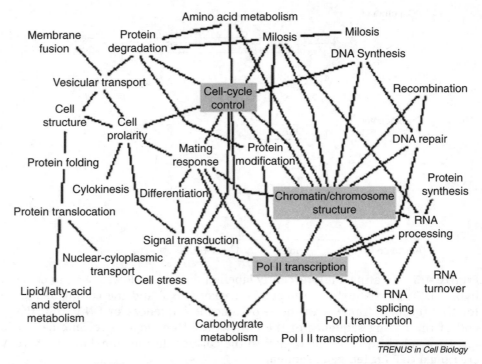

Figure 4.6 A diagram of the meta-network consisting of 32 functions (and their 70 or so associated interlinking connections) in place of the 1548 individual proteins and their 2700 links which constitute the yeast interactome.

yeast proteome. It found that many members of protein clusters associated with central roles (e.g. cell cycle) had interactions with proteins from many other cluster classes (e.g. DNA synthesis, transcription, polarity, etc.), thereby connecting biological functions into larger cellular processes. On the other hand, not many members of functional clusters for more peripheral cellular functions (e.g. RNA turnover) interacted with proteins from other clusters. This meta-analysis allowed a meta-network consisting of 32 functions (and their 70 or so associated interlinking connections) to be described in place of 1548 individual proteins and their 2700 links. The graphical representations of protein-interaction maps provide a rough outline of the complexity of protein associations.

This type of analysis only reveals protein–protein interactions arising under the artificial circumstances of being constitutively expressed at high levels in the same cellular compartment at the same time. However, a powerful complementary study provided a much more biologically relevant analysis of protein–protein interactions in this yeast: Gavin *et al.* (2002) exploited homologous recombination to integrate hundreds of ORFs fused to an affinity tag into their normal regulatory sequences. The transformed cells were then allowed to express the modified proteins and tandem affinity purification (TAP) was used to isolate the tagged protein and any associated proteins with it. The co-eluting proteins were then identified using standard MS methods. This approach identified 1440 distinct proteins within 232 multi-protein complexes in yeast. This study provided new information on 231 previously uncharacterized yeast proteins, and on a further 113 proteins to which the authors ascribe a previously unknown cellular role. Other types of global analysis in yeast serve to complement and support these studies. These include:

- Protein microarrays, a technology analogous to DNA microarrays, in which *target proteins* are immobilized on a solid support and probed with fluorescently labelled proteins. Proteins that bind to the target are thereby implicated as possibly interacting with them *in vivo*.

- Synthetic lethal screens, a genetic screening process which tests pairs of mutations together. This identifies mutations that alone are not lethal but together are incompatible with life and, therefore, provide strong evidence that the two gene products are functionally related.

- The identification of co-regulated mRNAs, which provides evidence that the encoded proteins are involved in related biological processes.

An overview of high-throughput systems designed to capture information on protein–protein interactions is provided in Figure 4.7.

Together, these global studies provide biological insights that are much deeper and dynamic than ever before. Moreover, comparisons of such data with similar information from other model organisms reveal that the basic constituents of cells, their interactions and processes have been conserved across aeons of evolutionary time and, therefore, are present in millions of organisms. By characterizing and annotating the constituent molecules and their interactions in one organism, it is often possible to extrapolate such information to other organisms. However, because the naming and description of genes and gene products varies widely in different organisms in this post-genomic era, it is even more vitally important for scientists to be able to communicate their findings to one another. That is why as far back as 1998, when there were still only three eukaryotic genomes available, genome researchers developed a shared species-independent controlled common vocabulary, and a carefully defined structural framework for organizing information, to allow communication across collaborating databases – the Gene Ontology (GO) project had been born. However, it is doubtful that the even the original consortium foresaw just how important GO was to become. Today, it is a key element in post-genomic research, in allowing scientists to produce standard annotations, improve computational queries, retrieve and analyse data from disparate sources and, even more critically, computationally extract biological insights from enormous data sets.

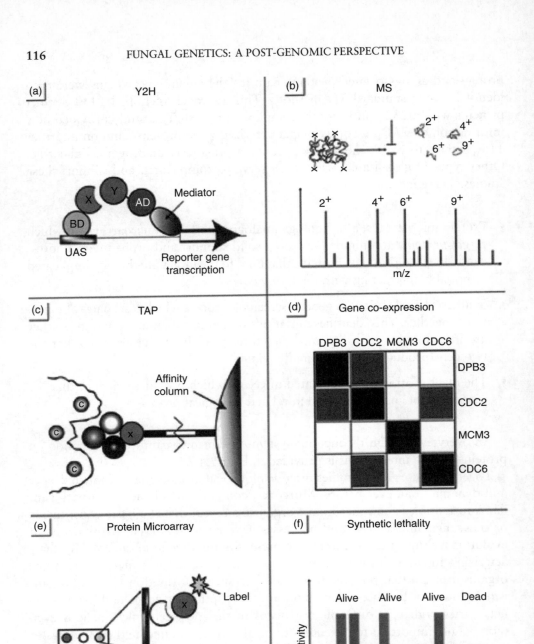

Figure 4.7 An overview of high-throughput systems designed to capture information on protein–protein interactions. (a) Yeast two-hybrid detects interactions between proteins X and Y, where X is linked to a DNA binding domain which binds to upstream activating sequence (UAS) of a promoter. (b) MS identifies polypeptide sequence. (c) TAP purifies proteins associated with the target

4.4.3 Functional Annotations are GO ...

In the field of informatics an ontology is essentially a graphical representation of knowledge (a knowledge map) consisting of nodes populated by terms from a strictly controlled vocabulary, joined by arrows, which represent relationships between the terms based on strictly defined logical statements. The GO actually consists of three independently constructed ontologies: the *molecular function* is constructed with a vocabulary that defines the biochemical activity of a gene product – that is, what it does (e.g. transporter, ligand, enzyme). The *cellular component* is constructed with a vocabulary that defines where in a cell the gene product is active (e.g. mitochondrion, cytoplasm, mitotic spindle). The *biological process* is constructed with a vocabulary that defines cellular activities accomplished by multiple molecular events (e.g. growth, translation, signal transduction). A very low resolution map of part of the *biological process* ontology is shown in (Figure 4.8) (a higher resolution map would have much more detail and complexity). The nodes in this example are populated by terms from the controlled vocabulary of biological processes joined by arrows representing the logical relationship statements 'is_a' or 'is_part'.

It is important to realize that the GO project is *not* a nomenclature for genes or gene products, but a means of querying and retrieving genes and proteins based on their shared biology. The power of the GO project lies in the fact that its vocabulary describes biological phenomena; that is, the *attributes* of biological objects (functions, processes and locations), not the objects themselves. Nevertheless, access to the objects is made possible by associating a GO term from one of the three ontologies with a gene or gene product stored in keyword databases such as Genbank, EMBL, MIPS and so on. The association must be further supported with a specific reference in the literature, plus an agreed evidence code, which is determined by the type of evidence presented in the reference (e.g. mutant phenotype, microarray expression, etc.), and the date that the annotation was assigned. The 'chromatin binding' node, in a subset of the

---

Figure 4.7 (Continued) (x). (d) Gene co-expression analysis produces the correlation matrix, where the dark areas show high correlation between expression levels of corresponding genes. (e) Protein microarrays (protein chips) can detect interactions between actual proteins rather than genes; target proteins immobilized on the solid support are probed with a fluorescently labelled protein. (f) Synthetic lethality method describes the genetic interaction when two individual nonlethal mutations result in lethality when administered together (a⁻ b⁻). (Shoemaker *et al.* (2007). This is an open-access article distributed under the terms of the Creative Commons Public Domain declaration which stipulates that, once placed in the public domain, this work may be freely reproduced, distributed, transmitted, modified, built upon, or otherwise used by anyone for any lawful purpose).

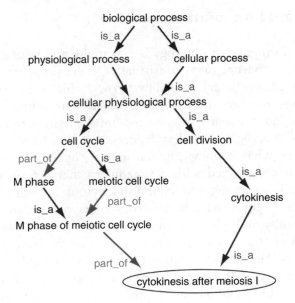

Figure 4.8 This low-resolution map of the *biological process* ontology is built by drawing up a carefully controlled vocabulary of terms about biological processes and then linking them with logical statements. This ontology is read from bottom to top. See http://www.geneontology.org/GO.ontology.relations.shtml for a full description of logical nomenclature for all three ontologies. http://www.yeastgenome.org/help/GO.html.

molecular function ontology, in Figure 4.9 is a case in point: It has been annotated with three columns of genes: six from *S. cerevisiae*, five from *Drosophila* and five from *Mus*. Their assigned gene designations reveal how difficult it is to search for/compare gene functions between these model organisms; the GO term 'chromatin binding', however, retrieves all related genes when applied across these three databases. The converse is also true: New genome sequences can be matched to GO terms by sequence similarity with sequences in databases that have already been annotated. The GO project allowed the computational transfer of biological annotations from the highly characterized and annotated yeast *S. cerevisiae* to a whole series of less well characterized fungi, including *Ashbya gossypii* and the fungal pathogens *Pneumocystis carinii*, *Sclerotinia sclerotiorum* and *C. albicans*. This affords biologists (fungal and others) the opportunity of understanding the biology of their favourite organism without necessarily developing technologies to undertake saturation analysis of its molecular interactions, such as those developed to explore the biology of *S. cerevisiae*.

However, even in the case of this highly characterized organism, identifying the 4-D arrangement of all the intricate cellular protein interactions, complete with biochemical kinetics, is still a long way off. Nevertheless, it is now possible to develop theoretical models of the interactions between collections of elements

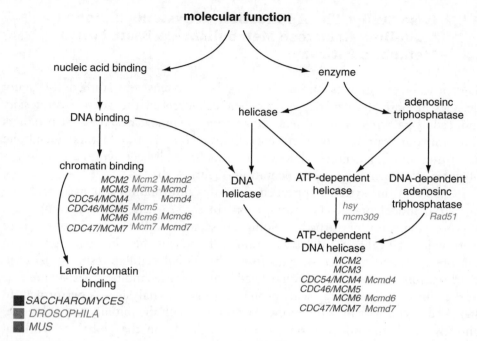

Figure 4.9   Using a subset of the *molecular function* ontology to illustrate the difference between the *annotation* and the *ontology*. The biological terms populating the nodes, and the arrows between the nodes, constitute the ontology. The gene designations for the three model organisms, linked to keyword databases such as Genbank, EMBL and MIPS, are the annotations. In the case of the single ontological term 'chromatin binding', there are 16 separate annotations arising from three different databases. (Ashburner *et al.* (2000)).

responsible for carrying out well-defined biological tasks. Such models can provide a rational framework in which to design a focused range of experiments – the data from which can, in turn, be used further to refine the model. Such attempts to reintegrate molecular details to reveal the secrets of the dynamic processes they mediate within the cell forms an experimental and theoretical approach referred to as *systems biology*.

4.5 Systems Biology

4.5.1 Establishing Cause and Effect in Time and Space

The post-genomics era has been dominated to date by gigantic data-accumulating exercises – data that allow us to analyse encoded information in time and space. Now the challenge is to reintegrate these data to provide meaningful insight into biological phenomena.

4.5.2 Case Study: The Application of Systems Biology to Modelling Galactose Metabolism – A Basic Fungal Metabolic Pathway

The process of galactose utilization has been extensively studied for many decades at both the genetic and biochemical levels in the yeast *S. cerevisiae*. Induced by galactose in the absence of glucose, the nine genes, the gene products (one transporter protein, four enzymes and four transcription factors), metabolic substrates and key gene regulatory networks have all been worked out during 30 years of pre-genomic experimentation (Figure 4.10).

In *step 1* of this systems approach all of this information was used to define an initial model of galactose utilization. In *step 2* each pathway component was systematically perturbed, yielding 20 separate cellular conditions. These were wild-type cells and deletion strains for each of the nine genes grown in the presence and absence of galactose. The global cellular response to *each perturbation* was detected and quantified using array technology. Also in the case of the wild-type cells, large-scale protein expression analysis was performed. In *step 3* all of this new information was integrated with the initially defined model and also with the information currently available on the global network of protein–protein and protein–DNA interactions in yeast.

Amazingly, this microarray analysis revealed that mRNA synthesis was significantly altered in the case of 997 yeast genes (i.e. approximately 20 % of all ORFs) in one or more of these perturbations. And this despite the fact that only nine gene

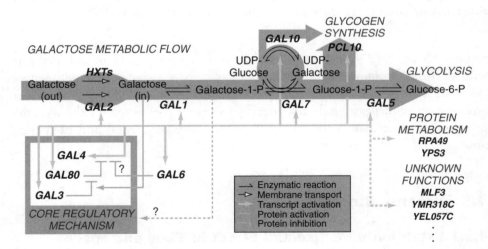

Figure 4.10 System analysis reveals new levels of regulation in a basic fungal metabolic pathway. Dotted interactions denote model refinements to galactose metabolism in yeast following a post-genomic global analysis of the yeast genome sequence, microarray-based mRNA analysis, databases of protein–protein interactions and rapid MS analysis of proteins.

products are needed for galactose metabolism. It would appear, therefore, that the 'great interconnectedness' of the yeast protein interaction network is extensively perturbed by a handful of perturbations to the galactose metabolic system.

In *step 4* the researchers formulated new hypotheses to explain the observations not predicted by the model, and then designed and executed additional perturbation experiments to test the new 'improved' model. For example, they detected an unexpected reduction in the expression levels in GAL enzyme genes (suggesting a regulatory effect) in cells carrying a deleted gal7 gene (encodes the enzyme responsible for converting galactose-1-P to glucose-1-P). This suggests the presence of an additional, heretofore unknown, regulatory circuit which down-regulates GAL gene expression when galactose-1-P accumulates in the cell. This new 'improved' model was tested by additionally deleting gal1 (encodes the enzyme that converts galactose to galactose-1-P). As predicted, this essentially reversed the effect.

This seminal systems biology paper was only made possible because the yeast genome had been sequenced, microarray analysis of yeast mRNA had been developed, databases of protein–protein interactions were available, rapid MS analysis of protein presence and abundance was possible and computing facilities to process all of the information were available. In short, systems biology is only possible in a post-genomic world.

4.6 Conclusion

From one gene – one enzyme in *Neurospora*, to cell cycle in yeast, fungal experiments have been at the forefront of altering the way biologists view their world. Yet the sum total of all the great discoveries in fungi from Beadle and Tatum to Hartwell and Nurse are set to pale into insignificance in the face of the potential offered by fungal systems biology in this, the *post-genomic era*. Led by *S. cerevisiae*, there is a paradigm shift towards *massively parallel comparative biology* in all types of organismal research. Even within the context of this book, such an approach offers rapid progress in every area of this field, including exploring native and transgenic fungal biology, identifying genomic differences for strain improvement, understanding the molecular consequences of fungal infection and identifying targets for the development of new antimycotics. The opportunities offered by post-genomic research are quite simply limitless.

Revision Questions

Q 4.1   Why is yeast a cornerstone of post-genomic research?

Q 4.2   What is the fundamental difference between pre- and post-genomics research?

Q 4.3   What is systems biology?

Q 4.4   What is an open reading frame (ORF)?

Q 4.5   How many 'new' ORFs were discovered when the yeast genome was analysed, how many of the 6000 ORFs were initially annotated with functions and how many have had functions determined since then?

Q 4.6   What does ORF redundancy reveal about the dynamics of genetic information in *S. cerevisiae*, *S. pombe* and *N. crassa*.

Q 4.7   Provide a brief description of the three basic approaches to identifying the functions of newly discovered ORFs.

Q 4.8   Explain how DeRisi *et al.* (1997) fabricated and used the first yeast microarray?

Q 4.9   Explain the meaning of: BLAST searches, complementary DNA (cDNA) libraries, expressed sequence tags (ESTs), serial analysis of gene expression (SAGE) and sequence alignments.

Q 4.10   Outline the challenges faced by proteomics in defining 4-D biology.

Q 4.11   How are fusion proteins made and used in yeast proteomics?

Q 4.12   Explain the 'systems biology' approach to understanding biological phenomena in yeast.

References

Ashburner, M., Ball, C.A., Blake, J.A. *et al.* (2000) Gene Ontology: tool for the unification of biology. *Nature Genetics*, **25**, 25–29. DOI: 10.1038/75556

DeRisi, J.L., Iyer, V.R. and Brown, P.O. (1997) Exploring the metabolic and genetic control of gene expression on a genomic scale. *Science*, **278**, 680–686.

Gavin, A.C., Bösche, M., Krause, R. *et al.* (2002) Functional organization of the yeast proteome by systematic analysis of protein complexes. *Nature*, **45**, 4–47.

Schwikowski, B., Uetz, P. and Fields, S. (2001) A network of protein–protein interactions in yeast. *Nature Biotechnology*, **18**, 1257–1261.

Shoemaker, B.A. and Panchenko, A.R. (2007) Deciphering protein–protein interactions. Part I. Experimental techniques and databases. *PLoS Computational Biology*, **3**, 337–344 (e42). DOI: 10.1371/journal.pcbi.0030042

Further Reading

Ansorge, W.J. (2009) Next-generation DNA sequencing techniques. *New Biotechnology*, **25**, 195–203.

Christie, K.R., Hong, E.L. and Cherry, J.M. (2009) Functional annotations for the *Saccharomyces cerevisiae* genome: the knowns and the known unknowns. *Trends in Microbiology*, **17**(7), 286–294.

Dujon, B. (2010) Yeast evolutionary genomics. *Nature Reviews Genetics*, **11**, 512–524.

Edwards, A.M., Kus, B., Jansen, R. *et al.* (2002) Bridging structural biology and genomics: assessing protein interaction data with known complexes. *Trends in Genetics*, **18**, 529–536.

Goffeau, A., Barrell, B.G. and Bussey, H. (1996) Life with 6000 genes. *Science*, **274**, 546–567.

Goffeau, A. (2000) Four years of post-genomic life with 6000 yeast genes. *FEBS Letters*, **480**, 37–41.

Hofmann, G., McIntyre, M. and Nielsenz, J. (2003) Fungal genomics beyond *Saccharomyces cerevisiae*? *Current Opinion in Biotechnology*, **14**, 226–231.

Huh, W.K., Falvo, J.V., Gerke, L.C. *et al.* (2003) Global analysis of protein localization in budding yeast. *Nature*, **425**, 686–691.

Ideker, T., Thorsson, V., Ranish, J.A. *et al.* (2001) Integrated genomic and proteomic analyses of a systematically perturbed metabolic network. *Science*, **292**, 929–934.

Kar-Chun, T., Ipcho, S.V.S. and Trengove, R.D. (2009) Assessing the impact of transcriptomics, proteomics and metabolomics on fungal phytopathology. *Molecular Plant Pathology*, **10**, 703–715.

Khurana, V. and Lindquist, S. (2010) Modelling neurodegeneration in *Saccharomyces cerevisiae*: why cook with baker's yeast? *Nature Reviews Neuroscience*, **11**, 436–449.

Kim, D.-U., Hayles, J., Kim, D. *et al.* (2010) Analysis of a genome-wide set of gene deletions in the fission yeast *Schizosaccharomyces pombe*. *Nature Biotechnology*, **28**, 617–628.

Magee, P.T., Gale, C., Berman, J. and Davis, D. (2003) Molecular genetic and genomic approaches to the study of medically important fungi. *Infection and Immunity*, **71**, 2299–2309.

Mannhaupt, G., Montrone, C., Haase, D. *et al.* (2003) What's in the genome of a filamentous fungus? Analysis of the *Neurospora* genome sequence. *Nucleic Acids Research*, **31**, 1944–1954.

Skinner, W., Keon, J. and Hargreaves, J. (2001) Gene information for fungal plant pathogens from expressed sequences. *Current Opinion in Microbiology*, **4**, 381–386.

Tunlid, A. and Talbot, N.J. (2002) Genomics of parasitic and symbiotic fungi. *Current Opinion in Microbiology*, **5**, 513–519.

Tylres, M. and Mannt, M. (2003). From genomics to proteomics. *Nature*, **422**, 193–197.

Uetz, P., Giot, L., Cagney, G. *et al.* (2000) A comprehensive analysis of protein–protein interactions in *Saccharomyces cerevisiae*. *Nature*, **403**, 623–627.

Winzeler, E.A., Shoemaker, D.D., Astromoff, A. *et al.* (1999) Functional characterization of the *S. cerevisiae* genome by gene deletion and parallel analysis. *Science*, **285**, 901–906.

Wood, V. Gwilliam, R., Rajandream, M.A. *et al.* (2002) The genome sequence of *Schizosaccharomyces pombe*. *Nature*, **415**, 871–880.

5
Fungal Fermentations Systems and Products

Kevin Kavanagh

5.1 Introduction

Fungi represent one of humanity's oldest domesticated organisms and are responsible for the production of some of our most enjoyable (e.g. alcohol), nutritious (e.g. bread) and medically useful (e.g. penicillin) products. Although not the subject of the material presented in this chapter, it should also be emphasized that fungi are extremely important decomposers of organic material, such as leaf litter. Many fungi engaged in decomposition are also the producers of enzymes and antibiotics of commercial importance.

The use of yeast to make bread and alcohol has been recorded for thousands of years. The Babylonians (circa 6000 BC) and Egyptians (circa 5000 BC) have left written accounts of their production of beers, wines and bread – all of which warranted the use of yeast. Indeed, the ancient Israelites and Egyptians recognized the difference between leavened (used yeast to 'raise' the dough) and unleavened bread from as early as 1200 BC. Yeast cells were first observed microscopically by Van Leeuwenhoek in 1680, and in 1838 the yeast involved in brewing was called *Saccharomyces cerevisiae*. Although the biochemical role of yeast in the fermentation process was not elucidated until 1863 by Louis Pasteur, the products of the process have been enjoyed for millennia!

Apart from their role in providing humans with food and alcoholic beverages for thousands of years, fungi have had a significant impact on the course of human history. The Roman Empire was one of the greatest empires the world has ever seen, and the reasons for its demise have been a source of much debate.

Fungi: Biology and Applications, Second Edition. Edited by Kevin Kavanagh.
© 2011 John Wiley & Sons, Ltd. Published 2011 by John Wiley & Sons, Ltd.

Recent evidence suggests that one factor that may have accelerated the decline of Rome was the reduced yield from cereal crops due to a series of warm, humid summers facilitating the growth of rusts and smuts. The reduced yield led to higher prices for bread with associated food riots. A similar scenario may have contributed to the French Revolution in 1789. The devastating famine of 1845–1848 in Ireland which killed a million people and forced a million to emigrate was due to potato blight caused by the fungus *Phythopthora infestans*. Fungi have also had beneficial effects on human history. The discovery of penicillin production by Alexander Fleming allowed the treatment of wound infections and subsequent improvement in patient survival. Prior to the D-day landings in June 1944, the Allies developed stocks of penicillin to treat wounded soldiers – without this antibiotic the deaths from the European invasion would have been far greater and the liberation of Europe might not have been attempted.

5.2 Fungal Fermentation Systems

Fungi are employed to produce a wide range of foods (e.g. bread, mycoprotein), alcoholic beverages (wine, beer), recombinant proteins, vitamins and antibiotics (Table 5.1). To achieve this productivity, a variety of fermentation systems are employed with fungi – the choice of which will depend upon the nature of the fungus (filamentous or yeast), the type of product that is required and the scale of the production (Table 5.2). Yeast cells have a typical mean generation time (time for the population to double) of 1.15–2 h, while filamentous fungi (moulds) divide every 2–7 h approximately; consequently, fermentation systems that are used with fungi may be of limited value for animal and plant cells, which have much longer doubling times. In addition, owing to the difference in growth morphology between yeast and filamentous fungi, some fermentation systems are suitable for use with the former cell type but not the latter. While a number

Table 5.1 Industrial applications of fungi.

Product	Example
Biomass	Production of baker's yeast
	Brewer's yeast tablets
	Single-cell protein (mycoprotein)
Cell components	Proteins (native or recombinant)
Products	Antibiotics (penicillin)
	Vitamins (B_{12})
Catabolite products	Ethanol
Bioconversion	Breakdown of range of carbohydrates

Table 5.2 Fungal fermentation systems and products.

Fermentation type	Example	Product
1. Solid	Cultivation of *A. bisporus*	Mushrooms
2. Batch	Brewing, winemaking	Beer, wine
3. Fed-batch	Cultivation of *S. cerevisiae*	Baker's yeast
4. Cell recycle batch	Wine making (Italy)	Summer wines
5. Continuous	Cultivation of *Penicillium graminerium*	Mycoprotein

of basic fermentation systems exist, modifications to these are often implemented to 'fine tune' a system to the production of a specific fungal product.

5.2.1 Solid Fermentation

Solid fermentation occurs where the fungus grows on a solid substrate (e.g. grain) in the absence of free water. In this case the fungus must be able to tolerate low water activity. This type of fermentation system has been used for the production of edible mushrooms (*Agaricus bisporus*) and for the growth of filamentous fungi for the isolation of antibiotics and enzymes. Solid fermentation is also employed to produce Soy sauce from soy beans using *Rhizopus* sp., and cheese may be produced from milk curd using *Penicillium* sp. The production of horticultural compost is also an example of a solid fermentation. A combination of yeast (*Candida lipolytica*) and filamentous fungus (*Chaetomium celluloyticum*) is used to degrade straw to produce fungal biomass. Solid fermentations have a number of applications, but they can be difficult to control and it may be expensive to sterilize the raw material.

5.2.2 Batch Fermentation

The most commonly employed system for the commercial cultivation of fungi is batch fermentation. This system has been used to produce alcoholic beverages and also for the production of industrial solvents such as ethanol and acetone. In batch fermentation the medium is inoculated at a low density with fungal cells. After a lag phase of minutes or hours, during which there is intense biochemical activity within the cells, the exponential phase of growth commences, which is accompanied by maximal utilization of nutrients. As the nutrient level declines, the rate of population growth slows until, in the stationary phase, there is no net increase in cell number and the population stabilizes. Prolonged incubation can result in a decrease in cell viability as cells die and the population declines.

Ethanol production by brewing yeast strains is maximal as the culture enters the stationary phase; consequently, brewers monitor the appearance of ethanol and halt the fermentation when levels have reached a peak. Continuation of the fermentation will ultimately result in depletion of the ethanol, as this is consumed by the yeast in the absence of the original carbohydrates. The culture will consequently display a biphasic growth curve.

Fed batch is very similar to batch fermentation, except that nutrients are added periodically during the fermentation process. This type of system is optimal for maximizing biomass and is employed for the production of baker's yeast. Another variation of batch fermentation is called the Melle–Boinot process or cell recycle batch fermentation (CRBF). In this system the cell mass at the end of a fermentation is used to inoculate the next fermentation, thus ensuring a high cell density throughout the process. By maintaining a high cell density more energy is available for ethanol production and the yeast cells become adapted to produce and tolerate ethanol at a higher level. CRBF typically proceeds for weeks or, in some cases, months, with each new cycle being inoculated with the cells from the previous cycle. The Melle–Boinot process is yeast strain- and substrate-specific, but is used for the production of summer wines in some regions of Italy and is an efficient system for the production of ethanol from xylose (a five-carbon sugar) by the yeast *Pachysolen tannophilus* (Figure 5.1). In this latter case the elevated ethanol production is accompanied by an enhanced ethanol tolerance.

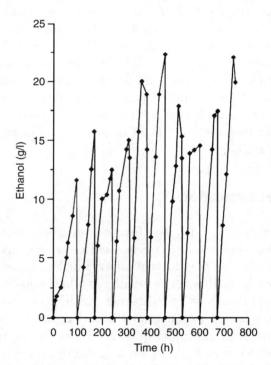

Figure 5.1 Cell recycle fermentation for the production of ethanol from xylose by the yeast *Pachysolen tannophilus*. (At the end of each fermentation cycle the cells are harvested and introduced into fresh fermentation medium.)

Interestingly, CRBF has no significant effect on ethanol production from glucose by *P. tannophilus*.

5.2.3 Continuous Cycle Fermentation

The other main fermentation system used with fungi is continuous cycle. In this system cells are grown under steady-state conditions and maintained at a particular stage of their growth cycle. Nutrients are added at a constant rate and biomass or spent medium is removed at the same rate in order to maintain a constant reaction volume. While continuous cycle fermentation has traditionally been difficult and expensive to establish for fungi, it is now routinely employed for producing mycoprotein and antibiotics.

5.2.4 Fungal Cell Immobilization

In addition to the different fermentation systems described above, there is also the possibility of immobilizing fungi in order to optimize productivity. Cell immobilization ensures that a high cell density is maintained, that cells are not washed out of the bioreactor, it reduces the opportunity for contamination and removes growth inhibition due to the production of a toxic metabolite such as ethanol. The main disadvantage of immobilization is that fungal cell viability decreases over time and that the immobilization system may degrade with continual usage.

Four basic immobilization systems are used with fungi. The first system involves the containment of the fungal cells in a membrane such as dialysis tubing, microfilters or other porous material (Figure 5.2a). One problem encountered with this type of system is membrane clogging, which reduces the efficacy of the system.

Yeast cells may also be attached electrostatically or covanently to solid surfaces (Figure 5.2b). In one such process, gluteraldehyde is employed to cross-link cells and attach them to a support which can be immersed into growth medium or used as a substrate over which medium is passed.

Fungal cells have frequently been immobilized by entrapment in alginate beads (Figure 5.2c). In this case the cells are resuspended in a 2–4 % (w/v) sodium alginate solution which is dropped into a calcium chloride solution where insoluble calcium alginate is produced in the form of spherical beads of varying diameters. Such beads are biochemically inert, porous and stable and entrap the fungal cells. Immobilization of yeast in alginate beads has led to an increase in ethanol productivity, but it has the disadvantage that the build-up of gas (especially carbon dioxide) within the bead may lead to disruption. Recently, other matrices, such as agar, gelatine and polyacrylamide, have been used for yeast entrapment.

The final form of immobilization involves inducing yeast cell flocculation or the formation of clumps. Flocculation is a natural process that normally occurs

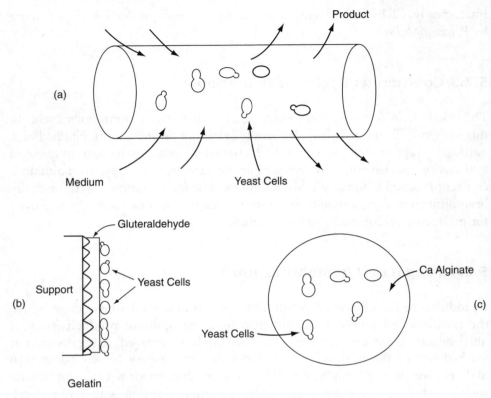

Figure 5.2 Fungal cell immobilization methods.

at the end of a fermentation. Inducing it at the end of a fermentation ensures a high cell density (potentially up to 10^{11} or 10^{12} cells/mL) and removes the need to centrifuge large volumes of culture medium to harvest cells.

Perhaps the most widely used system for immobilization is that which involves entrapping cells within membranes. This has a number of advantages, including the ability to ensure a high cell density, to raise ethanol productivity and to maintain high dilution rates in continuous cycle systems, since cell runoff is not a problem. Entrapped systems may present a diffusion barrier to the escape of metabolites through the surrounding matrix. In addition, there is the possibility that the metabolism of the yeast may change and affect the nature or quality of the desired product. Cell viability may be adversely affected in entrapped systems, since any toxins produced by the cells cannot diffuse away as easily and, hence, exert a direct impact on cells within the matrix.

5.2.5 Downstream Processing

Fungal fermentations can be employed to yield a range of products, but in many cases these must be concentrated before use. Typically, antibiotics and lipids are produced at a level of 10–30 g/L, single-cell protein at 30–50 g/L and ethanol at 70–120 g/L. Downstream processing (DSP) ensures that the product of interest is recovered in a reliable and continuous manner and that there is a large reduction in volume. The use of DSP will depend upon the nature of the product; for example, in the case of whisky, distillation is the main form of DSP. In the case of isolating proteins, the cell must be disrupted by sonication, enzymatic degradation of the cell wall or compression in a French press. Nucleic acids from the lysed cell are degraded using a nuclease. The enzymes of interest can be collected by ammonium sulfate precipitation and purified by running through a variety of columns containing Sephadex, DEAE-Sephacel or hydroxyapaptite. Once the product has been recovered it must be prepared for marketing, possibly by mixing with other agents and by packaging.

5.2.6 Factors Affecting Productivity of Fungal Fermentation

Irrespective of the nature of the fermentation system chosen or whether cells are free or immobilized, a number of parameters affect fungal productivity. Specific process requirements will dictate the choice of a fermentation system. For example, ethanol production by batch fermentation is well understood, but would be very poor if film fermentation was attempted. The nature of the fungal cell is critical, since yeast cells are amenable to growth in a bioreactor which uses spargers (paddles) to ensure nutrient mixing or oxygen transfer, whereas spargers would disrupt the mycelium of a filamentous fungus and adversely affect the fermentation. Agitation of fungal mycelium in bioreactors is usually achieved by bubbling air or gas (nitrogen) through the medium (Figure 5.3). Some fermentations require aerobic conditions, while others must be performed under semi-aerobic or anaerobic conditions. Certain fermentations require a specific initial inoculum in order to grow at an optimal rate. The nature of the growth medium must also be optimized for efficient fermentation to occur – many fungi are subject to catabolite repression; consequently, too high a concentration of glucose can inhibit the process. The temperature of the fermentation affects productivity, and a specific temperature can determine, for example, whether an ale or lager is produced during brewing.

Despite the wide range of fermentation systems available, the majority of commercially important fermentations are still either batch or fed batch. Part of the reason for this is that these systems are relatively easy to set up, the batch size can be varied to meet demand and, historically, most fermented beverages have been produced by batch fermentation. In addition, in the case of brewing the yeast cells are discarded at the end of the fermentation and a new inoculum is used, thus ensuring that strain-specific characteristics associated with a particular beer are not lost during prolonged fermentation. This is very important in ensuring the consistency of beers.

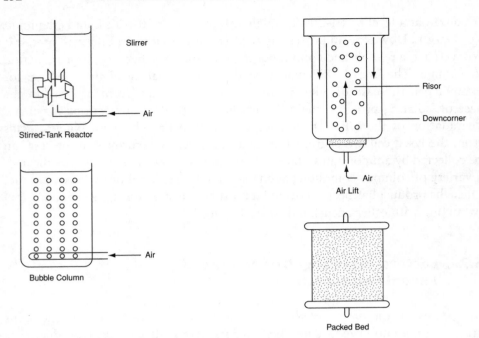

Figure 5.3 Means of aerating fungal cell cultures.

5.3 Commercial Fungal Products

5.3.1 Brewing

Yeasts are capable of producing ethanol by the semi-anaerobic fermentation of sugars, as in the case of beer, wine or cider manufacture. In addition to the production of ethanol, yeast also produces a range of compounds known as organoleptics, such as isoamyl alcohol, ethyl acetate and butanol, which give flavour to the final product. During the fermentation process the yeast converts glucose ($C_6H_{12}O_6$) initially into pyruvate through the process of glycolysis. Pyruvate is converted to acetaldehyde by the action of the enzyme pyruvate decarboxylase with the release of carbon dioxide. Acetaldehyde is subsequently converted to ethanol ($2C_2H_5OH$) by the action of alcohol dehydrogenase. The overall reaction can be summarized by the following equation:

$$C_6H_{12}O_6 \rightarrow 2C_2H_5OH + 2CO_2 \qquad \Delta G° \text{ (Fermentation)} = -58 \text{ kcal/mol}$$

Wine and cider are produced from grapes and apples respectively, where the fermentable carbohydrates are simple sugars such as glucose or fructose. In the case of beer production there is an added complication, in that the carbon source in barley, starch, is nonfermentable by most yeasts, which lack amylases and glucoamylases. In order to obtain a fermentable matrix, the barley is first

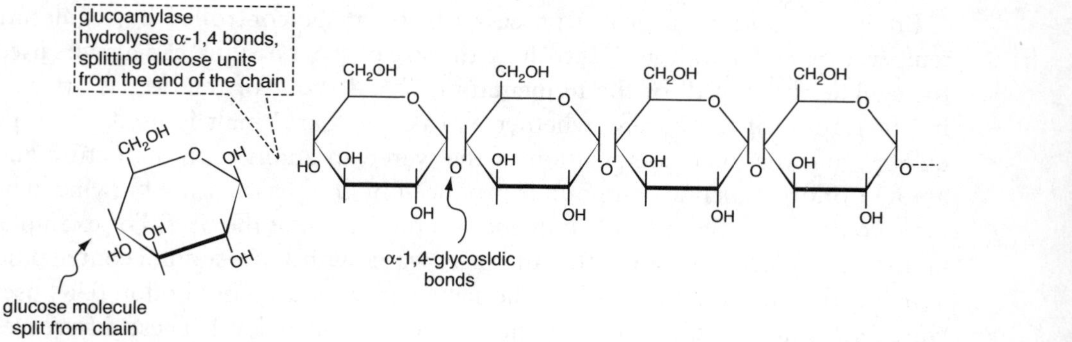

Figure 5.4 Breakdown of starch by amylases to yield a range of fermentable and non-fermentable sugars.

germinated in the process of malting. During malting, the endogenous α-, β- and gluco-amylases convert the starch into groups of fermentable sugars which the yeast can subsequently metabolize to ethanol (Figure 5.4). α-Amylase acts internally on the β-1,4 bonds in the starch molecule to form oligosaccharides of 7–12 glucose units. β-Amylase cleaves maltose units from the nonreducing ends produced on the oligosaccharide chains. The breakdown of starch to maltose and dextrins is known as saccharification. Once malting is complete, the barley seeds are dried and the nature of the drying affects the subsequent type of beer. A long, cool drying process produces a 'pale' malt which retains high enzymatic activity, whereas a short, hot drying process produces a 'dark' malt with low enzymatic activity. The dried seeds are crushed in the process of milling and the 'grist' is sprayed with hot water to extract the sugars and enzymes. At this point, nonfermentable carbohydrates (e.g. starch) represent the majority of sugars in the 'wort', but over a relatively short time the action of amylases converts such carbohydrates to fermentable compounds. After 2–3 h incubation, fermentable sugars represent approximately 75 % of the sugars in a typical wort, with non-fermentable carbohydrates constituting the remainder of the sugars. A typical wort at this stage contains the following fermentable sugars: fructose (2.1 g/L), glucose (9.1 g/L), sucrose (2.3 g/L), maltose (52.4 g/L) and maltotriose (12.8 g/L). Nonfermentable carbohydrates account for 23.9 g/L. The wort is boiled to sterilize it and reduce the pH. Hops are also added to the wort during boiling to impart flavour and tannins to the beer, which can give stability to the 'head' on the beer.

The yeast inoculum is added to the cooled and filtered wort and the brewing process commences. Traditionally, top-fermenting yeasts were used to make beer. These had an optimum temperature in the range 15–22 °C and rose to the top of the beer at the end of the process. Bottom-fermenting yeasts are now employed to produce the majority of beers and have a lower optimum temperature range than the top-fermenting yeast and settle to the bottom of the beer at the end of the fermentation in a process known as flocculation.

Brewing is a semi-anaerobic process which can be controlled by regulating temperature and agitation. Controlling the size of the inoculum can also be used to regulate the 'speed' of the fermentation. The flavour of the beer is affected by the nature of the barley, whether a 'pale' or 'dark' malt is used, the type of hops and the ionic composition of the water, in addition to the nature and amount of organoleptic compounds produced by the yeast during brewing. The ionic composition has a subtle influence on the nature of the beer. For example, Guinness produced in Dublin (Ireland) uses water with a low sodium content and a high carbonate content, whereas the same beer produced in London (UK) uses water with a high sodium content and a lower carbonate level. These differences lead to slight variations in the taste of the final products. At the end of the brewing process the beer is filtered to remove the yeast, pasteurized and packaged prior to dispatch. Some beers are left to mature prior to this final step (Figure 5.5).

Brewing can be used to give a range of beers depending upon the nature of the market to be supplied. Ales are typically produced using a fermentation temperature of 20–25 °C for 36 h, after which the wort is cooled to 17 °C for 72 h. In contrast, lager fermentations commence at 7–11 °C and can take up to 12–14 days to complete. Some beers, for example Budweiser, are produced using an extra source of amylases, such as rice. Human saliva also contains amylase and this has been used to produce 'native beers' from sorghum or millet. In this case the grain is ground-up and human saliva is added. The amylases in the saliva break down the starch, releasing fermentable carbohydrates which are fermented to ethanol by yeast. In recent years, 'light beers' have become commercially available. In these, the nonfermentable carbohydrates (dextrins, starch) are converted to fermentable carbohydrates by the addition of exogenous amylases and glucoamylases, which are subsequently fermented to ethanol. In January 2004 the Neuzeller Kloster brewery in Germany announced the launch of an anti-ageing beer! This beer is supplemented with vitamins and minerals 'designed to slow the ageing process'. Time will tell whether this new product really does delay the ageing process!

5.3.2 Recent Advances in Brewing Technology

Although traditional methods of producing alcohol for human consumption are still employed, much attention in recent years has focused on improving the yeast to give a higher ethanol yield, a faster fermentation time or to reduce the production of unwanted off-flavours (Table 5.3). Genetic engineering has been employed to generate brewing yeast that converts α-acetolactate directly to acetoin without the involvement of diacetyl – an organoleptic compound that is responsible for unpleasant flavours in beer. The gene for α-acetolactate decarboxylase (α-ALDC) has been introduced into yeast from bacteria and this allows the conversion of α-acetolactate directly to acetoin without the formation of diacetyl (Figure 5.6). Pilot studies using genetically modified yeast show

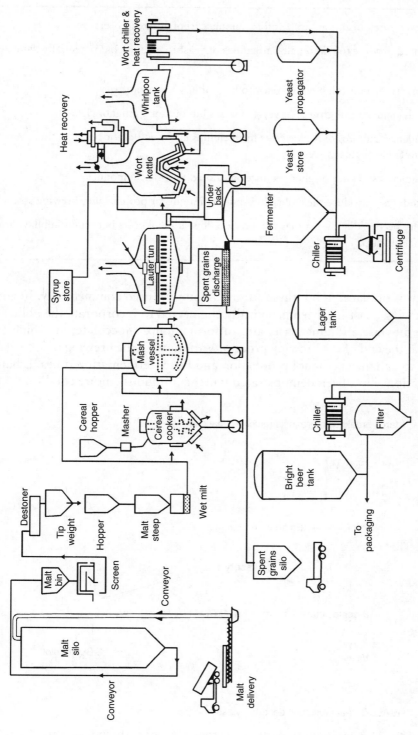

Figure 5.5 Flow chart of the stages in the production of beer.

Table 5.3 Reasons for attempting genetic manipulation of brewing yeast.

1. Carbohydrate utilization: utilize all remaining carbohydrate in beers

2. Increased product tolerance: alter tolerance of yeast to ethanol and potentially increase yield

3. Temperature: alter tolerance of yeast to high or low temperatures

4. Growth yield: curtail growth of yeast cultures but optimize ethanol yield

5. Flocculence: cultures of *S. cerevisiae* form clumps at the end of the fermentation. May need to enhance this process

6. Organoleptics: reduce or alter production of organoleptic compounds

7. By-products: introduce the ability to produce vitamins or proteins into brewing strains

8. Novelty beers: design yeast that will produce low-ethanol beers but retain full flavour of original product

lower levels of diacetyl, but these strains are not yet used commercially. In addition, brewing yeast has been genetically engineered to incorporate the ability to produce amylase and glucanase, and to form clumps (flocculate) at a high frequency at the end of the brewing process. Variations in the brewing process have also led to enhanced ethanol production and faster fermentation times. Ethanol fermentations have been demonstrated under continuous culture conditions and

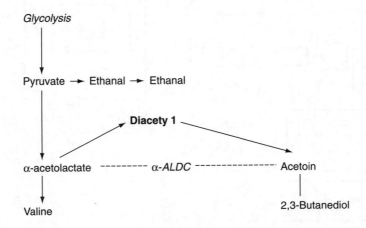

Starch breakdown products (glucose, maltose)

Glycolysis

Pyruvate → Ethanal → Ethanal

Diacety 1

α-acetolactate ---------- α-ALDC --------------- Acetoin

Valine

2,3-Butanediol

α-ALDC: α-acetolactate decarboxylase

Figure 5.6 The introduction of α-acetolactate decarboxylase into brewing yeast prevents diacetyl formation during ethanol production.

immobilized cells have been employed as the yeast inoculum. In this case a higher cell density is attainable, there is less production of diacetyl and the fermentation is faster than when free yeast cells are used. While genetically engineered yeast has been shown to be superior to conventional yeast in many respects, its use for the production of commercial products has not yet occurred due in part to the reluctance of the consumer to purchase genetically engineered produce.

5.3.3 Role of Yeast in Bread Making

Yeast plays a central role in the manufacture of bread. In this case the dough is infused with *S. cerevisiae* and aerated. The dough is left to stand for a short period in a warm environment during which time yeast cell respiration occurs and carbon dioxide is produced. Respiration can be summarized by the following equation:

$$C_6H_{12}O_6 + 6O_2 \rightarrow 6CO_2 + 6H_2O \qquad \Delta G° \text{ (Respiration)} = -686 \text{ kcal/mol}$$

The production of carbon dioxide gives bread its light, airy quality. While baking is essentially an aerobic process, some semi-anaerobic conditions develop in the dough and ethanol can be produced in small amounts. This, together with the carbon dioxide, is burnt off during the baking process and gives rise to the 'fruity' smell often associated with bakeries. In some parts of Japan the bread dough is left to stand for a few days, during which the oxygen is used up in respiration and fermentation commences. The dough soon contains appreciable levels of ethanol and is subsequently eaten!

5.3.4 Single-Cell Protein: A Novel Food Source

A range of fungi can be utilized to yield single-cell protein (SCP) from a variety of sources (Table 5.4). The filamentous fungus *Trichoderma viride* has been employed to obtain SCP from cellulolytic material, and the yeasts *Saccharomycopsis fibuliger* and *Candida tropicalis* are employed to produce SCP from starch in the 'Symba process'. In this process, starchy effluents from food-processing factories are inoculated with *Sacch. fibuligera*, which has the ability to hydrolyse and ferment soluble starch. The resulting mixture of oligosaccharides and reducing sugars are used to grow *C. tropicalis*, which can be used as a dietary supplement.

Although SCP may be produced from a variety of microorganisms, fungi have a number of advantages. Fungi have been well studied and characterized and the eating of fungi (e.g. mushrooms) and products containing fungi (e.g. Roqueforti cheeses) is well accepted. In addition, fungi have a low nucleic acid content and their filamentous morphology can be manipulated to give a fibrous appearance similar in appearance and texture to meat. In recent years fungi have been exploited for the production of SCP which is available commercially as a human food.

Table 5.4 Material and fungi used in SCP production.

Material	Fungi
Cellulose	*T. viride*
Ethanol	*Candida utilis*
Banana peels	*Pichia* species
Beef fat	*Saccharomycopsis* species and *C. utilis*
Sugars	Range of yeast species
Starch	*Saccharomycopsis fibuliger*

Fusarium graminearum is employed for the manufacture of mycoprotein and is marketed as Quorn™ mycoprotein. The search for a good producer of mycoprotein (also referred to as SCP) commenced in 1968 when Rank Hovis McDougal Ltd began screening over 3000 soil isolates for a fungus capable of growing on starch and giving a high yield of protein. Initial work concentrated on a *Penicillium* species, but an *F. graminearum* isolate (labelled A3/5) proved superior. Isolate A3/5 had an optimum growth temperature of 30 °C and a mean generation time of 3.5 h. When grown in medium containing 100 g glucose, 54 g of fungus results, of which 45 % (by weight) is protein.

The fermentation is now conducted in an air-lift continuous fermenter with cycle times of approximately 3000 h (Figure 5.7). The culture is maintained in the exponential phase of growth and the feedstock is starch. One problem associated with the long fermentation time is the appearance of a highly branched variant that arises spontaneously in the culture which adversely affects the texture of the final product. The appearance of the variant can be minimized by limiting nutrients and regulating pH.

Fungal mycelium is harvested on a continual basis, heat treated to reduce the RNA content from 10 % to 2 % and processed for marketing. The requirement to reduce the RNA content by thermal treatment is due to the fact that a high RNA intake can lead to a build up of uric acid in the body, leading to deposition in the kidneys and joints. Fungal mycoprotein has a net protein utilisation (NPU) of 75 compared with NPU values of 80 for beef and 83 for fish. The fungal mycelium can be processed to give an appearance and 'mouth-feel' of meat.

Quorn mycoprotein was launched on the market in 1985 and in 1999 over 790 t were produced. It is sold in ready-to-eat meals and also in the form of cubes of SCP which can be flavoured and used in home cooking in much the same way as chicken. It has the advantage of being suitable for vegetarians and those on low calorie diets, since it contains no animal ingredients and is low in cholesterol and fat. It is currently available in a range of European countries and the USA.

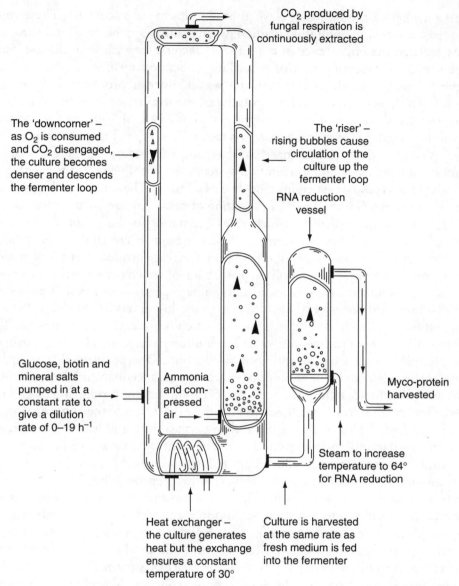

CO$_2$ produced by fungal respiration is continuously extracted

The 'downcorner' – as O$_2$ is consumed and CO$_2$ disengaged, the culture becomes denser and descends the fermenter loop

The 'riser' – rising bubbles cause circulation of the culture up the fermenter loop

RNA reduction vessel

Glucose, biotin and mineral salts pumped in at a constant rate to give a dilution rate of 0–19 h^{-1}

Ammonia and compressed air

Myco-protein harvested

Steam to increase temperature to 64° for RNA reduction

Heat exchanger – the culture generates heat but the exchange ensures a constant temperature of 30°

Culture is harvested at the same rate as fresh medium is fed into the fermenter

Figure 5.7 Continuous fermenter used in the production of Quorn™ myco-protein (Reproduced with permission from Trinci (1992).)

5.3.5 Antibiotic Production by Fungi

Perhaps one of the most important discoveries regarding the beneficial use of fungi for humans was the identification in 1928 by Sir Alexander Fleming that an isolate of *Penicillium notatum* produced a substance capable of killing Gram+ bacteria. This compound was subsequently identified as penicillin and was the first member of the β-lactam class of antibiotics to be discovered. These compounds function by inhibiting peptidoglycan synthesis in bacteria and their use has reduced the importance of the Gram+ bacteria as a cause of disease. Subsequent to the identification of penicillin production by *P. notatum*, a screen revealed that *Penicillium chrysogenum* was a superior producer. Following a series of mutagenic and selection procedures, the strain used in conventional fermentations is capable of producing penicillin at a rate of 7000 mg/L compared with the 3 mg/L of Fleming's *P. notatum* isolate. The type of fermentation system that is employed has also changed, so that most systems are submerged fermentations rather than solid fermentation systems. A typical penicillin fermentation yields three types of penicillin, namely F, G and V. The latter can be used directly; however, G is modified by the action of penicillin acylase to give a variety of semi-synthetic penicillins which show resistance to the action of bacterial penicillinases which are implicated in conferring anti-bacterial drug resistance.

The majority of antibiotics obtained from fungi are produced by fermentation and most are secondary metabolites, production of which occurs in the stationary phase and is linked to sporulation. Catabolite repression can inhibit antibiotic production, and one way to avoid this is to use low levels of glucose in the fermentation medium or to obtain a mutant which is not catabolite repressed. The chemical content of the medium must be monitored, since high levels of nitrogen or phosphate (PO_4) retard antibiotic production. One problem that seriously affects the productivity of antibiotic fermentations is feedback inhibition, where the antibiotic builds to high intracellular levels and retards production or kills the cell. One means of reversing this is to introduce low levels of the antifungal agent amphotericin B. This increases membrane permeability and leads to a decrease in intracellular antibiotic levels and a concomitant increase in production.

Antibiotic production can be maximized by optimizing production as a result of random mutagenesis and selection. Another approach has been to fuse or mate strains capable of producing high amounts of product with strains that are good secretors of product. Rational selection is a process where a chelating agent is introduced into the fermentation to complex all the metal ions present and, consequently, has a beneficial effect on antibiotic production. More recently, genetic cloning has been employed to express the genes for antibiotic production in another species – this has the possibility of producing hybrid antibiotics with potential novel targets.

5.3.6 Enzyme Production by Fungi

Fungi have been widely exploited as a source of industrially important enzymes for many years, and the principal fungi used in this regard are members of the *Aspergillus* and *Penicillium* genera (Table 5.5). Fungi have been the organism of choice for enzyme isolation, since their biology is well characterized and the ones of interest have been awarded 'generally regarded as safe' (GRAS) status.

Table 5.5 Fungal enzymes: sources and applications.

Fungus	Enzyme	Application
Yeast		
S. cerevisiae	Alcohol dehydrogenase	Ethanol assay
S. cerevisiae	Invertase	Confectionery
Kluyveromyces lactis	Lactase	Dairy industry
Mould		
A. niger	β-Glucanase	Brewing
A. oryzae	Protease	Meat tenderizer
A. oryzae	α-Amylase	Food industry
A. niger + A. oryzae	Lipase	Dairy industry
Aspergillus species	Pectinase	Fruit juice clarification
T. viride	Cellulase	Dehydrated foods
Mucor miehei	Rennin	Cheese manufacture

Catalase (EC 1.11.1.6) is used in cold sterilization and has been isolated from *Aspergillus niger*, lipase (EC 31.1.3. glycerol ester hydrolase) is used as a flavour enhancer and amylase (EC 3.2.1.1. α-1,4-glucan 4-glucanohydrolase), used for malting barley or improving the quality of bread, has been obtained from a number of *Aspergillus* species. Glucose oxidase (EC 1.1.3.4. β-D-glucose:O_2 oxidoreductase) is employed in glucose assays and can be isolated from *P. notatum*. Other enzymes that are obtained from filamentous fungi or yeast include cellulase (EC 3.2.1.4, β-1,4-glucan glucanohydrolase), which degrades cullulose, invertase (EC 3.2.1.26 β-D-fructofuranoside fructohydrolase), which converts sucrose to glucose and fructose and is used in jams and confectioneries, and pectinase (EC 3.2.2.15 polygalacturonide), which is used in the clarification of wine must and fruit juices.

Traditionally, fungi cultivated for the isolation of enzymes have been grown under solid fermentation conditions where the fungus is allowed to grow across a solid substrate in a low water environment. Typically, the substrate has been bran or other grain-based material, although recently other fermentation systems, such as fed batch, submerged or continuous cycle, using a wider range of carbon sources, have been utilized. Fungi are a good source of a number of enzymes and the isolation of these enzymes from fungi has many advantages over the use of animal or plant cells as sources of enzyme. Fungi display metabolic flexibility; that is, there are many variations in their metabolic pathways compared with animal cells. They can be grown readily using simple growth media and are amenable to genetic manipulation. Strain development to enhance enzyme production or stability can be achieved using mutagenesis or by applying a selection pressure for a specific enzyme. Fungi can also be transformed using a variety of vectors to produce recombinant proteins. Fungal enzymes, like cells, may be immobilized to increase their efficacy or to maintain them in a reactor. Immobilization of enzymes increases their stability but may also alter the pH at which they demonstrate optimum activity. One of the earliest examples of enzyme immobilization was performed by Tate and Lyle, who attached invertase to charcoal, which was subsequently used to hydrolyse sucrose.

In terms of fungal enzyme fermentation, one of the main problems is the control of bioreactor temperature, particularly when large volumes are used. Another problem encountered at the industrial level is the loss of strain-specific characteristics during prolonged fermentation, which can adversely affect the production of the desired enzyme. Owing to the nature of the fungal cell wall, intracellular enzymes must be released from the cell by some form of disruption. Methods such as mechanical breakage, alkali treatment and enzyme digestion of the cell wall have all been employed to release the enzyme(s) of interest. Following release of the enzyme there is a requirement to 'clean up' the material by removing nucleic acids, membranes and solids. The enzyme may be purified by affinity purification and used in its free state or in an immobilized state.

5.3.7 Mushroom Production

Mushrooms are widely consumed throughout the world and are produced in a multistage process. China, the USA and the Netherlands are the biggest producers of mushrooms, producing 442 000 t, 358 000 t and 240 000 t respectively. The biggest importer of fresh mushrooms is the UK (47 000 t), while the Netherlands and Poland are the two biggest European exporting countries. In western Europe the most commonly consumed mushroom is the edible mushroom (*A. bisporus*), but species such as *Pleurotus*, *Lentinus edodes* and *Auricularia* are the dominant cultivated mushrooms in the Far East.

The growth medium for the cultivation of *A. bisporus* is compost which consists of stable manure, straw and possibly chicken manure. The compost is spawned with a culture of *A. bisporus* (edible mushroom) grown on sterilized cereal grains and cultured at 25–28 °C and high humidity for 14 days to allow the mushroom mycelium to colonize the compost (Plate 5.1). The chitin-rich mycelium consists of bunches of septate hyphae which digest food externally and absorb digested material as the fungus grows throughout the compost. After this period the compost is 'cased' with a layer of neutralized peat, which has the effect of stimulating the formation of large primordia which will later develop into mushrooms. The exact function of casing is unclear, but it may provide biotic, chemical and/or physical factors which trigger mushroom formation. Under ideal conditions and in the absence of overt disease, flushes or crops appear at weekly intervals until the nutrients in the compost are depleted.

5.3.7.1 Fungal Diseases of Mushrooms

The conditions under which mushrooms are produced provide an ideal environment for the growth of a wide range of potentially damaging fungi. The fungi that affect mushroom crops can be divided into two broad classes: the 'weed moulds', which typically grow in the compost and compete with the mushroom mycelium for nutrients, water and space, and the fungal pathogens (or mycopathogens), which directly attack the mushroom. Weed moulds include such species as *Trichoderma*, *Chrysporium* and *Coprinus* and may affect the actual growth of the mushroom, but their impact on crop yield may be variable. Some species of *Trichoderma* (often referred to as green mould) (Plate 5.2) grow on the casing, while others, such as *Chrysosporium*, grow at the casing–compost interface and inhibit the growth of the mushroom mycelium.

The principal fungal pathogens of mushrooms are *Lecanicillium fungicola*, *Mycogone perniciosa* and *Hypomyces rosellus*. Each disease is highly destructive and the spores of each fungus are dispersed by air currents, water splashes and human contact. Dry bubble is caused by *Lecanicillium fungicola* and is characterized by a swelling of the mushroom stipe and the appearance of blue–grey spots on the cap from which a grey-coloured mycelium may develop (Plate 5.3). Wet bubble is caused by *Mycogone pernicosa* and infects cultivated mushrooms and probably wild mushrooms and toadstools also. Under humid conditions, infected mushroom stipes swell and the caps are small and misshapen. Amber-coloured droplets appear over the surface of the mushroom, which subsequently collapses into a wet bubble-like decaying mass. Cobweb disease of mushrooms is caused by *Hypomyces rosellus*, which colonizes the mushroom and also grows through and over the surface of the casing soil. Infected mushrooms become engulfed by the growth of the cobweb-like mycelium and the gills fail to develop.

The above pathogens are responsible for approximately 90 % of all fungal diseases of mushrooms and cause considerable economic loss to mushroom growers. Conventional control measures involve the use of fungicides and the adherence to strict hygiene standards. Under ideal conditions a grower would expect four crops from a batch of compost, but the presence of the above mycopathogens reduce the usual harvest to two or three flushes.

5.3.8 Soy Sauce

Fungi play an essential role in the production of Soy sauce, which is a very popular condiment for use with Asian food. Soy sauce is produced from the solid fermentation of crushed soy beans and wheat. Initially the soybeans are cooked to remove contaminants and then mixed with roasted wheat grains and pressed into cakes. These are then inoculated with *Aspergillus oryzae*, which, under aerobic conditions, grows over the surface of the cakes. The fermentation is allowed to continue for 24–48 h at 25 °C during which time invertases, amylases and cellulases produced by the fungus begin to degrade the matrix. After this period the cakes (now referred to as Koji) are mixed with brine (salt–water) and the yeast *Saccharomyces rouxii* and a range of lactobacilli are added. Anaerobic conditions develop which prevent the continued growth of *A. oryzae* and this is allowed to ferment for up to 6 months. After that time a dark, salty liquid is removed from the fermentation which has a high concentration of simple sugars, amino acids and vitamins. This can be left to mature for a longer period if required or alternatively it is pasteurized and bottled for sale.

5.3.9 Role of Fungi in Biofuel Production

There has been great interest recently in the production of fuel from renewable resources, especially as the reserves of fossil fuels begin to diminish and the costs of petroleum products escalate. Fungi may have a significant role in commercial biofuel production, especially since they are known to degrade lignocellulose material very effectively and many are good producers of solvents (e.g. *S. cerevisiae*). The fermentation of hexose sugars (e.g. glucose) by yeast is well established, but their ability to ferment pentose sugars (e.g. xylose) was only identified in 1980. Pentose sugars such as xylose are located in the xylan component of the hemicellulose of wood and plant materials. Approximately 175 million tons of potentially fermentable hemicellulose material are produced each year in the USA, and the efficient conversion of this to an economically beneficial fuel would be of enormous benefit. The fermentation of xylose to ethanol can be achieved using *S. cerevisiae*, but the xylose has to be converted to xylulose by the action of xylose isomerase in advance and the conversion requires a very high temperature (75 °C). In the early 1980s a number of xylose-fermenting yeasts were identified and these included yeasts such as *Pachysolen tannophilus*, *Candida shehatae* and *Pichia* spp. These achieve the fermentation of xylose to ethanol by first converting it to the pentitol xylitol by the action of xylose reductase. This is then converted to xylulose by xylitol dehydrogenase and this can then enter the pentose phosphate pathway of the cell. In contrast to glucose fermentation, xylose fermentation requires the presence of oxygen in determining the efficiency of the conversion to ethanol. It is believed that the requirement for oxygen lies in the ability of the cell to regenerate NAD^+ for xylitol dehydrogenase activity during respiration. Anaerobic conditions lead to an accumulation of xylitol and little ethanol production.

A number of filamentous fungi have been evaluated for their ability to produce biofuels. *Trichoderma reesei* is a cellulose-degrading fungus that is common in soils and decaying plant material throughout the world. This fungus was used to rapidly degrade plant biomass to produce easily fermentable sugars that could be subsequently converted to ethanol. In recent years much attention has been focused on an endophytic fungus isolated from trees growing in Patagonia. This fungus, *Gliocladium roseum*, can degrade cellulose and, interestingly, produce a wide range of chemicals (e.g. heptane, octane, benzene) that can be used as fuels. The possibility of utilizing this fungus or other closely related species to produce a type of diesel is being actively pursued.

While fungi will almost certainly play a very significant role in the production of biofuels, the continued development of this concept will only proceed if the economics of the process can be optimized and the adverse effects on food production can be reduced.

5.4 Conclusion

Fungi have been utilized for thousands of years for the production of various foods and beverages. While these applications are still important, fungi are now being used in novel ways for the production of SCP, antibiotics, enzymes and as expression systems for the production and secretion of foreign proteins. Consequently, the continued use of fungi on a large scale by humans is guaranteed.

Revision Questions

Q 5.1 Outline the main fermentation systems used commercially with fungi. Give examples of a product generated by each system.

Q 5.2 Describe the four methods for immobilizing yeast cells.

Q 5.3 What factors affect the productivity of a fermentation?

Q 5.4 Describe how recent advances in brewing technology could impact on the type of beer consumed in the future.

Q 5.5 Describe the processes involved in the production of Quorn mycoprotein.

Q 5.6 What factors affect the synthesis of antibiotics by fungi?

Q 5.7 What fermentation system is used to produce fungal enzymes?

Q 5.8 Outline the stages in the cultivation of *Agaricus bisporus*.

Reference

Trinci, A.P.J. (1992) Myco-protein: a twenty-year overnight success story. *Mycological Research*, 96(1), 1–13.

Further Reading

Books

An, Z. (ed.) (2005) *Handbook of Industrial Mycology*, Taylor & Francis Group.

Baltz, R.H., Davies, J.E. and Demain, A.L. (eds) (2010) *Manual of Industrial Microbiology and Biotechnology*, 3rd edn, American Society for Microbiology.

Boulton, C. and Quain, D. (2006) *Brewing Yeast and Fermentation*, Blackwell Publishing.

Cheung, P.K. (ed.) (2008) *Mushrooms as Functional Foods*, John Wiley & Sons, Ltd.

Fratamico, P.M., Annous, B.A. and Gunther, N.W. (eds) (2009) *Biofilms in the Food and Beverage Industries*, Woodhead Publishing Ltd.

Montville, T.J. and Matthews, K.R. (2008) *Food Microbiology: An Introduction*, 2nd edn, American Society for Microbiology.

Robson, G.D., van West, P. and Gadd, G.M. (eds) (2007) *Exploitation of Fungi*, Cambridge University Press.

Walker, G.M. (1998) *Yeast: Physiology and Biotechnology*, Wiley.

6

Pharmaceutical and Chemical Commodities from Fungi

Karina A. Horgan and Richard A. Murphy

6.1 Introduction to Pharmaceutical and Chemical Commodities

The economic significance of fungal biotechnology cannot be understated; indeed, as this chapter will outline, fungi have been exploited to yield a range of valuable products, some of which have proved extremely useful to mankind. Since the time of the pharaohs, fungi have been utilized for simple food processing; however, the last century has seen the development of fungal biotechnology for the subsequent production of valuable commodities such as antibiotics, enzymes, vitamins, pharmaceutical compounds, fungicides, plant growth regulators, hormones and proteins. As we move forward in the twenty-first century, this list will surely expand further. However, it is beyond the scope of this chapter to fully appreciate the enormous benefits and economic impact of fungi in the area of biotechnology. Instead, we will concentrate on a number of the more economically significant production processes, which have been developed through the utilization of fungi. The diverse natures of some of the economically important products produced by fungi are listed in Table 6.1.

Fungi: Biology and Applications, Second Edition. Edited by Kevin Kavanagh.

Table 6.1 Fungal products of economic importance.

Class of product	Typical example	Industrial/commercial application	Common production organism
Enzymes	Amylase	Starch processing Fermentation application	*Aspergillus niger* *Rhizopus oryzae.*
	Cellulase	Animal feed industry Brewing	*Trichoderma longibrachiatum*
	Protease	Meat/leather industry Cheese manufacture	*Aspergillus oryzae* *Rhizopus oligosporus*
Organic acid	Citric acid Itaconic acid Malic acid Fumaric acid	Soft drinks industry Chemical industry Beverage/food industry Food industry	*Aspergillus niger* *Candida/Rhodoturula* *Candida* *Candida*
Vitamins	Riboflavin Pyridoxine D-Erythro- ascorbic acid	Health industry Health industry Health industry	*Candida* *Pichia* *Candida*
Antibiotics	Penicillin Cephalosporin	Human/animal health Human/animal health	*Penicillium chrysogenum* *Cephalosporium acremonium*
Fatty acids	Stearic Dicarboxylic	Food industry Chemical industry	*Cryptococcus* *Candida*
Alcohol	Industrial alcohol Beverage alcohol	Fuel industry Beverage industry	*Saccharomyces* *Saccharomyces*
Pharmaceuticals	Lovastatin Cyclosporin	Human health Human health	*Monascus ruber* *Tolypocladium inflatum*
Amino acids	Lysine Tryptophan Phenylalanine	Health industry Health industry Health industry	*Saccharomyces* *Hansenula* *Rhodoturula*
Recombinant proteins	Insulin Phytase Hepatitis B surface antigen	Treatment of diabetes Phosphate liberation Vaccine preparation	*Saccharomyces cerevisiae* *Aspergillus niger* *Saccharomyces cerevisiae*

6.2 Fungal Metabolism

A common link between all fungi is their heterotrophic nature; they cannot manufacture their own food and depend on the organic material in other organisms for their survival. In a broad sense, however, it is possible to divide fungi into two main groups depending on how they obtain and assimilate nutrients. One group, the parasitic and mutualistic symbionts, obtains their nutrients in an effective manner from living organisms. The second group, saprotrophs, has the ability to convert organic matter from dead organisms into the essential nutrients required to support their growth. It is this second group that we are particularly interested in, as this collective of organisms gives rise to the production of the main bulk of the commodities commonly associated with fungi. However, regardless of this division, within the fungal life cycle one can clearly delineate the production of certain products or metabolites into two phases, namely primary and secondary metabolism.

Primary metabolites are those that are essential for growth to occur and include proteins, carbohydrates, nucleic acids and lipids. Indeed, the precursors of these primary products must be synthesized if they cannot be obtained from the growth medium. These primary metabolites have essential and obvious roles to play in the growth of the fungus. Typically, primary metabolites are associated with the rapid initial growth phase of the organism and maximal production occurs near the end of this phase. Once the fungus enters the stationary phase of growth, however, primary metabolites may be further metabolized. Examples of primary metabolites produced in abundance include enzymes, fats, alcohol and organic acids. Economically speaking, primary metabolites are easily exploited as the biochemical pathways involved in their production are widespread throughout the fungal kingdom, with common metabolites occurring in a wide range of fungi. This allows for the rapid screening of classes of fungi for such products and the easy development of production processes for their utilization. Primary metabolic processes have also been extensively usurped through the use of recombinant DNA technologies to the extent that heterologous proteins can be routinely produced by the host fungus as part of its primary metabolic phase.

In contrast to the primary metabolites, secondary metabolites are not essential for vegetative growth and, indeed, may have little or no primary function within the organism. Secondary metabolites are produced when the organism enters the stationary phase, once the initial phase of rapid growth has declined. The metabolites produced in this phase are often associated with differentiation and sporulation and can have profound biological activities, which in some instances have been exploited economically. A number of distinct differences are apparent between primary and secondary metabolites. In the first instance they have been shown to possess an enormous variety of biosynthetic origins and structures that are not in general found amongst the primary metabolites. Second, their occurrence tends to be restricted to a small number of organisms and, indeed, can vary between isolated strains of the same species. Finally, their production is characterized by the generation of groups of closely related compounds, which may have very different biological properties.

Important examples of secondary metabolites include medically important compounds such as antibiotics, statins, cyclosporins and ergot alkaloids. Agriculturally important secondary metabolites include strobilirubin, an antifungal compound, and plant hormones such as giberellic acid. Table 6.2 outlines some

Table 6.2 Example of primary and secondary metabolites.

Example	Production organism
Primary metabolites	
Enzymes	*Aspergillus* sp.
Industrial Alcohol	*Saccharomyces cerevisiae*
Organic acids	*Aspergillus/Candida*
Fats	*Candida*
Polymers	*Yarrowia*
Secondary metabolites	
Antibiotics	
Penicillin	*Penicillium*
Fusidic acid	*Fusidium coccineum*
Cholesterol lowering agents	
Lovastatin	*Monascus ruber*
Mevastatin	*Penecillium citrinum*
Immunosuppressing drugs	
Cyclosporin A	*Tolypocladium inflatum*
Plant hormones	
Giberellic acid	*Giberella fujikuroi*

of the more commonly found primary and secondary metabolites and the organisms from which they have been commercially exploited.

Fungal biotechnology has developed to allow the utilization of the metabolic processes inherent to the organisms in a commercially viable manner. In this chapter we will detail a number of the more important commercial commodities produced by fungi and outline the production processes for them.

6.3 Antibiotic Production

The most well-known and possibly best-studied secondary metabolites of fungi are a class of compounds known as antibiotics. These low-molecular mass compounds are so called because at low concentrations they inhibit the growth of other microorganisms. While many thousands of antibiotics have been discovered, their use has been limited to perhaps 60 at most due to the toxic properties they exhibit towards humans. Clinically speaking, the majority of antibiotics are produced by Actinomycetes, a bacterial order, and will not be dealt with here. Whilst several fungal genera produce antibiotics, only two do so to a commercially viable extent, and these include *Aspergillus* and *Penicillium*. The β-lactams, of which penicillin is the most infamous, not least because of its fortuitous discovery by Fleming in 1928, comprise a very large group of antibiotics and includes both the cephalosporins and penicillins. In 2000, the estimated world market for antibiotics was $28 billion, which underlies the importance both medically and economically of these metabolites.

At their core, all cephalosporins and penicillins possess a β-lactam (four-atom cyclic amide) ring on which side-chain substitutions and differences give rise to a series of antibiotics each with differing antibacterial activity. In addition to the so-called classical β-lactams, semi-synthetic varieties can be manufactured by the removal of the naturally occurring side chains and the subsequent chemical derivitization of the core β-lactam ring. Figure 6.1 illustrates the core β-lactam ring and the basic structures of penicillin and cephalosporin.

Gram-positive bacteria have on the outside aspect of the cell wall a layer which is composed of characteristic groupings of proteins and carbohydrates that comprise the antigenic determinants responsible for generating an immune response. Inside this outermost layer there is a polymeric structural layer known as peptidoglycan which is composed of repeating units of N-acetylglucosamine

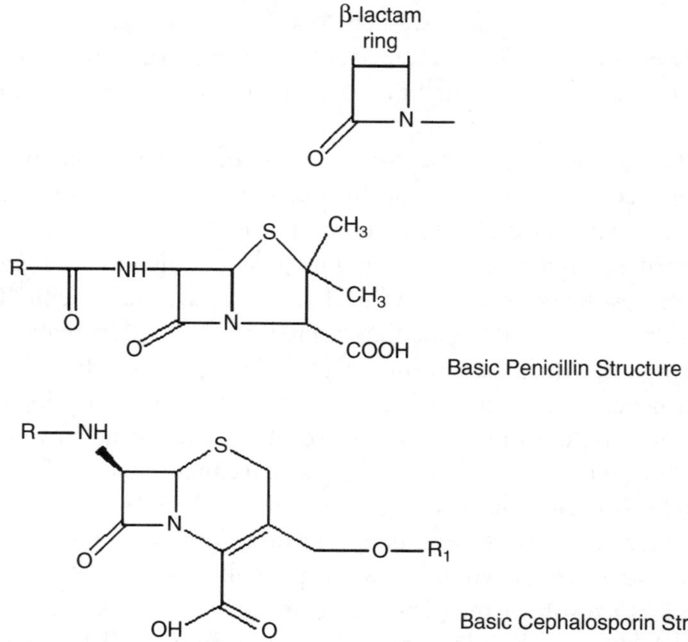

Figure 6.1 Structures of the core β-lactam ring and β-lactam antibiotics.

(NAG) and N-acetylmuramic acid (NAM). Associated with this cell-wall structure are a number of proteins known as penicillin binding proteins (PBPs), some of whose functions are as yet unclear. During cell-wall biosynthesis, a cross-linking process occurs whereby peptidoglycan strands become linked, leading to the structural stability of the wall. It is this cross-linking which is extremely sensitive to β-lactam antibiotics. For instance, various penicillins bind to the PBPs through their different side chains, leading to a variety of effects. Reaction with PBP-1 (a transpeptidase) produces cell lysis, while binding to PBP-2 (also a transpeptidase) leads to the generation of oval cells which are unable to replicate. Cephalosporins act in a very similar fashion to the penicillins and are also able to react with the PBPs by forming covalent bonds, thus leading to cellular lysis.

Gram-negative cells have a more complex cell-wall structure and usually contain an outer membrane and a complex periplasm comprised of lipopolysaccharides. Whilst the Gram-negative cell wall also contains a peptidoglycan layer, it is not as extensive as that of Gram-positive bacteria but is sensitive to β-lactam antibiotics owing to the presence of PBPs.

The word penicillin can be regarded as a generic term used to describe a large group of natural and semi-synthetic antibiotics that differ only by the structure of the side chains on the core aminopenicillanic acid ring. As a rule, the basic penicillin molecule consists of a β-lactam ring, a five-membered thiazolidine ring and a side chain. β-Lactams with non-polar side chains, such as phenylacetate and phenoxyacetate, are hydrophobic in nature and include penicillin G (benzylpenicillin) and penicillin V (methylpenicillin). The nonpolar penicillins are synthesized only by filamentous fungi.

Penicillins with polar side chains, such as D-α-aminoadipate, include penicillin N and possess hydrophilic characteristics. They are more widely synthesized by a range of microorganisms, including fungi, actinomycetes and unicellular bacteria.

The production of semi-synthetic penicillins is quiet easy and involves the removal of the side chain from naturally occurring penicillin and its subsequent replacement with a different side chain to yield a novel β-lactam derivative. Examples of semi-synthetic varieties include methicillin and ampicillin. Figure 6.2 illustrates the structure of natural and semi-synthetic penicillin.

One serious drawback with penicillins relates to the highly reactive nature of the β-lactam ring, which can result in their being susceptible to a variety of degradation processes. Factors which can affect their stability include their reactivity with hydroxide ions, which can result in the formation of inactive penicilloic acid, or their acid-sensitive nature, which can lead to their degradation at low pH. Acid sensitivity can be overcome clinically by use of the compounds in a buffering solution. A more serious limitation to their use, however, is their susceptibility to a group of enzymes known as penicillinases, which are produced by bacteria and can result in the generation of antibiotic resistance. The most common of these enzymes is β-lactamase, which cleaves the β-lactam ring and thus inactivates the antibiotic. A variety of acylases have also been identified, whose mode of action is to cleave the acylamino side chain of the antibiotic, thus

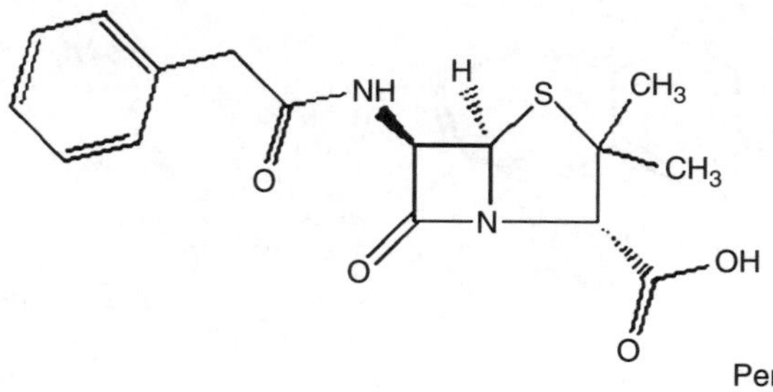

Penicillin G (Natural)

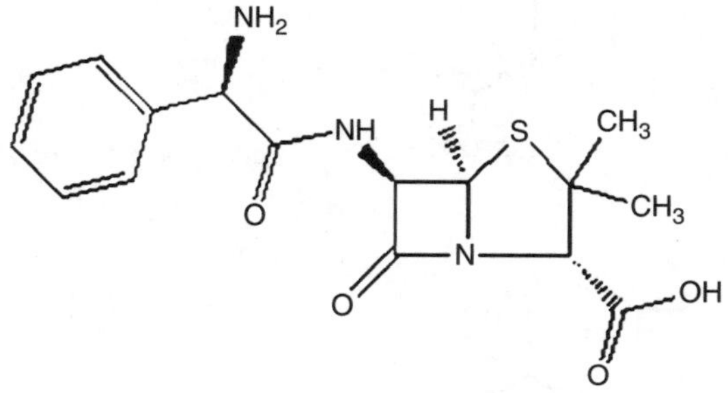

Ampicillin (Semi-Synthetic)

Figure 6.2 Natural and semi-synthetic penicillins.

rendering them inactive. To combat these enzymes, a number of compounds, such as clavulinic acid or the sublactams, have been developed that when given in combination with the susceptible antibiotic result in the permanent inactivation of the antibiotic-degrading enzymes. Of more importance, however, has been the development of the semi-synthetic penicillins, many of which are resistant to β-lactamase and other penicillinases. For instance, methicillin is completely resistant to these enzymes, though it does have the disadvantage that it is less effective. Almost all β-lactamase-resistant penicillins are less potent than the parent molecules.

Cephalosporins are very closely related to the penicillins and were initially discovered soon after; however, unlike the penicillins, their use was limited for a long period until a clinically useful agent was found. Cephalasporin C is regarded as the prototypical cephalosporin, and following its structural elucidation it was found to be a β-lactam with a six-membered dihydrothiazine ring instead of the five-membered thiazolidine ring characteristic of the penicillins. Chemical removal of the side chain of cephalosporin C results in the generation of 7-aminocephalosporonic acid (7-ACA), which can be used as a synthetic starting point for most of the cephalosporins available today. Indeed, it is more economically feasible to produce 7-ACA from penicillin G by a series of synthesis reactions rather than to incur the prohibitive costs of fermentation to produce the antibiotic.

One notable feature of the structure of the cephalosporins is their reduced chemical reactivity relative to the penicillins. However, some β-lactamases are more efficient at cleaving cephalosporins than penicillins, and this has led to the

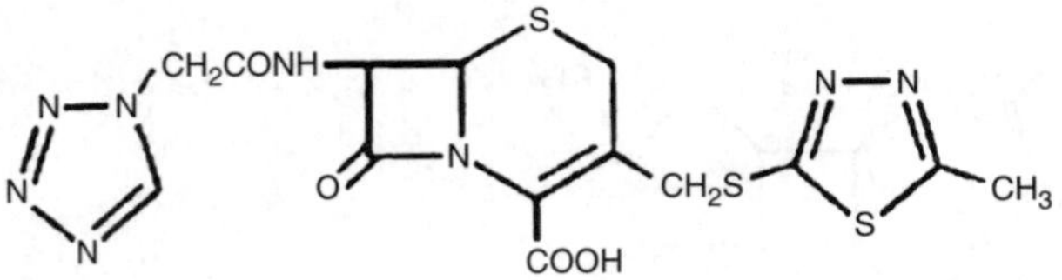

Cefazolin (1st generation)

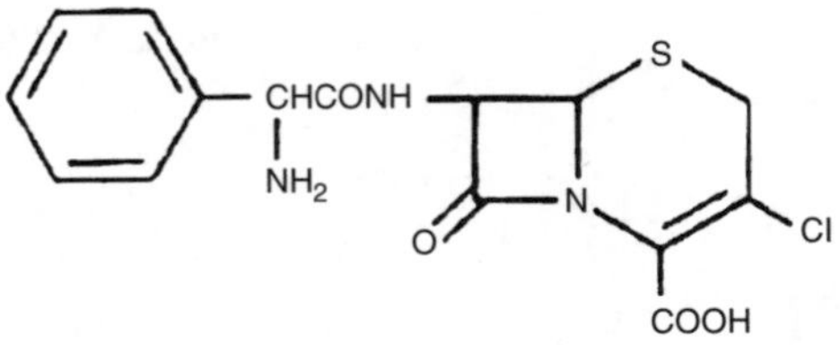

Cefachlor (2nd generation)

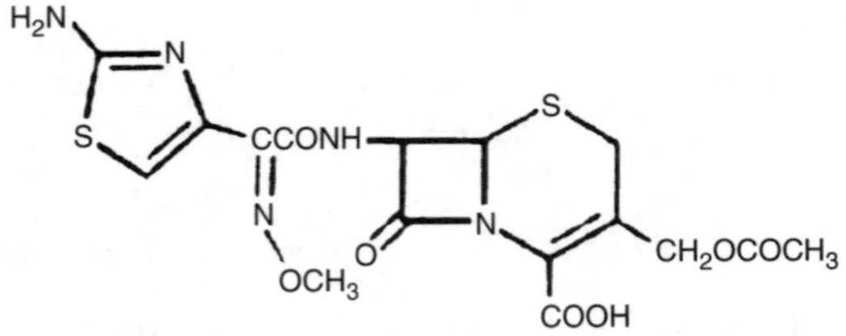

Cefotaxime (3rd generation)

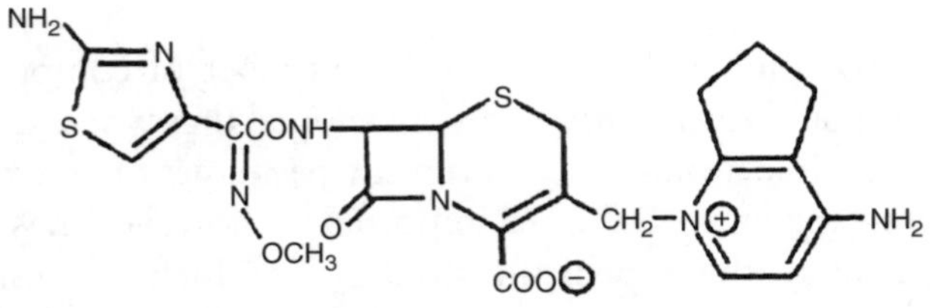

Cefpirome (4th generation)

Figure 6.3 Cephalosporin structures.

development of so-called second-, third- and fourth-generation cephalosporins. These compounds all differ in their antimicrobial properties, susceptibility to microbial resistance, absorption, metabolism and side effects. Examples of first-generation cephalosporins include cephalothin and cephazolin, second-generation cephalosporins include cefamandole and cefachlor, third-generation cephalosporins include cefotaxime and cefixime and fourth-generation examples include cefapime (Figure 6.3).

6.3.1 Antibiotic Production Cycles

At a cellular level, the production pathways for the cephalosporins and penicillins share some similarities; indeed, the first two steps are common to both classes of antibiotic. Initially, a tri-peptide known as ACV is formed from the amino acids L-cysteine, L-α-aminoadipic acid and L-valine. This key intermediate is then converted to isopenicillin N (IPN) by the enzyme IPN synthase. It is this intermediate which gives rise to both the penicillins and the cephalosporins. In the case of penicillin formation, IPN is hydrolysed to 6-amino penicillanic acid (6-APA), which can be used subsequently to give rise to specific penicillins. Alternatively, in the formation of the cephalosporins, IPN is epimerized to penicillin N, which is further reacted enzymatically to yield deoxycephalosporin C. This last molecule can then undergo further modification to give rise to cephalosporin C and cephamycin C.

As discussed earlier, the generation of semi-synthetic varieties of penicillins and cephalosporins is a simple process. By reacting the core penicillin compound 6-APA with a variety of organic acids, numerous penicillins can be duly formed. Indeed, the production of 6-APA is now carried out through the removal of the side chain from penicillin G and then reacted to yield a range of antibiotics. Similarly, the removal of the side chain from cephalosporin C to yield 7-amino cephalosporonic acid (7-ACA) can lead to the generation of numerous cephalosporins through the reaction of this compound with a variety of acids.

6.3.2 Industrial Production of Antibiotics

Industrially speaking, penicillin production is a relatively inefficient process, where it is estimated that only 10 % of the carbon source utilized in the fermentation ends up as antibiotic. Production of β-lactam antibiotics occurs best under conditions of carbon, nitrogen and phosphorus limitation and at low growth rates. Each manufacturer uses a different production process, the details of which are closely guarded. Overall, though, the basics of the production process are similar in nature. Production starts with the inoculation of a primary culture from a preserved culture stock. Typically, the culture stock can be in the form of lyophilized spores or spores preserved in liquid nitrogen. Numerous other preservation methods exist and will not be outlined here.

Primary culturing can utilize either agar slants or liquid culture, with agar slants being the most common. The primary culture is then used to inoculate a secondary culture, which in the case of antibiotic production is aimed at the generation of spores. Secondary culturing can take place in agar-coated bottles or on particulate material, both of which result in the generation of a large quantity of spores.

A spore suspension prepared from the secondary culture is subsequently used to inoculate liquid media as part of an inoculum build-up process. It should be pointed out that stringent aseptic techniques are used throughout the process to prevent the contamination of the antibiotic-producing culture. Practically speaking, industrial strains of antibiotic-producing fungi are less robust than naturally occurring fungi due to the aggressive mutation and selection pressures placed on them when they were originally isolated.

Depending on the size of the process, the scale-up procedure can have as many as three or four stages. Typically, the initial seed culture produced from the secondary spore suspension is less than 10 L. Following a defined period of growth, this can be used to inoculate a culture of less than 20 000 L, which in the final stages of the production cycle will serve to start a culture of up to 300 000 L. One important point is the nature of the product that the production cycle is centred on. Antibiotics are secondary metabolites, and in order to obtain the maximum productivity from the final-stage culture it is necessary to ensure that growth of the organism is limited and the organism enters its secondary metabolism phase. This is usually achieved by designing the growth medium to ensure that a key nutrient becomes limiting at the right time to effect the change in metabolism necessary for antibiotic production. In the case of penicillin production this is usually achieved by limiting the supply of glucose.

At the end of the fermentation it is necessary to separate the antibiotic material from the fungal mycelia, medium constituents and any other metabolites produced during the process. This is known as downstream processing, and the types of steps involved will depend on the antibiotic in production and also on the production process. Typically, it will involve some form of centrifugation or filtration to remove the fungal biomass and additional steps such as solvent extraction, ultrafiltration, chromatography and drying to produce a relatively pure antibiotic which can then be used for the manufacture of pharmaceutical preparations. It is estimated that over 10 000 t of penicillin G alone are produced by fermentation each year.

6.3.3 Additional Fungal Antibiotics

Fungi also produce a number of other antibiotics that are structurally unrelated to the β-lactams. Griseofulvin, a natural organic compound containing chlorine, is produced by *Penicillium griseofulvin*. This compound is interesting, as it inhibits the growth of fungi by preventing the assembly of fungal microtubules and, thus, mitosis. Another unrelated antibiotic is the steroidal compound fusidic acid, which is produced by *Fusidium coccineum*. This antibiotic is active against Gram-positive bacteria and has clinical use against β-lactam-resistant strains of bacteria.

6.4 Pharmacologically Active Products

In addition to antibiotics, fungi, as illustrated in Table 6.2, produce a range of other secondary metabolites. Some of these compounds are very significant in terms of their medical importance, including cyclosporin A and a group of compounds with cholesterol-lowering properties known as the statins. Other compounds, which will be discussed, include the alkaloids and the gibberellins.

6.4.1 Cyclosporin A

Immunosuppressive drugs have transformed modern transplant surgery by vastly reducing the incidence of organ rejection. The discovery and exploitation of the powerful immunosuppressant cyclosporin A has relied almost completely on fungal biotechnology. Attempts at chemically synthesizing the drug have served to illustrate the complexity of fungal secondary metabolism. Cyclosporin A is produced by the fungus *Tolypocladium inflatum* and was initially isolated from a Norwegian soil sample. The compound inhibits the production of interleukin-2 by T-lymphocytes and in so doing inhibits any potential immune response stimulated by antigens produced against transplanted organs. Cyclosporin A has also found use in the treatment of medical conditions such as psoriasis and eczema, owing to the role of interleukin-2 in mediating inflammatory responses.

The structure of cyclosporine A has shown it to be a heavily methylated cyclic peptide. In a similar fashion to other secondary metabolites, a range of over 25 cyclosporin analogues is produced by *T. inflatum*, and while 17 have antifungal activity, only two are immunosuppressants. Following a series of strain improvements using mutagenesis and culture optimizations, gram-quantity yields per litre have been achieved under optimized fermentation conditions. Despite the best efforts at chemical synthesis of the drug, production of cyclosporine A is still only economically feasible by natural means.

6.4.2 Statins

The so-called statins are a group of compounds which act as potent competitive inhibitors of 3-HMG-CoA-reductase, a key enzyme in the biosynthesis of cholesterol. These organic acids interact with the enzyme through their acidic side groups and in doing so effect a reduction in plasma cholesterol levels. The most important statins commercially are the mevinic acids, with the most notable being lovastatin from *Monascus ruber* and mevastatin from *Penicillium citrinum*. Both mevastatin and lovastatin can be converted into the compounds ML-236A and monacolin J respectively by chemical means or by microbial transformation. Each of these compounds differs in their affinity for 3-HMG-CoA-reductase and, thus, in their effectiveness. Research and development over the last two decades has shown that a number of fungi produce a range of similar compounds with cholesterol-lowering effects. One interesting aspect of their commercial production lies in the ability to produce these compounds by liquid fermentation (e.g. lovastatin from *Aspergillus terreus*) or by using solid-state fermentation, the so-called Koji process (e.g. lovastatin from *Monascus rubber*).

6.4.3 Alkaloids

Members of the genus *Claviceps*, a parasitic fungus which grows on a wide variety of grains, synthesize numerous secondary metabolites known as alkaloids.

These compounds are produced in the sclerotia of the fungus, the resting structure with which the fungus ensures its survival over winter. Many of these alkaloids are pharmaceutically important and can act in a variety of ways, but they particularly affect the central nervous system, causing hallucinations or convulsions. The common core of these compounds is the tetracyclic alkaloid ring structure (ergoline nucleus) in which the nitrogen atom at position 6 is usually methylated. This core structure is derived from tryptophan and mevalonic acid and can be modified with varying degrees of complexity to give rise to a multitude of alkaloids, each differing in their potency and toxicity. These derivatives can then be used in the chemical synthesis of additional pharmacologically active compounds. For instance, lysergic acid diethylamide (LSD) is synthetically derived from lysergic acid, an alkaloid produced by *Claviceps paspali*. Medically speaking, many of these alkaloids are important owing to their negative effects; indeed, alkaloid-contaminated bread caused numerous outbreaks of ergotism until the eighteenth century. There is also evidence to suggest that the Salem Witch trials were brought about following outbreaks of ergotism.

Medically useful alkaloids have, however, been isolated, and the most useful of these are the alkaloids ergometrine and its methylated derivative methyl ergometrine. Both of these compounds stimulate contractions of uterine smooth muscle and can also be used as vasoconstrictors to control excessive bleeding after childbirth.

6.4.4 Gibberellins

The gibberellins are a group of diterpenoid compounds containing 19 or 20 carbon atoms and are capable of promoting numerous developmental processes in plants. Examples of effects that may be mediated by these compounds include the induction of bolting, production of hydrolytic enzymes and stimulation of both cell elongation and cell division. These secondary metabolites are biosynthesized from mevalonic acid by *Gibberella fujikora*, though they have also been isolated from *Sphaceloma manihoticola*, *Neurospora crassa*, *Rhizobium phaseoli* and *Azospirillium lipoferrum*. They have numerous agricultural applications, including their use in malting, fruit ripening and improving fruit set and size. On an industrial scale, the most important production organism is *G. fujikora*, from which mutated strains secrete gibberellins at gram per litre quantities.

6.4.5 Endophytic Compounds

The development of drug resistance in infectious microorganisms (e.g. species of *Staphylococcus*, *Mycobacterium* and *Streptococcus*) to existing antibiotic compounds has led to a requirement for new chemotherapeutic agents that are highly effective, possess low toxicity and have a minor environmental impact. In the continual search by both pharmaceutical and agricultural industries for new products, natural selection has been found to be superior to synthetic chemistry for discovering novel substances that have the potential to be developed into new industrial products. Since natural products are adapted to a specific function in nature, the search for novel secondary metabolites should concentrate on organisms that inhabit novel ecosystems.

In the past few decades, plant scientists have begun to realize that plants may serve as a reservoir of untold numbers of organisms known as endophytes. Endophytic fungi, a polyphyletic group of highly diverse, primarily ascomycetous fungi that are defined functionally by their occurrence within tissues of plants without causing any immediate negative effects, are found in liverworts, hornworts, mosses, lycophytes, equisetopsids, ferns and seed plants from the arctic tundra to the tropics. Once inside their host plant, endophytes usually assume a dormant state either for the whole lifetime of the infected plant tissue or for an extended period of time – that is, until environmental conditions are favourable for the fungus or until the ontogenetic state of the host changes to the advantage of the fungus, which may then turn pathogenic. Colonization of host plants by endophytic fungi is believed to contribute to the host plant an adaptation to biotic and abiotic stress factors. It is of special interest that in many cases host plant tolerance to biotic stress has been correlated with fungal natural products.

Although work on the utilization of this vast resource of poorly understood microorganisms has been initiated, the enormous potential for the discovery of a wide range of extremely beneficial products holds exciting promise. This is witnessed by the discovery of a wide range of products and microorganisms (Table 6.3). There are other characteristics of endophytic fungi that also render them desirable for manipulation in an industrial screening programme. Most screening has focused on soil-dwelling fungi; little attention has been directed toward endophytes. As a consequence, they have not been subjected to intensive screening programs, which suggests thst the vast majority remain largely undiscovered.

An array of natural products has been characterized from endophytes, which include anti-cancerous, anti-fungal, anti-bacterial, anti-viral, insecticidal and immunosuppressants agents. The discovery of the paclitaxel (Taxol)-producing endophytic fungus *Taxomyces andreanae* from the yew plant *Taxus brevifolia* has led to a more comprehensive examination of other *Taxus* species and plants for the presence of paclitaxel-producing endophytes. Paclitaxel, a multi-billion dollar anti-cancer compound produced by the yew plant, has activity against a broad band of tumour types, including breast, ovarian, lung, head and neck cancer. Taxol, a tetracyclic diterpene lactam (shown in Figure 6.3), was first isolated from the bark, roots and branches of western yew, *Taxus brevifolia*, in the late 1960s (Figure 6.4). The main natural source of Taxol is found in the bark of yew (*Taxus*), where it exists in low concentrations of 0.01–0.05 %; however, *Taxus* species are endangered and grow very slowly. Traditional methods of extracting taxol from the bark of *Taxus* species are inefficient and environmentally costly.

Table 6.3 Isolated endophytic fungi and their bioactive products.

Endophyte	Host plant	Metabolite	Bioactivity
Colletotrichum gloeosporiodes	*Artemisia mongolica*	Colletotric acid	Antimicrobial
Muscodor roseus	*Erythrophelum chlorosachys*	Volatile antibiotic	Antimicrobial
Muscodor albus	*Cinnamomum zeylanicum*	1-Butanol, 3-methyl acetate	Antimicrobial
Phoma sp.	*Taxus wallachina*	2-Hydroxy-6-methyl benzoic acid	Antibacterial
Fusarium sp.	*Selaginella pallescens*	CR377	Antifungal
Cryptosporiopsis quercina	*Tripterigeum wilfordii*	Cryptocandin	Antifungal
Nodulisporium sp.	*Bontia daphnoides L*	Nodulisporic acids	Anti-insecticidal
Semiatoantlerium tepuiense	*Maguireothamnus speiosus*	Taxol®	Anti-cancer
Taxomyces andreanae	*Taxus brevifolia Nutt*	Taxol	Anti-cancer
Tubercularia sp.	*Taxus mairie*	Taxol	Anti-cancer
Pestalotiopsis microspora	*Taxus wallachina*	Taxol	Anti-cancer
Sporomia minima	*Taxus wallachina*	Paclitaxel	Anti-cancer
Rhinocladiella sp.	*Tripterygium wilfordii*	22-Oxa-[12]-cytochalasin	Anti-tumor
Pestalotiopsis guepinii	*Wollemia nobilis*	Taxol	Anti-tumor
Fusarium subglutinans	*Tripterygium wilfordii*	Subglutinols A & B	Immunosupressive

Adapted from Tejesvi *et al.* (2007).

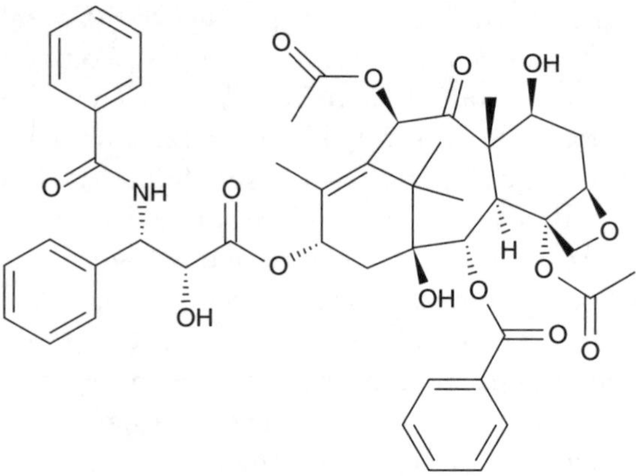

Figure 6.4 Chemical structure of Taxol.

Microbial fermentation has demonstrated that the isolation and identification of Taxol-producing endophytic fungi is a new and feasible approach to the production of Taxol. Presently, the development and utilization of Taxol-producing fungi have made significant progress worldwide. For industrial production purposes, optimization of fermentation conditions is necessary to increase the yields of Taxol-producing fungi. In addition to reducing costs and increasing yields, fungal fermentation as a way of producing Taxol is also beneficial to protecting natural *Taxus* tree resources. Taxol produced by fungi on large-scale industrial fermentation has attractive development prospects, and it will have enormous market and social benefits.

Natural products from endophytic microbes have been observed to inhibit or kill a wide variety of harmful disease-causing agents, including, but not limited to, phytopathogens, as well as bacteria, fungi, viruses and protozoans that affect humans and animals. *Cryptosporiopsis quercina*, a fungus commonly associated with hardwood species in Europe, has been isolated as an endophyte from *Tripterigeum wilfordii*, a medicinal plant native to Eurasia. On Petri plates, *C. quercina* demonstrates excellent antifungal activity against some important human fungal pathogens – *Candida albicans* and *Trichophyton* sp. Since infections caused by fungi are a growing health problem, especially amongst AIDS patients, and those who are otherwise immunocompromised, new antimycotics are needed to combat this problem. A unique peptide antimycotic, termed cryptocandin, has been isolated and characterized from *C. quercina*. This compound contains a number of hydroxylated amino acids and a novel amino acid (3-hydroxy-4-hydroxy methyl proline). The bioactive compound is related to known antimycotics, such as the echinocandins and the pneumocandins. Cryptocandin is also active against a number of plant pathogenic fungi, including *Sclerotinia sclerotiorum* and *Botrytis cinerea*, and is currently being tested and developed by several companies for use against a number of fungi causing diseases of skin and nails.

Endophytes produce substances that can influence the immune system of animals. Subglutinols A and B are immunosuppressive compounds produced by *Fusarium subglutinans*, an endophyte of *T. wilfordii*. The compounds both have IC50 values of 0.1 μM in the mixed lymphocyte reaction assay. In the same assay, cyclosporin is roughly as potent as the subglutinols. These compounds are being examined more thoroughly as immunosuppressive agents. Their role in the endophyte and its relationship to the plant are unknown.

In the continuous search for novel drug sources, endophytic fungi have proven to be a largely untapped reservoir of bioactive products, with great chemical diversity. These compounds have been optimized by evolutionary, ecological and environmental factors. The development of drugs from endophytes with high potency will offer much needed new remedies for acute and chronic human diseases. As so many bioactive compounds have been isolated from endophytes which only occupy a small portion of total endophyte species, it is obvious that there is a great opportunity to find reliable and novel bioactive products in endophytes which may be used as clinically effective compounds in the future.

6.5 Chemical Commodities

Several industrially important chemicals are produced via biological processes using moulds and yeasts. In terms of world production volume, the most important of these is citric acid. From 1978 to 1984, the average rate of increase of total consumption of citric acid in western countries was about 3.5–6 % per year; in 1995 the world market for citric acid was at 500 000 t.

6.5.1 Citric Acid

Citric acid is the principal organic acid found in citrus fruit. To meet with increasing demands, it is produced from carbohydrate feedstock by fermentation with the fungus *Aspergillus niger* and yeast of the genus *Candida*. The initial commercial production of citric acid was achieved using *A. niger* in a surface fermentation process. The development of the process of submerged fermentation in the 1950s was a major turning point in citric acid production. Citric acid's main use is as an acidulant in soft drinks and confectionary. A more recent application of citric acid is as a metal complexing agent, to reduce oxidative metal deterioration and for metal cleaning. With the increasing requirement for citric acid, its production by fermentation is increasing continually with about 500 000 t of it being produced annually.

A number of fungi and yeasts have been used over the years for the production of citric acid, but *A. niger* remains the preferred fermentation organism for commercial production. The main advantages for using this organism are its ease of handling, its ability to ferment a wide variety of cheap raw materials and its high yields.

A variety of raw materials, such as molasses, starchy materials and hydrocarbons, have been employed as substrate for the production of citric acid. Sucrose, cane molasses or purified glucose syrup from maize are sometimes used according to availability and price. Molasses has been acclaimed as a low-cost raw material and it contains 40–55 % of sugars in the form of sucrose, glucose and fructose. There are considerable variations in the culture conditions reported in the literature for citric acid production by *A. niger*. To ensure high productivity, it is essential that the media contain major nutrients like carbon, nitrogen, phosphorous and also trace elements. The fermentation process is also influenced by aeration temperature and pH.

The use of different carbon sources has been shown to have a marked effect on yields of citric acid by *A. niger*. *A. niger* can rapidly take up simple sugars such as glucose and fructose. Sucrose is usually the sugar of choice; at industrial scale, the fungus possesses an extracellular mycelial-bound invertase which, under the acidic conditions of citric acid fermentation, hydrolyses sucrose to its monomers. A sugar concentration of 14–22 % is considered the optimal level for maximum production yields. Lower sugar concentrations lead to lower yields of citric acid, as well as the accumulation of oxalic acid.

The nitrogen sources for citric acid production by *A. niger* are generally ammonium sulfate, ammonium nitrate, sodium nitrate, potassium nitrate and urea. The presence of phosphorus in the fermentation medium has a profound effect on the production of citric acid. Too high a level of phosphorus promotes more growth and less acid production. Potassium dihydrogen phosphate (0.1 %) has been reported to be the most suitable phosphorus source. Maintenance of a low pH is essential for production; generally, a pH below 2 is required for optimal fermentation. Citric acid fermentation is an aerobic process and increased aeration rates have resulted in enhanced yields and reduced fermentation times. Trace elements are also a major factor in the yields obtained in citric acid fermentation. When trace elements are growth limiting, citric acid accumulates in larger quantities.

6.5.2 Production by Filamentous Fungi

A number of different fermentation processes exist for the production of citric acid.

6.5.2.1 Surface Fermentation

Surface culturing was the first process employed for the large-scale production of microbial citric acid. Despite the fact that more sophisticated fermentation methods (submerged process) have been developed, surface culturing techniques are still employed, as they are simple to operate and install. Another advantage of this culturing method is energy costs for surface fermentation are lower than those of submerged fermentation. The mycelium is grown as a surface mat in shallow 50–100 L stainless steel or aluminium trays. The trays are stacked in stable racks in an almost aseptic fermentation chamber.

The carbohydrate sources (usually molasses) for the fermentation medium is diluted to 15 % sugars, the pH is adjusted to 5–7 and any required pretreatment is carried out. After the addition of the nutrients, the medium is sterilized, cooled and pumped into the trays. Inoculation is performed by introducing spores, either by generating a spore suspension or by blowing spores over the surface of the trays along with air. Spores subsequently germinate and form a mycelial mat. The temperature is maintained at 28–30 °C and the relative humidity between 40 and 60 %.

During fermentation, considerable heat is generated, necessitating high aeration rates. Air provides oxygen to the organism and also controls the fermentation temperature and the relative humidity. As the fermentation progresses, the pH decreases to below 2.0. If the fermentation pH rises to 3.0, oxalic acid and gluconic acid may be formed in considerable amounts. Fermentation progresses for 8–12 days, after which time the fermented liquid is poured out of the pans and separated from the mycelium for further processing. Fermentation yields are in the range of 70–75 %.

6.5.2.2 Submerged Fermentation

This process is now more popular for the commercial production of citric acid. It requires less space, is less labour intensive and higher production rates are obtained. With submerged fermentation, a stirred tank reactor or a tower fermentor may be used (Figure 6.5).

In view of the low pH level which develops during fermentation and the fact that citric acid is corrosive, the use of acid-resistant bioreactors is desirable. An important consideration with bioreactors designed for citric acid production is the provision of an aeration system, which can maintain a high dissolved oxygen level. With both types of bioreactor, sterile air is sparged from the base, although additional inputs are often used in tower fermentors.

The medium preparation in submerged fermentations involves appropriate dilution of the carbon source, pretreatment addition of the appropriate nutrients and sterilization in line or in the bioreactor. Inoculation is performed either by the addition of a suspension of spores or precultivated mycelia. When spores are used, they need to be dispersed in the medium; therefore, addition of a surfactant is usually necessary. With precultivated mycelia the inoculum size is usually about 10 % of the fresh medium. Air is sparged through the medium at a rate of 0.5–1.5 VVM (volume of air per volume of culture medium per minute) throughout the

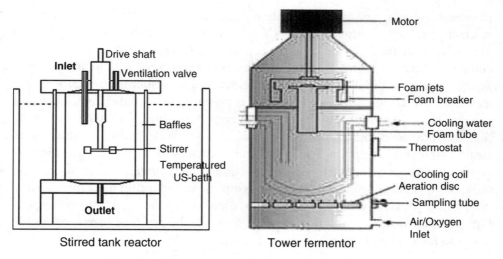

Figure 6.5 Types of reactor for citric acid production.

fermentation. Under optimal conditions, fermentation is completed in 5–10 days. Submerged fermentation can be performed by continuous and fed-batch modes, but generally it is carried out by the batch feed mode.

6.5.2.3 Solid-State Fermentation

The Koji process or solid-state fermentation, developed in Japan, is the simplest process for production of citric acid. This process is the solid-state equivalent of surface fermentation. The raw materials used are sweet potato fibrous residues, rice or wheat bran and fruit wastes. The carbohydrate source is moistened with water to about 70 % moisture. The moist carbohydrate is then steamed for sterilization, placed in trays and inoculated using conidia of *A. niger*. The pH at the start of fermentation is 5.5. The starch is hydrolysed by amylase produced by the fungus and subsequently converted to citric acid. The fermentation is complete in 4–5 days. The main problem with this process is the presence of trace elements, which cannot be removed by standard methods.

6.5.3 Production by Yeast

Yeasts are also commercially employed in the commercial production of citric acid from various carbon sources. Yeast strains that are used in the production of citric acid include *Saccharomyces lipolytica*, *Candida tropicalis*, *Candida olephila*, *Candida guilliermondii*, *Candida citroformans* and *Hansenula anomala*. There are a number of advantages when using yeast in comparison with filamentous fungi. Yeasts can tolerate high initial sugar concentration, they are insensitive to trace metals and can thus ferment crude carbon sources without any treatment, they also have a great potential for being used in continuous culture and have a high fermentation rate. For commercial production of citric acid by yeast, tower fermentors with efficient cooling systems are employed. To inoculate the fermentation, an inoculum is prepared in a smaller fermentor and is subsequently transferred into the production fermentor. The temperature of the fermentation is maintained between 25 and 37 °C, depending on the type of strain employed. The pH is generally >5.5, but can fall during fermentation. A continuous process for citric acid production using *Candida* cultured on cane molasses has been developed. Processes employing *n*-paraffin as carbon source with strains of *Candida* as the fermenting organisms have been fully developed, but they have become uneconomic with the rise in the prices of petroleum products and have never been run on a large scale.

6.5.3.1 Citric Acid Metabolic Pathways

The exact mechanism for citric acid production is not clearly understood, but involves an incomplete version of the tricarboxylic acid cycle. Possibly during the initial metabolism of glucose there is an increase in cellular oxaloacetate levels, which decrease the catabolism of citrate by α-ketoglutarate dehydrogenase and simultaneously increase the rate of citrate synthetase. The condensing enzyme citrate synthetase brings about the biosynthesis of citric acid by the condensation of acetyl-CoA and oxaloacetic acid. This condensation of C2 and C4 compounds is the major route of citrate synthesis (Figure 6.6). Citrate synthetase has been shown to have allosteric regulation. Oxaloacetic acid for citric acid formation is achieved by way of the citric acid cycle and by the anaplerotic reaction for the high yield of citric acid. Once the concentration of citric acid in the cells is high enough, the acid has to be excreted.

6.5.3.2 Citric Acid Recovery

Following the 5–10 days' fermentation the microbial cells are separated from the fermented liquor by centrifugation or filtration. Yeast-based fermentation liquors kept neutral with $CaCO_3$ or lime need to be acidified at this point with mineral acid before this step. Citric acid is then precipitated from the filtrate or supernatant as insoluble calcium citrate tetrahydrate by the addition of lime. The filtered, washed calcium salt is then treated with sulfuric acid in an acidulator. A precipitate of calcium sulfate is formed and filtered off. The remaining citric acid solution is treated with active carbon, passed through cation exchangers and concentrated by evaporation before it crystallizes. Alternatively, citric acid may be extracted from the filtered broth using either tributyl phosphate or long-chain secondary or tertiary amines. The acid is extracted into solvent at low

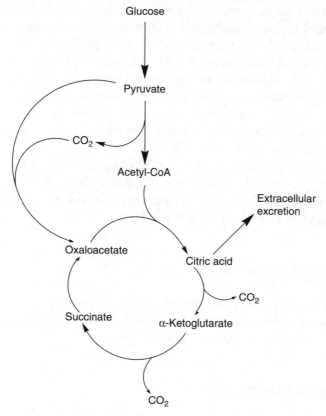

Figure 6.6 Citric acid biosynthetic pathway.

temperatures and re-extracted into water at a higher temperature. The purified solution is concentrated and crystallized.

6.5.4 Itaconic Acid Production

Itaconic acid is used to alter the dyeing characteristics of vinyl polymers and also in the manufacture of polymers used in emulsion paints. Itaconic acid accumulation was originally observed in *Aspergillus itaconicus*; *A. terreus* and mutants of this strain are now more widely used. The main carbon sources used in the commercial production of itaconic acid include glucose together with inorganic salts, purified molasses or media containing a portion of beet molasses. Calcium and zinc are also essential in the growth medium.

6.5.4.1 Fermentation Process

Once the medium is prepared and sterilized, inoculation is performed either by the addition of a suspension of spores or pre-cultivated mycelia. Though surface fermentation in trays has been used, the submerged method is more widely preferred. Aerated and agitated stainless steel tanks are employed and provision for cooling is necessary. Fermentation temperatures for itaconic acid production are quite high: approximately 40 °C. The pH of the media must be reduced to 2 to initiate production; once the accumulation of itaconic acid is well under way, higher yields are obtained if the medium is partially neutralized. The fermentation is highly aerobic and continuous aeration is required to decrease production losses. Following 72 h of fermentation, yields of 60 % can be obtained based on carbohydrate source supplied.

Carbohydrate is metabolized, in *A. itaconicus* cells, by glycolysis to pyruvate, which is further converted through the citric acid cycle to aconitic acid. Aconitic acid is then converted to itaconic acid by the enzyme aconitic acid decarboxylase, an enzyme that is extremely oxygen dependent.

6.5.4.2 Itaconic Acid Recovery

The mycelium is separated from the fermentation medium by filtration and the resultant liquor clarified. The itaconic acid is then recovered by evaporation and crystallization, by ion exchange or solvent extraction.

6.5.5 Vitamin Production

Vitamins, essential nutrients required in small quantities, have a well documented and accepted value to the health of humans and animals. There is a large need for extra vitamins, other than those derived from plant and animal food sources, due to unbalanced food habits or processing, food shortage or disease. Added vitamins are either prepared chemically or biotechnologically via fermentation or bioconversion processes (Table 6.4). Several vitamins are at the moment only produced via organic chemical synthesis. However, for many of these compounds, microbiological processes for their production are rapidly emerging and some are already taking over. Compounds such as riboflavin (B_2), ergosterol (provitamin D_2), cyanocobalamin (B_{12}), orotic acid (B_{13}), vitamin F group and vitamin C are now produced exclusively via fermentation.

6.5.5.1 Vitamin B$_2$ (Riboflavin)

Riboflavin is commonly used in animal feed and human nutrition. It is produced by both synthetic and fermentation processes, with the latter recently increasing in application. The first fermentations were initially carried out in 1965, but were shut down after 3 years as they proved more expensive than chemical processes.

Table 6.4 Industrial production of vitamins.

Vitamin	Organic extraction	Chemical synthesis	Microbial synthesis			
			Bacterial	Fungal	Algal	World production (tons year^{-1})
Vitamin B$_1$	+					2 000
Vitamin B$_2$	+		+	+		2 000
Niacin (B$_3$,PP)	+		+			8 500
Pantothenic acid (B$_5$)	+					4 000
Vitamin B$_6$	+					1 600
Biotin (B$_8$)	+		+			3
Folic acid (B$_9$)	+					300
Vitamin B$_{12}$			+			10
Vitamin B$_{13}$			+			100
Vitamin C	+		+			70 000
Vitamin A	+					2 500
Provitamin D$_2$				+		
Provitamin D$_3$	+	+				25
Vitamin E	+	+		+	+	6 800
Vitamin F		+		+	+	1 000
Vitamin K2	+					2

With improvement in producer organisms, the microbial process was revived in 1975 and is now increasingly used for riboflavin production. Although bacteria (*Clostridium* sp.) and yeasts (*Candida* sp.) are also good producers, currently the two closely related ascomycete fungi *Ashbya gossypii* and *Eremothecium ashbyi* are considered the best riboflavin producers. *A. gossypii* is the preferred strain for production, as *E. ashbyi* is genetically unstable.

Soya bean oil and soya bean meal are the substrates most commonly used in *A. gossypii* fermentations. Riboflavin production occurs during the late phase of growth when all the glucose in the medium is exhausted. While glucose remains in the medium, during the early phase of growth it is converted to lipid droplets, which are used later in riboflavin production. Supplementation of the culture medium with glycine or ribitol stimulates riboflavin formation. Both compounds are precursors in the riboflavin synthetic pathway and their effects suggest a limitation of central metabolites.

An alternative biotechnological process for the commercial production of riboflavin is through the fermentation of yeast; that is, *Candida famata*. Mutants of this strain also exist which can produce up to 200 g of riboflavin per litre following 8 days fermentation.

6.5.5.2 Vitamin D

Vitamins D_2 and D_3 are used as antirachitic treatments and large amounts of these vitamins are also used for fortification of food and feed. Vitamin D_2 (ergocalciferol) is obtained by the UV irradiation of yeast ergosterol (provitamin D_2) (Figure 6.7). Efficient fermentation processes for ergosterol accumulation have been established. *Saccharomyces cerevisiae* is known to accumulate high levels of sterols. Of about 20 sterols encountered in *S. cerevisiae*, ergosterol, ergosta-5,7,22,24(28)-tetraen-3β-ol, zymosterol and lanosterol are considered to be the major sterols, of which ergosterol makes up over 90 %.

Yeast cells consume carbohydrate as energy and carbon sources by aerobic and anaerobic metabolism. The concentration of carbohydrate and the supply of oxygen determine which metabolic pathway yeast cells utilize. To overcome the repression caused by insufficient nutrients or an oversupply of carbohydrate, fed-batch methods have been used in the process of ergosterol fermentation. In the yeast culture process, glucose is preferred, and when the glucose concentration reaches a low level the cell growth is confined. Then a short period of adaptation occurs as cells continue to grow by consuming the ethanol, produced in the first phase, as the carbon source. The whole process appears to be a two-phase process, with the ergosterol content increasing when the specific growth rate is decreased.

6.5.6 Fungal Pigments

Fungi have the potential for use in production processes that are themselves less polluting than traditional chemical processes. For instance, many fungi produce pigments which have application in both the textile and food industries, as evidenced by the way in which mildew growth can lead to permanent staining of textiles and plastics. They could, therefore, be used for the direct production of textile dyes or dye intermediates, replacing chemically synthesized forms which have inherent environmental effects during their production and waste disposal. The recent approval of fungal carotenoids as food colourants by the European Union has served to strengthen the prospects and global market for use of non-carotenoid fungal polyketide pigments. Fungal production of colourants not only

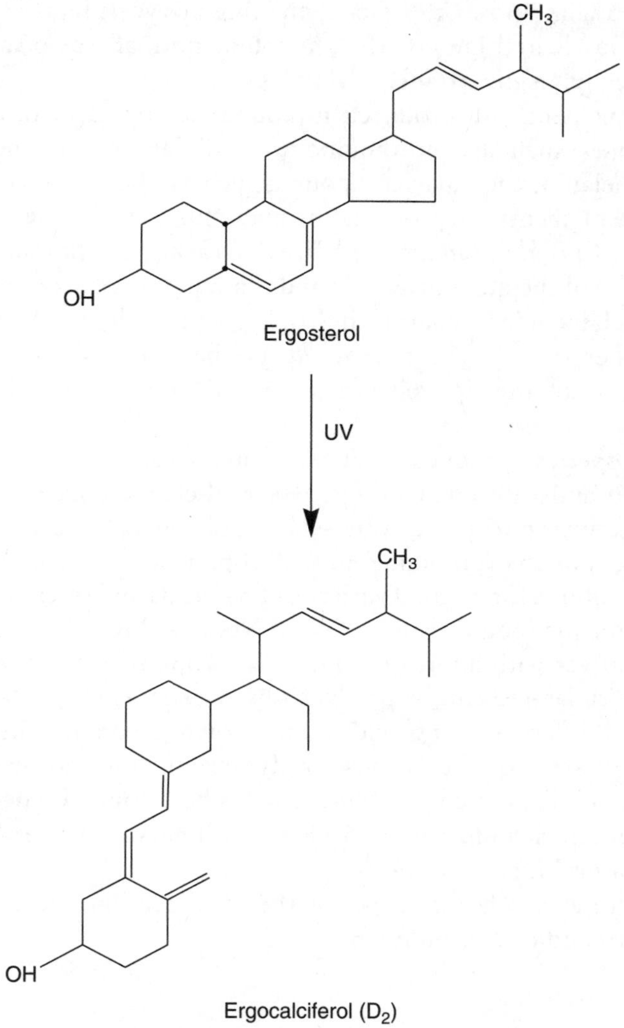

Figure 6.7 Vitamin D biosynthesis.

has environmental and safety benefits, but also confers on the manufacturer the benefit of making the production process independent of the seasonal supply of raw materials and also minimizes batch-to-batch variations. Fungal pigments are known to exhibit unique structural and chemical diversities and have an extraordinary range of colours.

Carotenoids such as β-carotene are produced by a wide range of *Mucorales* fungi and are suitable for addition to a variety of foods. The yeast *Phaffia rhodozyma* has become the most important microbial source for the production of the carotenoid pigment astaxanthin and is responsible for the orange–pink colour of salmonid flesh and the reddish colour of boiled crustacean shells.

Feeding farmed salmonids with a diet containing this yeast induces pigmentation of the white muscle and imparts the red colour normally associated with wild fish. Economically, its importance is very high.

Polyketide pigments are produced in abundance by filamentous fungi, and include quinones such as anthraquinones and naphthaquinones, dihydroxy naphthalene melanin, and flavin compounds such as riboflavin. Yellow-coloured anthraquinone pigments are produced by many fungi, including *Eurotium* spp., *Fusarium* spp., *Curvularia lunata* and *Drechslera* spp. Anthraquinone is an important member of the quinone family and is a building block of many dyes, examples of which include catenarin, chrysophanol, cynodontin, helminthosporin, tritisporin and erythroglaucin. *Emericella* spp. have been shown to produce alternative yellow-coloured pigments such as the epurpurins, falconensins and falconensones.

Monascus species produce orange, water-insoluble pigments such as monascorubrin and rubropunctatin. These well-characterized compounds can be converted to high-purity red, water-soluble pigments by reaction with amino acids, yielding monascorubramine and rubropunctamine. These pigments are suitable as colourants for a broad variety of foodstuffs and often serve as suitable replacements for the food dyes FD&C Red No. 2 and Red No. 4. Interestingly, pigment derivatives with improved functional properties in the colour range of orange–red to violet–red can be produced by *Monascus* fermentations through the inclusion of different amino acids in the growth medium. One concern with the use of *Monascus* for the purpose of dye production, however, is that the fermented rice substrate used in the process has been found on occasion to contain the mycotoxin citrinin. For the present, this limits the use of *Monascus* as a producer of natural food colourants.

Clearly, there is significant scope for the identification and development of food colourants and dyes from fungi.

6.6 Yeast Extracts

Maintenance of an adequate food supply challenged humans during much of their early existence. Since ancient times, both western and oriental cultures have used microorganisms to transform or produce food. Fungi have a crucial role to play in the processing of many foods, improving the texture digestibility, nutritional value, flavour or appearance of the raw material used. The first industrial production of micoorganisms for nutritional purposes took place in Germany, when Torula yeast was produced and incorporated into soup and sausages. Yeast cells may be solubilized either partially or completely by autolysis and several other techniques. On further processing, the slurries can be converted into a variety of preparations and products which are useful in the laboratory and as ingredients in food, feeds and fermentation media. Amongst the principal products are concentrates of yeast invertase and β-galactosidase, soluble yeast components in liquid, paste, powder or granular form, and isolated fractions of yeast cell constituents, such as protein and cell walls (glucan, mannooligosaccharide, emulsifiers mannoprotein) liberated by cells fractured mechanically. The major commercial products are clear water-soluble extracts known generally as yeast extract, autolysed yeast extract and yeast hydrolysate.

During the early part of the last century, studies on spent yeast from breweries led to the development of yeast extracts for the food industry. At present, this technology has been extended to several other types of yeast, thus providing a much wider range of yeast extract to the food industry. As natural flavourings approved by the FDA, yeast extracts are used as condiments in the preparation of meat products, sauces, soups, gravies, cheese spreads, bakery products, seasonings, vegetable products and seafoods. A reliable economical source of peptides, amino acids, trace minerals and vitamins of the B complex group, yeast extracts are nutritional additives in health-food formulations, baby food feed supplements and for enrichment of growth and production media for microorganisms and other biological culture systems.

6.6.1 Yeast Extract Production

There are three distinct manufacturing practices for yeast extract production: autolysis, plasmolysis and hydrolysis. Autolysis is a process by which the cell components within the cells are solubilized by activation of enzymes, which are inherently present in the cell. This is achieved by carefully controlling temperature, pH and time, with the careful addition of enzymes or reagents to stimulate degradation and release of the cell contents into the medium. The amount of free amino acid present in the extract can serve as a rough guide to indicate the degree of hydrolysis. Free amino acids are known to directly or indirectly exert major influences on food flavour.

A yeast extract manufacturing process, outlined in Figure 6.8, that has gained more acceptance in Europe than in the USA is plasmolysis. In this process, yeast cells are treated with salt and begin to lose water; the cytoplasm separates from the cell wall. When this happens, cells die and the degradative process begins. The advantages of this process are the fact that no specialized equipment is required and salt is relatively cheap to purchase and readily available. One drawback of this process is the high salt content of the extract. The hydrolytic process utilizes the action of hydrochloric acid on yeast at specific temperatures and pressure. Hydrolysis is carried out until the required concentration of free nitrogen is achieved; this usually takes 6–12 h, and shorter, more efficient hydrolysis can be achieved at higher temperature and pressure. The hydrolysate is neutralized with sodium hydroxide; the extract is then filtered and concentrated.

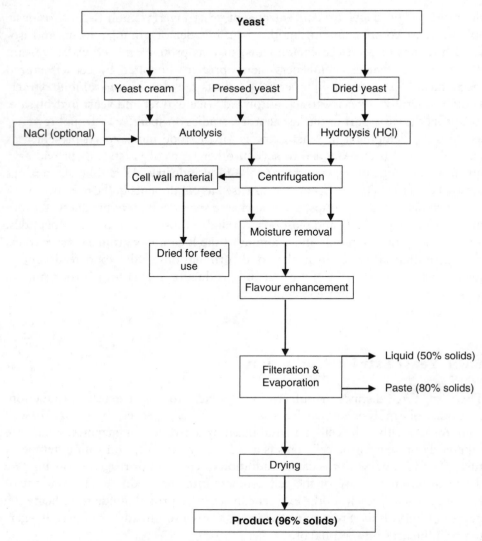

Figure 6.8 Yeast extract manufacturing process.

6.7 Enriched Yeast

With increasing demand for organic and non-genetically modified additives in human and animal feeds, interest in the production of 'organic vitamins and minerals' has increased in recent years. One of the major concerns with the production of these organic products is economic viability. Thus, to overcome this concern, the incorporation into or enrichment of yeast (*S. cerevisiae*) with minerals has developed as a common industrial process. One of the most commonly available of these yeast-type products is selenium-enriched yeast, such as Sel-Plex®.

6.7.1 Selenium Yeast Production

Since selenium and sulfur exist in the same group within the periodic table and have many similar chemical characteristics, microbes such as the bakers yeast, *S. cerevisiae*, have been shown to be unable to distinguish between either atom. Indeed, *S. cerevisiae* has the ability to metabolize selenium and incorporate it into molecules where sulfur would normally exists as the native atom. When propagated in a nutrient-enriched medium containing reduced levels of sulfur but enriched with selenium as inorganic selenite or selenate salts, *S. cerevisiae* can utilize selenium as it does sulfur, resulting in the biosynthesis of various organic selenium compounds. The majority of the selenium in selenium yeast exists as analogues of organo-sulfur compounds, such as selenomethionine, selenocysteine and selenocystine. Small amounts are also thought to exist as selenohomocystenine, selenocystathione, methylselenocysteine, *S*-adenosyl-selenomethionine, the selenotrisulfide, selenoglutathione and various seleno-thiols, some of which are represented in Figure 6.9. The biosynthesis of such organo-selenium compounds is thought to be achieved through the biochemical pathways of organo-sulfur biosynthesis, which have been well characterized. These selenoamino acids can then be utilized by the protein synthesis machinery of the yeast cell and incorporated into protein molecules.

Crude protein, shown to account for approximately one half of dry weight of yeast, was characterized and was shown to be comprised of approximately 80 % amino acids, 12 % nucleic acids and 8 % ammonia. Examination of these results shows the total sulfur-containing amino acid content of *S. cerevisiae* to be 1.99 % (w/w) with 1.21 % as methionine and 0.78 % as cysteine. This represents 2600 and 2080 ppm of organically bound sulfur in these forms. Studies have shown that up to 50 % of the methionine moieties in proteins can be replaced with selenomethionine while retaining biological activity. These findings would indicate that well in excess of 2000 ppm of yeast sulfur could be replaced by selenium with possibly no adverse affects on protein synthesis or growth characteristics of the yeast cells. Therefore, if propagation conditions are carefully controlled, a growth pattern can be induced which allows the incorporation of selenium in the yeast to levels in excess of 2000 ppm, over 50 000 times the normal level of 0.04 ppm; the production process for selenium yeast is outlined in Figure 6.10.

Selenium yeast has many advantages over traditional inorganic selenium. These include a lessening of environmental concerns pertaining to the toxicity of selenium, as selenium yeast displays reduced toxicity over inorganic selenium forms. Another advantage associated with increased bioavailability is a reduction in the level of unabsorbed selenium excreted by the faecal route. This would prevent toxic selenium build-up by concentration in the faeces and is of significance where intensive farming techniques are employed. Economically, too, benefits are to be seen, with increased bioavailability, lower quantities of selenium would be required to supplement selenium-deficient diets, resulting in less expensive feeds.

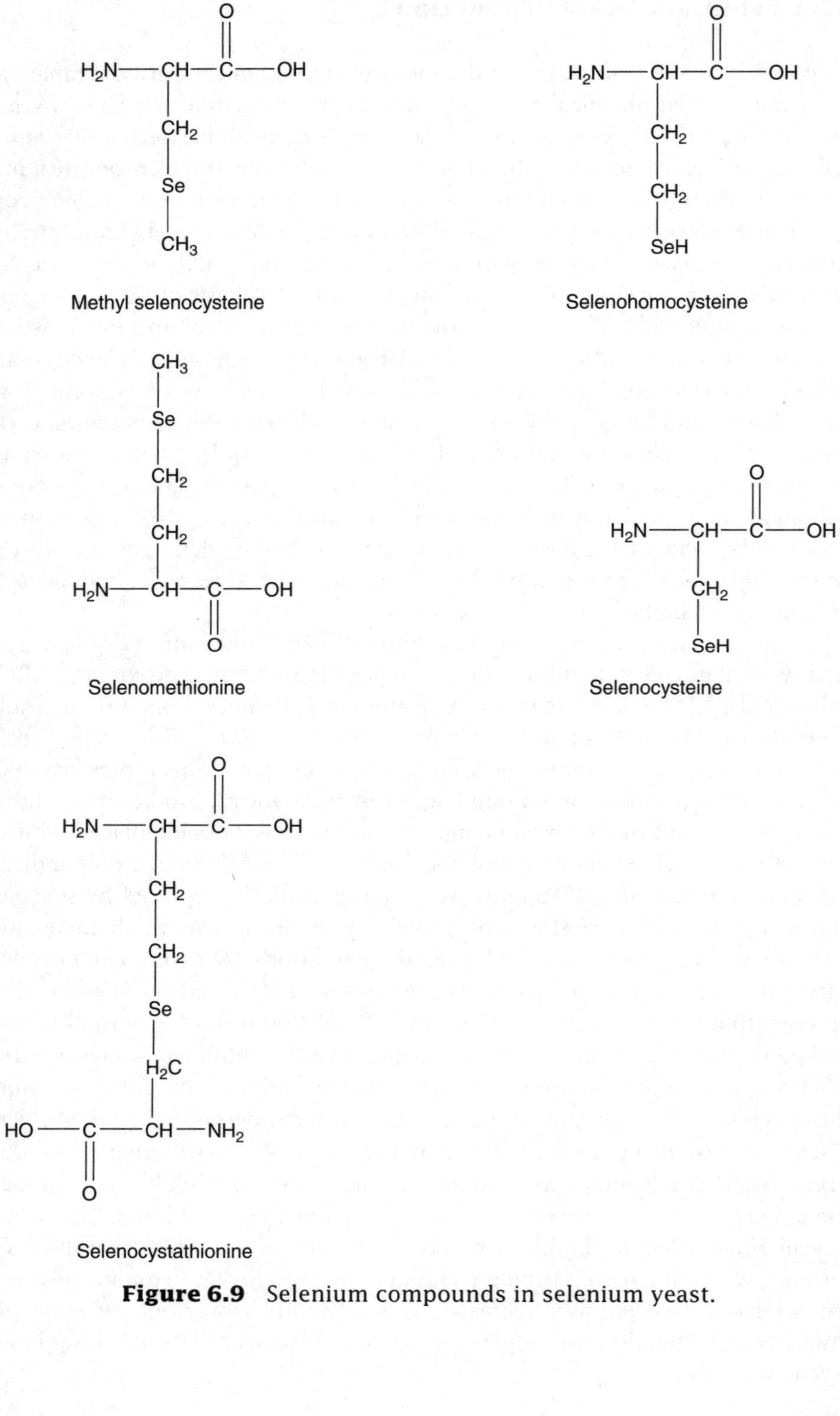

Figure 6.9 Selenium compounds in selenium yeast.

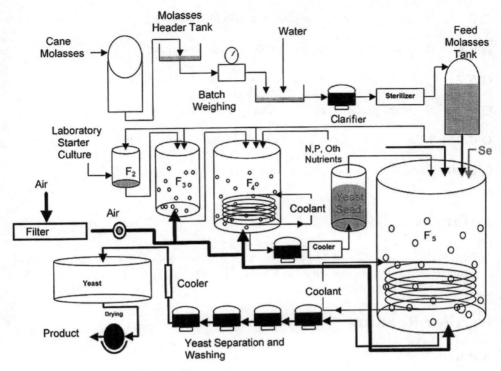

Figure 6.10 Selenium yeast production.

6.8 Conclusions

Within the fungal life cycle one can clearly delineate the production of certain products or metabolites into two phases, namely primary and secondary metabolism. Fungal biotechnology has developed to allow the utilization of these metabolic processes in a commercially viable manner. The economic significance of this cannot be understated when one considers the use of fungi in the production of valuable commodities such as antibiotics, enzymes, vitamins, pharmaceutical compounds, fungicides, plant growth regulators, hormones and proteins.

Revision Questions

Q 6.1   Give three examples of both primary and secondary metabolites.

Q 6.2   Name the main cell-wall component on which penicillins act.

Q 6.3   How does cyclosporine A function as an immunosuppressant?

Q 6.4   How do the so-called statins act to prevent cholesterol biosynthesis?

Q 6.5 What fungus is the main producer of cephalosporins?

Q 6.6 How does *Saccharomyces cerevisiae* utilize selenium?

Q 6.7 Name the yeast responsible for the production of astaxanthin.

Q 6.8 Name the preferred fungal source of vitamin B_2 (riboflavin).

Q 6.9 What is the medically useful alkaloid ergometrine used for?

Q 6.10 Name three medically important roles that endophytes can play.

References

Tejesvi, M.V., Nalini, M.S., Mahesh, B. *et al.* (2007) New hopes from endophytic fungal secondary metabolites. *Boletín de la Sociedad Química de México*, 1(1), 19–26.

Further Reading

Books

Anke, T. (ed.) (1991). *Fungal Biotechnology*, Chapman and Hall, London.

Arora, K.D. (ed.) (2005) *Fungal Biotechnology in Agricultural, Food and Environmental Applications (Mycology)*, Dekker, New York.

Bu'lock, J. and Kristiansen, B. (eds) (1987) *Basic Biotechnology*, Academic Press, London.

Carlile, M.J., Watkinson, S.C. and Gooday, G.W. (2001) *The Fungi*, 2nd edn, Academic Press, London.

Tkacz, J.S. and Lange, L. (eds) (2004) Advances in Fungal Biotechnology for Industry, Agriculture, and Medicine, Kluwer Academic/Plenum Publishers, New York.

Walsh, G. (1998) *Biopharmaceuticals: Biochemistry and Biotechnology*, John Wiley and Sons, Inc., New York.

Walsh, G. and Headon, D. (1994) *Protein Biotechnology*, John Wiley and Sons, Inc., New York.

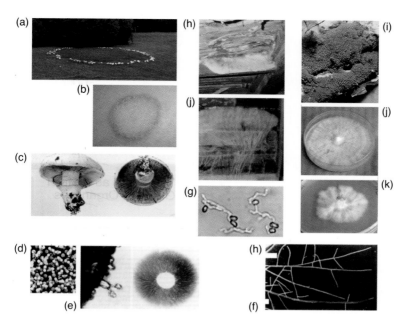

Plate 1.1 Radiating circular pattern of (a) *Chlorophyllum molybdites* fruit-bodies forming a fairy ring in grassland and (b) *Tinea corporis* infection of the skin. (c) Fruiting bodies of *Agaricus bitorquis*, (d) scanning electron micrograph of *Agaricus bisporus* gill surface (Carlie *et al.* (2001). Permission obtained for First Edition), (e) light micrograph of basidium bearing two spores (Carlie *et al.* (2001). Permission obtained for First Edition), (f) spore print from underside of fruit-body (Carlie *et al.* (2001). Permission obtained for First Edition), (g) germ-tubes emerging from spores, (h) branching hyphae growing on an agar plate (Carlie *et al.* (2001). Permission obtained for First Edition); (all from Carlile *et al.* (2001), Academic Press). The dry-rot fungus *Serpula lacrymans* (i) colony decaying timbers in a wall void, and forming a red–brown fruiting structure, (j) close-up of underside of a basidiocarp, (k) exploratory fan-shaped mycelium with connected rhizomorphs, (l) fast- and (m) slow-growing colonies growing on agar media.

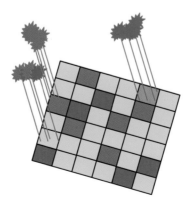

Plate 4.1 Two-colour microarray technology; 6400 distinct ORFs are arrayed on the slide (only 36 are shown here). Two separate cDNA samples, one from yeast growing in glucose at the beginning of the fermentation labelled with a green fluorescent dye and the other from later in the fermentation labelled with the red dye. Red colour indicates gene expression increased relative to the reference, green colour indicates gene expression decreased relative to the reference and yellow colour indicates no change in expression level. Three ORFs are shown in detail here.

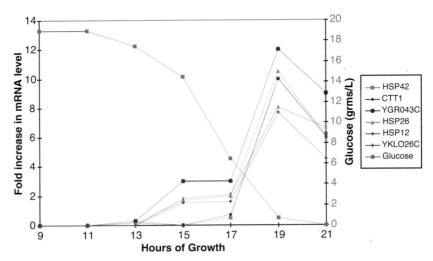

Plate 4.2 Pattern interpretation of microarray (after DeRisi et al. (1997)). This shows how the expression levels of genes known to be stress-induced via STREs in their promoters were identical to one another. (The promoter sequences of other genes with the same profile but not previously identified as stress-inducible were than examined - many contain one or more recognisable STRE sites).

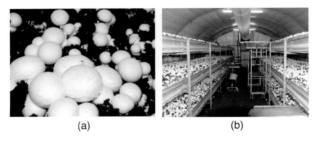

(a)   (b)

Plate 5.1 (a) Cultivated mushrooms. (b) A mushroom tunnel with shelves containing compost and casing for growth of mushrooms.

Plate 5.2 *Trichoderma* inhibiting the growth of mushrooms.

Plate 5.3 Mushroom showing evidence of *Lecanicillium* dry bubble infection.

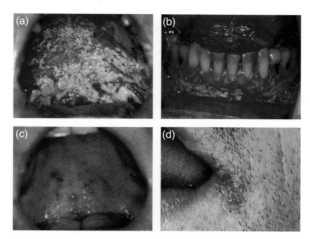

Plate 10.1 Photographs showing different forms of oral candidosis. Pseudomembranous candidosis of the palate (a) and oral mucosa (b), erythematous candidosis of the palate (c) and angular cheilitis (d).

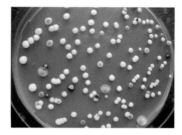

Plate 10.2 Colonies of *Candida* species grown from a clinical sample taken from the oral cavity of a patient with oral candidosis and grown on a CHROMagar Candida™ plate. Most *Candida* species can be identified on the basis of the colour of the colonies produced (e.g. *C. albicans* colonies are green–blue, *C. glabrata* colonies are pink and *C. parapsilosis* are pale cream).

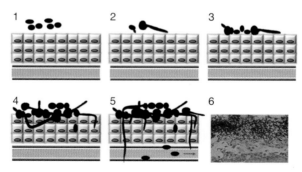

Plate 10.3 A schematic diagram indicating the stages of candidal infection. (1) *Candida* cells (black) come into contact with host epithelium (blue). (2) *Candida* cells, particularly germ tubes and hyphae, adhere to the epithelial cell surface via specific interactions between candidal adhesins (e.g. Hwp1) and host ligands. (3) Once they have bound to the tissue the *Candida* cells begin to proliferate and to produce hyphae. (4) Growth continues by budding and production of extensive levels of hyphae (growth is often associated with formation of a biofilm). (5) Fungal hyphae and yeast cells eventually penetrate through the epithelial layer to the tissues below, ultimately reaching the bloodstream (red), via which they disseminate throughout the body to cause disseminated candidosis. (6) A photograph of reconstituted human epithelial tissue infected with *C. albicans*.

Plate 12.1 The disease triangle.

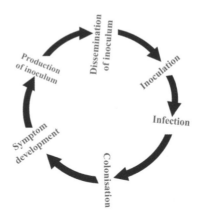

Plate 12.2 Variation in the head blight susceptibility of two barley cultivars to *Fusarium* head blight disease. Both heads were inoculated with *Fusarium culmorum*, but only the cultivar on the left shows disease symptoms (premature bleaching of head spikelets).

Plate 12.3 A generalized fungal disease cycle.

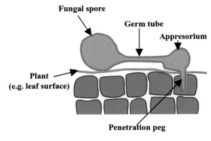

Plate 12.4 Illustration of fungal penetration of a host plants by means of a specialized appresorium and penetration peg.

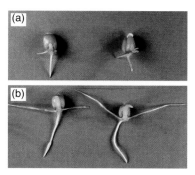

Plate 12.5 Effect of deoxynivalenol, a non-host-specific fungal trichothecene toxin, on the germination of wheat seedlings. Seeds germinated on toxin solution (a) show reduced root and coleoptile growth compared with those germinated on water (b).

Plate 12.6 Clubroot of brassicas caused by *Plasmodiophora brassicae.* Roots swell and develop tumours due to increased hormone (e.g. cytokinin) production.

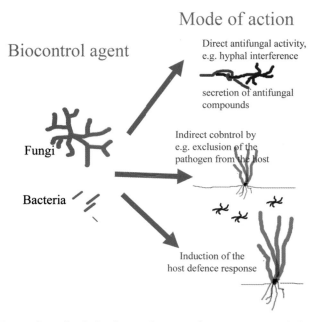

Plate 12.7 Means by which biological control agents can inhibit the development of a fungal disease.

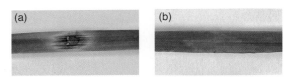

Plate 12.8 Biological control of net blotch disease of barley (caused by *Perenophora teres*) by bacteria (*Bacillus* spp.) originating from a cereal field. The blotch disease symptom severity caused by this pathogen on inoculated leaves in the absence (a) and presence (b) of the biocontrol agent.

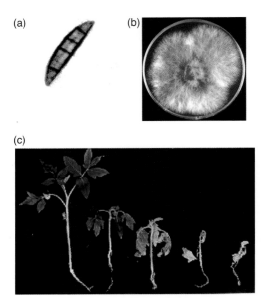

Plate 12.9 *Fusarium* wilt. Microscopic analysis of typical asexual macroconidia (a) and mycelial growth on potato dextrose agar (b) of *Fusarium* species (*Fusarium oxysporum* also produces microconidia and chlamydospores). (c) *Fusarium oxysporum* f.sp. *lycopersici* wilt of tomato: symptoms range from healthy (left) to severe wilting and stunting (right) (courtesy Dr Antonio Di Pietro, Cordoba, Spain; Di Petro et al. (2003)).

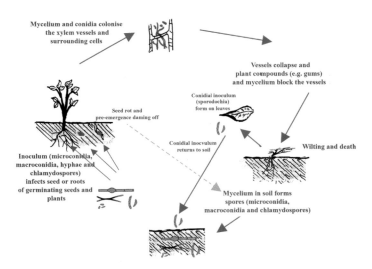

Plate 12.10 Simplified *Fusarium* wilt disease cycle (adapted from Agrios (1988)).

Plate 12.11 Yellowing/browning of leaves of an elm branch due to Dutch elm disease. The entire tree subsequently wilted and died.

Plate 12.12 *Phytophthora infestans* late blight of potatoes. Microscopic analysis of asexual sporangia containing zoospores (a) and sexual oospore (b). Disease symptoms: water-soaked dark lesions on leaves (c) and diseased tubers exhibit brown/black blotches on their surface, and internally they exhibit water-soaked dark-brown rotted tissue (d).

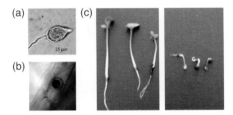

Plate 12.13 *Pythium* damping-off disease. (a) Asexual sporangia containing zoospores; (b) sexually produced oospore formed on the surface of a tomato seedling; (c) seedling of cress that are healthy (left) and other exhibiting damping-off symptoms (right).

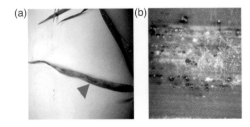

Plate 12.14 *Septoria* leaf blotch of wheat caused by *Mycosphaerella graminicola*. (a) Oval disease lesions (surrounded by a chlorotic halo) on an infected leaf, running parallel to the leaf blade. (b) Black pycnidia develop on mature lesions and cirri may form on the lesions if the weather is dry for prolonged periods.

Plate 12.15 Sunken lesions on broad bean pod due to anthracnose disease caused by *Colletotrichum lindemuthianum.*

Plate 12.16 Black stem rust of mature wheat stems caused by *Puccinia graminis* f.sp. *tritici.* The three stems possess black pustules of teliospores; these pustules arise from red urediospore pustules that erupt through the stem epidermis earlier in the growing season.

(a)   (b)

Plate 12.17 Smut diseases of cereals. Covered smut of oats caused by *Ustilago hordei* (a) and covered smut or 'bunt' of wheat caused by *Tilletia* species (b). Grain are replaced with black spore masses and hence destroyed.

Plate 12.18 Powdery mildew disease of wheat caused by *Blumeria* (= *Erysiphae*) *graminis.* Symptoms of powdery mildew disease include chlorotic or necrotic leaves, stems and fruits covered with mycelium and fruiting bodies of the fungus, often giving a white 'woolly' appearance.

7

Biotechnological Use of Fungal Enzymes

Shauna M. McKelvey and Richard A. Murphy

7.1 Introduction to Enzymes

The roots of modern enzymology may be traced back to the nineteenth century, when it was shown that an alcohol precipitate of malt extract contained a thermolabile substance which converted starch into fermentable sugars. The enzyme responsible was termed diastase because of its ability to yield soluble dextrins from insoluble starch granules. By the mid-ninteenth century the existence of several additional enzymes, including polyphenol oxidase, peroxidase and invertase, were recognized. An amylolytic preparation, termed Taka-diastase, was the first enzyme preparation to be patented for industrial use. It was produced by the filamentous fungus *Aspergillus oryzae* when grown on rice. The patent was lodged in 1884 by Dr Jokichi Takamine, a Japanese immigrant to the USA. Since then, the technology to identify, extract and produce enzymes on a commercial scale has progressed dramatically, and they are now used in many industrial processes.

Of all the enzymes discovered, approximately half are of fungal origin. Numerous factors favour microbially sourced enzymes over those derived from plants or animals, including the wide variety of catalytic activities, higher yields, ease of genetic manipulation and rapid production in inexpensive media. Other appealing factors include the fact that they are biodegradable and are active under mild conditions with respect to temperature and pH. Enzymes have long played a central role in mankind's attempts to utilize biological systems for a variety of purposes. Examples of harnessing their power include cheese making, brewing, baking, the production of antibiotics and in the manufacture of commodities

Fungi: Biology and Applications, Second Edition. Edited by Kevin Kavanagh.
© 2011 John Wiley & Sons, Ltd. Published 2011 by John Wiley & Sons, Ltd.

such as leather, indigo and linen. They also find applications in areas such as detergent and paper production, the textile industry and in the food and drinks industry in products ranging from fruit juice to coffee, tea and wine.

7.2 Enzymes in Industry

The first serious attempt at the turn of the twentieth century to use enzymes for industrial purposes met with limited success. Disappointing results were due largely to a lack of understanding of enzyme activity. Only in the last 30–50 years have enzymes been characterized and their kinetics understood. This increased knowledge has in turn facilitated the application of enzymes in a variety of industrial processes.

The majority of enzymes currently used in industry may be described as hydrolytic depolymerases (i.e. pectinases, lipases, carbohydrases, etc.) and the single most significant industrial application of enzymes is the inclusion of proteases and amylases in detergent preparations. Carbohydrases (in particular, amylases and pectinases) account for a significant portion of the remaining market. The glucose isomerase enzyme, which is utilized industrially in the production of high-fructose syrup from glucose syrups, is one of the few non-depolymerases to command significant industrial volume.

Rapid developments have occurred in the enzyme supply market over the past few years due to the evolution of the biotechnology industry. Volume growth for this thriving sector is currently between 4 and 5 % of the annual growth rate and annual sales figures have crossed the US$2 billion mark. Current advancements in protein engineering and recombinant gene technology have revolutionized enzyme production and commercialization by extending the list of enzymes now available (Table 7.1).

Three main segments dominate the industrial enzyme market: technical enzymes, food enzymes and animal feed enzymes. Technical enzymes, including protease, amylase and cellulose, account for the highest percentage of sales. These enzymes are extensively used in the detergent, starch, textile, leather, paper and pulp, and personal care industries. The second largest segment in the market is for food enzymes. Included in this group are enzymes employed in the dairy, brewing, wine and juice, fats, oils and baking industries, such as lipase and pectinase. The final area is that of feed enzymes, comprising enzymes such as phytase, xylanase and β-glucanase.

7.3 Current Enzyme Applications

Given the wide and varied applications for industrial enzymes, discussion will be limited to the most important in financial terms.

Table 7.1 Sources and applications of industrial enzymes.

Enzyme	Organism	Main application areas
α-Amylase	*Aspergillus niger, Aspergillus oryzae*	Starch processing and food industry
Chymosin	*A. niger*	Food industry
Cellulase	*Trichoderma viride, Trichoderma reesei*	Textile, pulp and paper industry
Cellobiohydrolase	*T. viride, T. reesei*	Textile, pulp and paper industry
Glucoamylase	*Aspergillus phoenicis, Rhizopus delemar*	Starch processing industry
Glucose oxidase	*A. niger, A. oryzae*	Textile industry, biosensor
Laccase	*Trametes versicolor*	Textile, pulp and paper industry
Lipases	*A. niger, A. oryzae*	Food and detergent industry
Pectin lyase	*T. reesei*	Food industry
Proteases	*A. niger, A. oryzae, R. delemar*	Food and detergent industry
Phytase	*A. niger, A. oryzae*	Food industry
Rennin	*Mucor miehei*	Food industry
Xylanases	*T. reesei, Trichoderma konignii, A. niger*	Textile, pulp, paper and bakery industry
Amyloglucosidase	*A. niger*	Starch syrups, dextrose, foods
Invertase	Yeast species	Confectionary industry

Adapted from Meyer (2008).

7.3.1 Detergent Industry

The use of enzymes originates from 1913, when pancreatic enzymes were used in a pre-soaking product. Shortly after this the first enzymatic detergent, Brunus, was marketed. Since then the application of enzymes in detergents has become widespread, with different enzymatic formulations continuously emerging, such as mannanase, which aids in the removal of various food stains containing guar gum. Detergent additives still represent the largest application of industrial enzymes, and protease is the most dominant enzyme, accounting for 30 % of the market. Cellulases have also been employed in household washing powders; however, they are considered a more expensive enzyme source. They function by allowing the removal of small, fuzzy fibrils from fabric surfaces such as cotton and improve the appearance and colour brightness.

The increased usage of these enzymes as detergent additives is mainly due to their cleaning capabilities in environmentally acceptable, non-phosphate detergents.

Important factors to be considered when including enzymes in detergents include:

- they should have a high activity and stability over a wide pH and temperature range;
- they should be effective at low levels (0.4–0.8 %);
- their compatibility with various detergent components;
- having a long shelf life.

7.3.2 Bioethanol and Biodiesel

Owing to recent increased environmental concerns, rising oil prices and fast-diminishing resources of fossil fuels, there has been a surge of interest in the production of biofuels such as bioethanol and biodiesel. Ethanol is the most widely used biofuel today, and intense efforts are currently underway to develop a technically viable process for conversion of cellulose and lignocellulose into bioethanol. As an incentive to reduce the price of cellulase enzymes used in processing cellulosic materials, the US Department of Energy awarded a research grant of US$32 000 000 to Genencor and Novozyme in an attempt to make bioethanol production more economically feasible.

Biodiesel is an ester derived from oils such as soybean, rapeseed, and vegetable and animal fats. Vegetable-derived biodiesel is much cleaner than that of petroleum, as it does not produce sulfur dioxide and the soot particulate is minimized by one-third. Through trans-esterification with lipase enzymes, organically derived oils are combined with ethanol or methanol to form fatty esters and these esters can then be blended with conventional diesel fuel or used as 100 % biodiesel.

7.3.3 Tanning Industry

Traditionally, lime and sodium sulfide mixtures have been used in the leather industry for the de-hairing of skins and hides. This method causes pollution and is unpleasant. However, through the advent of biotechnology, safer, more environmentally friendly means of treating animal skins have been developed. The use of a cocktail of proteases and lipases extracted from fungal strains such as *Aspergillus* have now become common practice. The enzyme mixture causes the swelling of hair roots, which allows the hair to be removed easily. Elastin and keratin are then degraded and hair residues removed in a process called bating. The end product is of a higher quality than leather manufactured using traditional methods.

7.3.4 Effluent and Waste Treatment

With respect to the waste treatment industry, enzymes are now playing a significant role. For instance, lipases are used in activated sludge and other aerobic waste processes, as they aid in the breaking down of solids and prevention of fat blockages. Proteases are employed in the processing of waste feathers from poultry slaughterhouses. Approximately 5 % of the body weight of poultry is feathers, which can be considered a high protein source of feed provided their keratin structure is completely degraded. Total solubilization of feathers can be achieved through enzymatic hydrolysis along with sodium hydroxide addition and mechanical disintegration.

Vast amounts of wastewater are produced each year in the textile industry, with China producing in the region of 0.65 billion tonnes per annum, and disposal of this effluent causes environmental pollution. Currently, textile effluent is mainly treated using biological methods or a combination of biological and chemical methods. However, enhancements of biological treatment techniques, such as bio-augmentation, immobilized microorganisms and microorganism activity enhancements, are currently being studied with a view to improving the efficiency of this system.

7.3.5 Food Processing

The food industry is one of the most significant utilizers of enzymatic activities, ranging from catalases and lipases in cheese making and cheese ripening, to pectinases in wine and fruit juice clarification, α-amylases in dextran liquefaction and proteases in meat tenderization. The main reasons for the increased interest in the use of enzymes in food-related processes include:

- They are very specific and controllable chemical catalysts; as such, the production of undesirable by-products may be avoided.

- Catalysis takes place under mild conditions of temperature, pH and pressure compared with chemical methods of conversion.

- Enzymes are biodegradable and chemical toxicity problems are greatly reduced.

- Their immobilization on solid supports can provide technological advantages in processing and also avoid the presence of the enzyme in the final product, thus reducing the possibility of allergenicity.

Lipases find application in the production of leaner fish, refining rice flavour and modifying soybean milk. Other applications include the flavour enhancement of cheeses and the production of cheese products such as soft cheeses. Proteases play a vital role in meat tenderization, as they possess the ability to hydrolyse connective tissue proteins as well as muscle fibre proteins. Enzymes are applied directly to the meat or injected directly into the bloodstream. Bread making is a technique that has been around for thousands of years. Currently, amylase, protease and cellulase are being used as dough improvers. Their application results in improved texture, volume, flavour and freshness of the bread, as well as improved machinability of the dough. In addition, current efforts are being made to further understand bread staling and the mechanisms behind the enzymatic prevention of staling when using α-amylases and xylanases.

7.3.6 Fruit Juice Maceration

A combination of pectinases, cellulases and hemicellulases (collectively called macerating enzymes) are used in the extraction and clarification of fruit and vegetable juices. In addition, α-amylases, amyloglucosidase and laccase have been used to prevent haze formation in starch-containing fruits such as apples. Treatment of fruit pulps with pectinase also showed an increase in fruit juice volume from bananas, grapes and apples. The demand for enzymes such as these is bound to increase as the range of applicable fruits and vegetables rises.

A process called vacuum infusion has been developed using pectinase to ease the peeling of citrus fruits. This process was developed from the observation that the infusion of certain impure naringinase preparations into grapefruit peel in an attempt to reduce bitterness also softened the albedo (the white portion of citrus peel). This application may also apply to the pickling process, where excessive softening may occur during fermentation and storage. Thus, enzyme infusion to alter the sensory attributes of fruits, vegetables and other foods has enormous potential in food biotechnology.

7.3.7 Animal Feed

The animal feed industry is an extremely important part of the world's agro-industrial activities, with an annual production of >624 million tonnes of feed, worth in excess of US\$50 000 000. It is an industry that has gone through many changes in the past few years, as consumers and industry have looked into how compound animal feeds are produced, how the animals are reared and how the systems of animal husbandry in use today affect the environment. The incorporation of enzymes, including β-glucanase, cellulase, amylase and protease, into feedstuffs has been demonstrated to contribute to enhanced production performance. Perhaps the best example is the addition of β-glucanase and xylanase to barley-based diets in the poultry industry. Noted positive effects included improvements in digestion and absorption of feed components, as well as weight gain by broiler chicks and egg-laying hens.

Other positive effects of these enzymes are evident in the digestion of non-starch polysaccharides (NSPs) by barley-fed piglets. Table 7.2 lists enzyme activities which may contribute to enhanced animal performance when incorporated into animal feedstuffs. The target substrates of the enzymes are also listed, as are the recorded beneficial effects.

Supplementation of the diet with selected enzyme activities may also promote a decrease in the overall pollutive effect of animal excreta. This is particularly true in the case of dietary phosphorus, a large proportion of which remains unassimilated by monogastrics. The inclusion of suitable, microbially derived phytases in the diet can improve the efficiency of nutrient utilization and reduce waste. Current research has identified new fungal enzymes with 4–50-fold higher specific activities than previously reported.

Phytic acid is an organic compound of phosphate and is the main storage form of phosphorus in plants (60–65 % of phosphorus present in cereal grains). Phytate-bound phosphorus is poorly hydrolysed, as broilers have a low capacity

Table 7.2 Enzyme activities and associated supplementational effects.

Enzyme(s)	Substrate	Beneficial effects
β-Glucanases	Mixed linked glucans	Reduction in viscosity of digesta
		Reduction in incidence of pasted vents
		Improvements in litter characteristics
		Reduction in dirty egg problems
Xylanases	Pentosans	Reduction in viscosity of digesta
Cellulases	Cellulose	Promotes a more comprehensive digestion of vegetative matter
Phytases	Phytic acid	Removal of anti-nutritional effects of phytic acid
Proteases	Proteins	Supplementation of endogenous proteolytic and
Amylases	Starch	amylolytic capacity to benefit young or sick animals whose digestive function may not be operating maximally.

Adapted from Walsh (1998).

for its hydrolysis. Phytic acid also hinders the assimilation of other nutritionally important proteins and metals, such as calcium, zinc and magnesium, as it binds tightly to these. The majority of this form of phosphorus, therefore, is excreted in the manure, causing environmental problems in areas of intensive livestock production. Inclusion of phytate-degrading enzymes such as phytase has yielded many dramatic beneficial results on an environmental basis and the addition of such enzymes to animal diets is well documented. Use of phytase has a threefold beneficial effect: the anti-nutritional properties of phytic acid are destroyed, a lesser requirement for feed supplementation with inorganic phosphorus results and reduced phosphate levels are present in the faeces, which leads to less loading of the environment with phosphorus. Currently, phytase utilization is restricted to locations where there is a considerable pollution load, but if increases in efficiency can be achieved, then the application of this enzyme will become more widespread.

7.4 Future Direction of Industrial Enzymes

There is little doubt that in the near future the use of industrial enzymes will expand dramatically in areas such as biopulping, food processing, carbohydrate conversions, chemical conversions, food and animal feed additives, cleaning and detoxification of environmental toxins. Prior to the advent of the tools of molecular genetics, the use of enzymes as industrial catalysts was limited to perhaps 20 enzymes that could be produced inexpensively in large amounts. The use of genetic engineering to clone the genes of non-abundant enzymes for subsequent overexpression in heterologous hosts promises to expand the opportunities for using enzymes in industrial applications greatly.

7.5 Specific Enzymes

In a commercial sense the main enzymes include protease, cellulase, xylanase, lipase, amylase and phytase and can be produced by many different genera of microorganism, including fungal strains of *Aspergillus*, *Rhizopus* and *Penicillium*. The following deals specifically with these enzymes.

7.5.1 Protease

Proteases have been widely described as a diverse class of enzymes that are known to hydrolyse the peptide bond (CO-NH) in a protein molecule. They vary with respect to their pH optima and, in general, alkaline proteases are produced by bacteria and acid proteases by fungi. Proteolytic enzymes occur naturally in all living organisms and represent 2 % of all the proteins present. They have been applied in numerous industrial processes and the great diversity of proteases has attracted worldwide attention in attempts to exploit their physiological and biotechnological applications.

Proteases are physiologically necessary for living organisms and are found in a wide diversity of sources, such as plants, animals and microorganisms. However, microbes possess certain qualities that see them as a preferred choice, including their broad biochemical diversity and susceptibility to genetic manipulation. Microbial proteases, therefore, have been extensively studied and their molecular properties are understood in detail.

Filamentous fungal strains such as *Aspergillus* have been widely used for industrial production of protease. Compared with bacteria, fungi produce a wider variety of protease with broad pH activity ranges.

7.5.1.1 Types of Protease

Proteases are subdivided into two major groups, depending on their site of action: exopeptidases and endopeptidases. Exopeptidases, also known as peptidases, are classified according to whether they split off single amino acids from the N-terminus or C-terminus of peptide chains and are specific for dipeptide substrates. Additional exo-acting peptidases cleave dipeptide units from the N-terminus or the C-terminus of proteins. Endopeptidases, also known as proteinases, cleave peptide bonds internally within a polypeptide. Based on the functional group present at the active site, endoproteinases are further classified into four prominent groups, namely serine, cysteine, aspartic and metalloproteases (Table 7.3).

Aminopeptidases are exo-acting peptidases and catalyse the cleavage of amino acids from the N-terminus of protein or peptide substrates. They are widely distributed throughout the animal and plant kingdoms and are found in a variety of microbial species, including bacteria and fungi.

As the name implies, carboxypeptidases cleave amino acids from the C-terminal of the polypeptide chain and liberate a single amino acid or a dipeptide. They can be divided into three groups based on the nature of the amino acid residue at the active site of the enzyme: serine carboxypeptidases, metallo-carboxypeptidases and cysteine carboxypeptidases.

Serine proteases are characterized by the presence of a serine group in their active site and are characterized as endopeptidases with a strong proteolytic activity coupled with low specificity. The basic mechanism of action of serine proteases involves transfer of the acyl portion of a substrate to a functional group of the enzyme. They are widespread amongst bacteria, viruses and eukaryotes, suggesting that they are vital to all organisms. Serine proteases have also been found in the exopeptidase, oligopeptidase and omega peptidase groups. Based on their structural similarities, they have been further divided into 20–30 families.

Proteases with low pH optima are abundant in filamentous fungi, and a large proportion of these have been shown to have properties consistent with aspartic proteinases. These molecules contain an aspartic residue at the active site and

Table 7.3 Classification of proteases.

Protease	Mode of action[a]	EC no.
Exopeptidases		
Aminopeptidases	●↓-o-o-o-o—	3.4.11
Dipeptidyl peptidases	●-●↓-o-o-o—	3.4.14
Tripeptidyl peptidase	●-●-●↓-o-o—	3.4.14
Carboxpeptidase	—o-o-o-o-o↓-●	3.4.16–3.4.18
Serine-type protease		3.4.16
Metalloprotease		3.4.17
Cysteine-type protease		3.4.18
Peptidyl dipeptidases	—o-o-o-o↓-●-●	3.4.15
Dipeptidases	●↓-●	3.4.13
Omega peptidases	*-●↓-o-o—	3.4.19
	—o-o-o↓-●-*	3.4.19
Endopeptidases	—o-o-o↓-o-o-o—	3.4.21–3.4.34
Serine protease		3.4.21
Cysteine protease		3.4.22
Aspartic protease		3.4.23
Metalloprotease		3.4.24
Endopeptidases of unknown catalytic mechanism		3.4.99

Adapted from Rao *et al.* (1998).
[a]Open circles represent the amino acid residues in the polypeptide chain. Solid circles indicate the terminal amino acids and stars signify the blocked termini. Arrows show the sites of action of the enzyme.

are unaffected by chelating agents, thiol-group reagents or by serine protease inhibitors. They share similarities with the animal digestive enzymes, pepsin and rennin and, therefore, are employed as replacements for animal proteases and in cheese manufacture.

Cysteine proteases occur in both prokaryotes and eukaryotes and about 20 families have been recognized; however, reports of their occurrence in fungi are very limited. The activity of all cysteine proteases depends on a catalytic dyad consisting of cysteine and histidine, and the order of Cys and His residues differs amongst the families. Cysteine proteases have a highly similar mechanism of action to that of serine proteases.

Metalloproteases are the most diverse of the catalytic types of protease and about 25 families have been recognized. The metalloproteases typically contain an essential metal atom and show optimal activity at neutral pH. Ca^{2+} is essential for stability and the molecules, therefore, are inactivated and destabilized by chelating agents.

7.5.1.2 Application of Proteases

The use of proteases in the food industry dates back to antiquity; for example, the ancient Greeks utilized enzymes from microorganisms for various purposes such as baking, alcohol production and meat tenderization. The importance of proteases in industry has been widely investigated and applications have been found in sectors such as detergent, feed, pharmaceutical, animal feed, diagnostics and fine chemical industries. Of these industries, the detergent and feed sectors are the highest exploiters of proteolytic activity.

There is a wide application for proteolytic enzymes in the detergent industry due to a number of attractive characteristics, including stability over a broad temperature range and optimal activity in the alkaline pH range. Proteases added to laundry detergents enable the release of proteinaceous stains such as keratin, blood, milk and gravy. In response to the current energy crisis and awareness for energy conservation, researchers are continually screening for new proteases. Attempts have been made to produce protease enzyme, using various substrates such as shrimp and crab shell powder, soybean meal and fish waste.

With respect to the dairy industry, the main application for proteases is in the manufacture of cheese, as they have high milk clotting properties. Calf rennet, extracted from the fourth stomach of calves, has traditionally been used in the manufacturing of cheese; however, increased demand resulted in a shortage of rennet and the search began for alternative sources. Recombinant chymosin has successfully replaced calf rennet and is now produced from fungal strains, including *Mucor miehei* and *Mucor pusillus*. In addition, fungal protease from *Penicillium roqueforti* and *Penicillium caseicolum* play central roles in the cheese ripening process. *A. oryzae* proteinases are widely employed in the baking industry to help control break texture and gain dough uniformity.

The wide diversity and specificity of proteases are used to great advantage in developing effective therapeutic agents. Collagenases are increasingly used for therapeutic applications in the preparation of slow-release dosage forms. Proteolytic enzymes from *A. oryzae* have been used as digestive aids to correct lytic enzyme deficiency. Overall, the wide diversity of proteases allows for further exploitation of these microbial powerhouses.

7.5.2 Cellulase

Cellulose is the principal component of plant cell walls and is the most abundant, renewable polymer on Earth. Its structure consists of a linear polymer of 1,4-β-linked glucose residues. Individual cellulose molecules are linked together by hydrogen bonds to produce larger, crystalline structures. Owing to the complexity of this structure, crystalline cellulose is not amenable to attack by single enzymes. As a consequence, cellulolytic microorganisms have been utilized to secrete mixtures of cellulolytic activities which degrade cellulose.

It is widely documented that complete degradation of cellulose requires the coordinate action of three main enzymes, including endoglucanases, exoglucanases and β-glucosidases. While cellulase is an endoglucanase that hydrolyses cellulose randomly, producing numerous oligosaccharides, cellobiose and glucose, exoglucanases hydrolyse β-1,4-D-glucosidic linkages in cellulose, releasing cellobiose from the non-reducing end. On the other hand, β-glucosidases hydrolyse cellobiose to glucose (Table 7.4). As it is a major waste by-product both in nature and from man's activities, cellulose holds tremendous potential as a renewable energy source.

Table 7.4 Components of aerobic fungal cellulases and their mode of action on the cellulose chain.

Enzyme type	EC code	Synonym	Mode of action
endo-(1,4)-β-D-Glucanase	EC 3.2.1.4	Endoglucanase or endocellulase	—G—G—G—G— ↑ ↑ Cleaves linkages randomly
exo-(1,4)-β-D-Glucanase	EC 3.2.1.91	Cellobiohydrolase or exocellulase	G—G—G—G—G— ↑ Releases cellobiose either from reducing or non-reducing end
exo-(1,4)-β-D-Glucanase	EC 3.2.1.74	Exoglucanase or glucohydrolyase	G—G—G—G—G— ↑ Releases glucose from non-reducing end
β-Glucosidase	EC 3.2.1.21	Cellobiase	G—G, G—G—G—G ↑ ↑ Releases glucose from cellobiose and short-chain xylooligosaccharides

Adapted from Bhat and Bhat (1997).

7.5.2.1 Sources of Cellulase

Cellulolytic enzymes are produced by a wide variety of bacteria and fungi, aerobes and anaerobes, mesophiles and thermophiles. The most detailed studies have been on the cellulolytic enzyme systems of the aerobic fungi *Trichoderma reesei*, *Trichoderma viride*, *Penicillium pinophilium*, *Sporotrichum pulverulentum* and *Fusarium solani*.

7.5.2.2 Application of Cellulase

Active research on cellulases and related polysaccharides began in the early 1950s owing to their enormous potential to convert lignocellulose to glucose and soluble sugars.

Biotechnology of cellulases and hemicellulases began in early 1980s, first in animal feed, followed by food applications. Subsequently, these enzymes were used in the textile, laundry and the pulp and paper industries.

Cellulose is one of the most abundant polymers on Earth with exceeding good potential as a renewable energy source. During World War 2 the US army was alarmed at the rate of deterioration of cellulosic materials in the South Pacific, including clothing, tents and sand bags. Several organizations within the army set up laboratories to find an immediate solution to this problem. As a result, a parent strain QM6A was isolated and identified as *T. viride* and later recognized as *T. reesei*. The immediate benefit of the army's programme led to further research on selection and characterization of hyper-cellulolytic *T. reesei* strains. These projects not only improved the production of cellulase by *T. reesei*, but also aroused worldwide activity.

Fungal enzyme systems involved in cellulose degradation have been extensively studied owing to their potential value in biotechnology. Applications are found in food sectors such as juicing, in which cellulases, along with numerous other enzymes, are used in juice extraction and clarification. In the farming industry, cellulolytic activities are used as supplements for livestock, which in turn increases yield and performance. In the textile industry, cellulases have achieved success because of their ability to modify cellulosic fibres in a controlled and desired manner. Cellulases have only been applied to the textile industry in the past 10 years, but are now the third largest group of enzymes used. Combinations of cellulase and hemicellulase have been used in the pulp and paper industry in areas such as biomechanical pulping and biobleaching.

One of the main applications for cellulolytic enzymes is in the conversion of biomass to ethanol for fuel production. Unlike fossil fuels, ethanol is a renewable energy source produced through fermentation of sugars. Ethanol is widely used in the USA as a partial gas replacement, with approximately 1 billion gallons of fuel ethanol manufactured from cornstarch using a traditional yeast-based process. This process, using a high-value grain, is made possible by the production of animal feed and other co-products. Fuel ethanol that is produced from corn has been used in gasohol or oxygenated fuels since the 1980s; however, the cost of ethanol as an energy source still remains relatively high compared with fossil fuels. As this industry is quite mature, there is limited opportunity for major process improvements. A potential source for low-cost ethanol production is to utilize lignocellulosic materials such as crop residues (straws, hulls, stems, stalks) grasses, sawdust, wood chips and solid animal waste. Development and implementation of such technologies would provide employment, reduce oil imports, improve air quality and provide a natural solution to the disposal of solid wastes. An improvement in the organisms and processes for the bioconversion of lignocellulose to ethanol offers the potential to increase efficiency and reduce the production cost of fuel ethanol to that of petroleum.

Cellulase costs are a critical factor with respect to improving the economics of ethanol production. One approach to increase the effectiveness of cellulase utilization is to develop recombinant microorganisms, which provide some of the enzymes necessary for cellulose solubilization, minimizing the accumulation of soluble inhibitory products and reducing the requirement for fungal cellulase.

7.5.3 Xylanase

Xylan is a major structural polysaccharide in plant cells and one of the most abundant organic substances in nature with a high potential for degradation to useful end products. It is a heteroglycan composed of a linear chain of xylopyranose residues bound by $\beta(1 \rightarrow 4)$ linkages, with a variety of substituents linked to the main chain by glycosidic or ester linkages. Owing to the structural heterogeneity of xylan, complete degradation requires the synergistic action of different xylanolytic enzymes such as endoxylanase, β-xylosidase, α-glucuronidase, α-arabinofuranosidase and esterase. *endo*-1-4-β-Xylanase is the most important, as it initiates the degradation of xylan into xylooligosaccharides and xylose. Xylan is found in large quantities in hardwoods from angiosperms (15–30 % of the cell-wall content) and softwoods from gymnosperms (7–10 %), as well as in annual plants (<30 %), and is typically located in the secondary cell wall of plants.

7.5.3.1 Sources of Xylanase

Xylanases are widespread amongst a wide array of organisms, including bacteria, algae, fungi, protozoa, gastropods and anthropods. Some of the most important fungal sources include strains of *Aspergillus* and *Trichoderma*. The habitats of these microorganisms are diverse and typically include environments where plant materials accumulate and deteriorate, as well as in the stomach of ruminants.

Although numerous organisms produce this class of enzyme, filamentous fungi are the preferred choice for commercial production owing to their high specificity, mild reaction conditions and high level of enzyme production compared with other sources. Fungal xylanases are generally active at mesophilic temperature (40–60 °C) with a slightly acidic pH; however, xylanases have also been reported to be active in extreme environments. The majority of these have been found to be members of families 10 and 11. Psychrophilic fungi such as *Penicillium* sp., *Alternaria alternata* and *Phome* sp. have been isolated from the Antarctic environment. Common features of these xylanases are a low temperature optimum, high catalytic activity at low and moderate temperatures and poor stability.

7.5.3.2 Application of Xylanase

Xylanases are produced by a large number of different fungal strains; however, commercial production is more or less restricted to *Trichoderma* sp. and *Aspergillus* sp. Advancements such as increased activity, thermostability and stability under acidic and alkaline conditions may enhance the repertoire of fungal xylanases utilized in industry. The potential biotechnological applications of xylan and xylanases have been of major interest to researchers over the past 10 years. The major end products of xylan, namely furfural, derived from agricultural residues, and xylitol, obtained from wood residues, are of major importance.

Inclusion of xylanolytic activities in industry has been widely documented and they have found application in the food and beverage industries; for example, to improve the properties of dough and the quality of baked products, as well as in the clarification of juice and wines. In addition, they are used in the poultry industry to increase feed efficiency and improve the nutritional value of silage and greenfeed. Other less well documented applications include in brewing, to increase wort filterability in addition to reducing haze in the final product, and in coffee extraction, in the preparation of soluble coffee.

The widest application for xylanase is in the paper and pulp industry, where environmental regulations have put a restriction on the usage of chlorine in the bleaching process. The by-products formed during chemical processing are toxic, mutagenic, bioaccumulating and cause numerous harmful disturbances in biological systems. Biobleaching has now replaced traditional methods and involves using microorganisms and enzymes to bleach pulp. Microbial xylanase cleaves the xylan backbone and enhances the accessibility of lignin in wood fibres to bleaching chemicals, such as chlorine dioxide (ClO_2), and can reduce the amount of bleaching chemicals required to produce pulps of desired brightness value. Besides bleaching, the use of xylanase helps increase pulp fibrillation, reduces the beating times, hence reducing energy consumption, and increases the freeness in recycled fibres.

The number of xylanase enzymes in production is on the increase; for instance, the United States Patent and Trademark Office has listed 468 patents since 2001. Stringent environmental regulations and awareness to reduce the emission of greenhouse gases have added an incentive for future research developments in the study of xylanases.

7.5.4 Amylase

Amylases, including α-amylase, β-amylase and glucoamylase, are perhaps the most important enzymes in present-day biotechnology owing to their wide-ranging application in numerous industrial processes, including the food, fermentation, textiles and paper industries. Such is their success that microbial amylases have successfully replaced the chemical hydrolysis of starch in starch-processing industries.

The starch biopolymer consists of α-D-glucose joined together in two polymeric units, amylose and amylopectin.

Amylose can be considered as a linear molecule with glucose units linked through α-1,4 bonds with a double helical crystalline structure containing six D-glucose molecules per turn. In contrast, amylopectin is a highly branched structure with an average branch chain of 20–25 glucose units in length and with 4–6 % α-1,6 bonds at branch points.

Amylopectin may have a molecular weight in excess of 10^8, making it one of the largest molecules in nature. Starch granules are round or irregular in shape and in the raw or unprocessed state are between 1 and 100 µm long, being held together with internal hydrogen bonds.

Amylolytic enzymes occur widely and are produced by many species of fungi. While many different microorganisms produce amylases, obtaining a strain capable of producing commercially acceptable yields remains challenging. Commercial production of amylolytic enzymes is carried out by two main methods: submerged fermentation and solid-state fermentation. In recent times, solid-state fermentation has gained in popularity for the production of starch-saccharifying enzymes.

Thermostability of amylase enzymes has become a feature of most of the enzymes with industrial application. As a consequence, thermophilic microorganisms are of special interest for the production of thermostable amylases.

7.5.4.1 α-Amylase

α-Amylases (also known as *endo*-1,4-α-D-glucan glucohydrolyase, 1,4-α-D-glucan glucanohydrolase or glycogenase) are extracellular enzymes which randomly cleave 1,4-α-D-glucosidic linkages that occur between adjacent glucose units in linear amylase chains. These endo-acting enzymes hydrolyse amylase chains in the interior of the molecule and are classified according to their action and end products. Those that produce free sugars are termed 'saccharogenic' and those which liquefy starch without producing free sugars are known as 'starch-liquefying'. By hydrolysing randomly along the starch chain, α-amylase breaks down long-chain carbohydrates such as starch, yielding maltotriose and maltose from amylose, or maltose, glucose and limit dextrin from amylopectin. Owing to its ability to cleave starch indiscriminately, α-amylase tends to be faster acting than β-amylase.

α-Amylase is produced commercially from bacterial and fungal sources, including strains of *Aspergillus* and *Bacillus*. In general, bacterial α-amylase is preferred over fungal amylase owing to the characteristic thermostability that the enzyme has. Typically, *Bacillus amyloliquefaciens* and *Bacillus licheniformis* are employed for commercial applications. The use of thermostable α-amylase can minimize contamination risk in addition to reducing reaction time and also minimizing the polymerization of D-glucose to iso-maltose. Filamentous fungi have also been assessed for their ability to produce α-amylase. The thermophilic fungus *Thermomyces lanuginosis* has been shown to produce high levels of α-amylase with interesting thermostability properties. A strain of *Aspergillus kawachii* has been used to produce high levels of acid-stable α-amylase using solid-state fermentation. *Pycnoporus sanguineus* has been identified as a source of α-amylase and cultivation in the solid state resulted higher enzyme production than in submerged fermentation. Finally, a strain of *Rhizopus* has been found which produces a thermostable α-amylase under solid-state conditions.

7.5.4.2 β-Amylase

β-Amylases (also known as 1,4-α-D-glucan maltohydrolase, glycogenase or saccharogen amylase) are typically of plant origin, but a few microbial strains which produce this enzyme have also been identified. Being an exo-acting enzyme, it cleaves non-reducing chain ends of amylose, amylopectin and glycogen molecules through hydrolysis of alternate glycosidic linkages, yielding maltose. As this enzyme is unable to cleave α-1,6-glycosidic linkages in amylopectin, it results in incomplete degradation of the molecule, producing only 50–60 % maltose and a β-limit dextrin. Unlike other amylase enzymes, only a small number of β-amylases of microbial origin are produced commercially. The enzyme has primarily been characterized from bacterial strains, although fungal strains such as *Rhizopus* have been reported to synthesize β-amylase.

7.5.4.3 Glucoamylase

Glucoamylase (also known as glucan 1,4-α-glucosidase, amyloglucosidase, *exo*-1,4-α-glucosidase, lysosomal α-glucosidase or 1,4-α-D-glucan glucohydrolase) hydrolyses single glucose monomers from non-reducing ends of amylose and amylopectin. Unlike α-amylase, however, most glucoamylases are also able to cleave the 1,6-α-linkages at the branching points of amylopectin, albeit at a reduced rate. Glucose, maltose and β-limit dextrins are the end products of glucoamylase hydrolysis. The majority of glucoamylases are multidomain enzymes consisting of a catalytic domain connected to a starch-binding domain. Glucoamylase enzymes have been isolated from numerous microbial sources. Filamentous fungi, however, constitute the major source of glucoamylase, which is widely used in the manufacture of glucose and fructose syrups. Commercially speaking, glucoamylase is typically produced from solid-state cultures of *Aspergillus niger* using numerous agro-industrial waste residues, such as wheat bran, rice bran, rice husk, gram, wheat and corn flour, tea waste, copra waste and so on. In addition, strains of *Aspergillus awamori* have been frequently utilized for glucoamylase production, as have numerous *Rhizopus* isolates. Interestingly, *Rhizopus* strains can be classified into four groups based on their soluble-starch-digestive glucoamylase (SSGA) and raw-starch-digestive glucoamylase (RSGA) activities. Glucoamylase is also produced by yeast cultures such as *Saccharomyces cerevisiae*, *Saccharomycopsis fibuligera*, *Candida fennica*, *Candida famata* and *Endomycopsis fibuligera*.

7.5.4.4 Application of Amylolytic Enzymes

Starch-containing agricultural biomass can be used as a potential substrate for the production of gaseous or liquid fuels, feed proteins and chemicals through microbial conversion. These substrates include corn (maize), wheat, oats, rice, potato and cassava, which on a dry basis can contain around 60–75 % (wt/wt) of starch.

Starch liquefaction is achieved by dispersion of insoluble starch granules in aqueous solution followed by a partial hydrolysis using thermostable amylases. Industrially speaking, the starch suspension for liquefaction is generally in excess of 35 % (w/v) and the viscosity is extremely high following gelatinization. Thermostable α-amylase is used to effect a reduction in viscosity and the partial hydrolysis of starch. Before the introduction of thermostable amylases, starch liquefaction was achieved by acid hydrolysis (hydrochloric or oxalic acids, pH 2 and 140–150 °C for 5 min). The introduction of thermostable amylases has brought about the use of milder, more environmentally friendly processing conditions. The formation of by-products is reduced and refining and recovery costs are lowered.

Starch represents a high-yielding ethanol resource and production has been reported from numerous materials, such as corn, wheat, potatoes, cassava and corn stover. Ethanol production is reliant on the use of fungal saccharifying enzymes, as the carbohydrates in the raw materials are not directly fermentable by most yeast. Starch is first hydrated and gelatinized by milling and cooking, and then broken down by amylolytic enzymes which pretreat the starch and hydrolyse it into simple sugars. These sugars can then be converted by yeast into ethanol for use in many different applications.

Starch conversion into high-fructose corn sweeteners or syrups is of major industrial importance. Owing to their high sweetening property they can be used to replace sucrose syrups in foods and beverages. High-fructose corn syrup is produced by first milling corn to produce corn starch. This is then treated with α-amylase to produce shorter oligosaccharide sugars. Glucoamylase from strains of *Aspergillus* breaks the sugar chains down even further to yield glucose. This is subsequently converted by xylose isomerase into a fructose-rich syrup containing approximately 42 % fructose. A final purification step yields a high-fructose corn syrup with a fructose content of 90 %.

Amylase enzymes also find use in bread making. The main component of wheat flour is starch, from which amylases can produce smaller weight dextrins for the yeast to utilize, imparting flavour and causing the bread to rise. Amylase enzymes are added into the bread improver, thus making the overall process faster and more practical for commercial use.

7.5.5 Lipase

Lipases (triacylglycerol acylhydrolases) are a class of hydrolytic enzyme which are widely distributed amongst microorganisms, plants and animals. Their principal function is to hydrolyse triglycerides into diglycerides, monoglycerides, fatty acids and glycerol, but they can also catalyse esterification, interesterification and transesterification reactions in nonaqueous media. Approximately 35 different lipases are commercially available, and one of the main advantages of lipases is the fact that they can be used not only in water, but also in water–organic solvent mixtures or even in pure anhydrous organic solvents. This versatility enables lipases to have potential applications in the food, detergent, pharmaceutical, leather, textile, cosmetic and paper industries.

7.5.5.1 Sources of Lipase

Lipases from a large number of fungal sources have been characterized and shown to have a wide range of properties, depending on their source, with respect to positional specificity, fatty acid specificity, thermostability and activity optima.

Strains of fungi producing lipase of commercial interest include *Rhizopus* sp., *Rhizomucor miehei*, *Geotrichum candidum*, *Pichia burtonii* and *Candida cylidracae*. *Rhizomucor miehei* lipase is probably the most used lipase obtained from fungi, even being used as a model for the determination of the structure of some other lipases.

7.5.5.2 Application of Lipase

Microbial lipases are an important group of biotechnologically valuable enzymes and it is anticipated that the market for lipase will continue to grow, mainly in detergent and cosmetics markets. At present, lipases are considered to be one of the largest groups based on their commercial use in a billion-dollar business that comprises a wide variety of different applications. Lipases are mainly used in two different ways: either as biological catalysts to manufacture food ingredients or in the manufacture of fine chemicals. Novel biotechnological applications for lipases have been demonstrated, such as in the synthesis of biopolymers and biodiesel and in the production of enantiopure pharmaceuticals, agrochemicals and flavour compounds. The potential for the manufacture of industrially important chemicals which are typically manufactured from fats and oils by chemical processes and that could be generated by lipases with greater rapidity and better specificity under mild conditions also exists.

Their value as biocatalysts lies in their ability to act under mild conditions, their high stability in organic solvents, broad substrate specificity and a high degree of regio- and/or stereo-selectivity in catalysis. These enzymes are active under ambient conditions and this reduces the impact of reaction conditions on sensitive reactants and products.

Lipases have found extensive application in the dairy industry, where they are used for the hydrolysis of milk fat. Current uses include enhancement of flavour in cheese, acceleration of cheese ripening and lipolysis of butterfat and cream. The use of lipase in milk fat hydrolysis endows many dairy products, such as soft cheese, with specific flavour characteristics. For example, the addition of lipases that primarily release short-chain (mainly C4 and C6) fatty acids promotes a sharp, tangy flavour, while the release of medium-chain (C12, C14) fatty acids imparts a soapy taste to the product.

The promotion of free fatty acid release by lipolytic enzymes enables their participation in simple chemical reactions and can initiate, for instance, the synthesis of flavour ingredients such as acetoacetate, beta-keto acids, methyl ketones, flavour esters and lactones. A whole range of fungal lipase preparations have been developed for the cheese manufacturing industry, including enzymes from *M. miehei*, *A. niger* and *A. oryzae*.

In the food industry, some fats have greater value because of their structure. However, lipase-catalysed trans-esterification of cheaper oils can enable cheaper fats to be used, a good example of which is the production of cocoa butter from palm mid-fraction. Lipase-catalysed trans-esterification in organic solvents is an emerging industrial application and has been used for the production of cocoa butter equivalent, human milk fat substitute, pharmaceutically important polyunsaturated fatty acids and in the production of biodiesel from vegetable oils.

M. miehei and *Candida antarctica* lipase have been successfully employed in the esterification of free fatty acids in the absence of organic solvent and in the trans-esterification of fatty acid methyl esters in hexane with isopropylidene glycerols. Inter-esterification using an immobilized *M. miehei* lipase has been utilized for the production of vegetable oils such as corn oil, sunflower oil, peanut oil, olive oil and soybean oil containing omega-3 polyunsaturated fatty acids.

The use of lipases to carry out industrial hydrolysis of tallow has a number of advantages; significantly, the heat requirement is reduced, thereby decreasing the consumption of fossil fuels and obviously the environmental impact. Additionally, there is less degradation of unsaturated fatty acids due to the lowered reaction temperature and, as a consequence, pure, natural fatty acids can be obtained from highly unsaturated oils. With regard to their nutritional value, un-degraded polyunsaturated fatty acids may be important to preserve in the production of food additives such as mono- and di-glycerides.

Lipases play a number of roles in the textile industry. They have been used, for instance, to assist in the removal of size lubricants, and commercial preparations used for the desizing of cotton fabrics contain both α-amylase and lipases. The most commercially significant application of lipases is in their addition to detergents, both in household and industrial products. Nowadays, detergents typically contain one or more enzymes, such as protease, amylase, cellulase and lipase. These can reduce the environmental impact of detergent products, since they not only save energy by enabling a lower wash temperature to be used, but they also allow the content of other chemicals in detergents to be reduced. Additionally, lipase and other enzymatic components are biodegradable, leave no harmful residues, have no negative impact on sewage treatment processes and do not present a risk to aquatic life.

The first commercial lipase, Lipolase, originated from the fungus *Thermomyces lanuginosus*. Owing to the low yield of enzyme, the gene encoding the protein was cloned and subsequently expressed in *A. oryzae*. This enzyme has been extensively used in the detergent and textile industries. One of the most commercially successful lipase preparations is a recombinant *Humicola* enzyme which has been heterologously expressed in *A. oryzae* and which has application in the detergent sector.

7.5.6 Phytase

Phytase is an enzyme that has found widespread use as a feed additive in the animal industry owing to its absence in the gastrointestinal tracts of monogastric species such as pigs and poultry. The enzyme hydrolyses an anti-nutritional factor known as phytate or phytic acid to liberate inositol and inorganic phosphorus.

As an essential element, phosphorus is essential for the growth and development of all organisms, playing key roles in skeletal structure and in vital metabolic pathways too numerous to mention. The negative effects of phosphorus-deficient diets on livestock performance are manifold and are well documented, including reduced appetite, bone malformation and lowered fertility. To counteract this, an external source of phosphorus must be supplied in sufficient quantity to meet the daily requirements of the animal. This can result in the environment becoming overloaded with phosphorus in areas of intensive livestock production. The development of enzyme technology based on supplementing diets with sources of microbial phytase has proven to be a practical and effective method of improving phytate digestibility in monogastric animal diets.

The principal storage form of phosphorus in feedstuffs of plant origin is the hexaphosphate ester of myoinositol, more commonly known as phytic acid, and it accounts for up to 80 % of the phosphorus in grains and seeds. Phytic acid and its salts and esters are considered to be anti-nutritional factors owing to their ability to bind essential minerals such as calcium, zinc, magnesium and iron. They may also react with proteins, thereby decreasing the bioavailability of protein and other nutritionally important factors. Considerable interest has been focused on the supplementation of animal feeds with exogenous (mainly microbial) sources of phytases with the specific aim of liberating phytate-bound phosphorus in the gastrointestinal tract.

7.5.6.1 Sources of Phytase

A broad range of microorganisms produce phospohydrolases or phytases capable of catalysing the hydrolysis of phytate. These include yeast, such as *S. cerevisiae*. and fungi such as *Aspergillus ficuum*, which produces a highly thermostable, highly active phytase enzyme. Phytases have also been found in bacteria; however, they tend to display pH optima which are in the neutral to alkaline range and which, therefore, are relatively inactive at monogastric stomach pH values.

Over 200 fungal isolates belonging to the genera *Aspergillus*, *Mucor*, *Penicillium* and *Rhizopus* have been tested for phytase production. An additional survey of 84 fungi from 25 species for phytase production indicated that the incidence of phytase production is highest in aspergilli. Of all the organisms surveyed, *A. niger* NRRL 3135 produces the most active extracellular phytase. This fungus produces two different enzymes: one with pH optima at 5.5 and 2.5 and one with a pH optimum of 2.0, which are designated PhyA and PhyB respectively. In light of these considerations, the favoured microbial source of exogenous phytase is filamentous fungi such as *A. niger*.

7.5.6.2 Application of Phytase

The addition of microbial phytase to the feedstuffs of monogastric animals was described over 35 years ago. Published research shows that enzymatic treatment of feed using microbial phytase sources increases the bioavailability of essential minerals and proteins and provides levels of growth performance as good as or better than those with phosphate supplementation.

Obviously, there is significant potential for the use of fungal phytase in view of the pollution caused by inorganic phosphorus supplementation and the legislative regulations to reduce this by at least 50 % in the EU alone. In Europe, livestock production, and in particular pig production, has received much of the blame for phosphorus pollution and has been targeted in an attempt to relieve the phosphorus burden of the land. Legislation against phosphorus pollution has been adopted in many countries; farmers must operate within a legal limit per unit of land. For instance, in the Vendée region in France, the amount of phosphorus may not exceed 44 kg/ha (100 kg P_2O_5/ha).

Current global estimates are that animal feed producers with combined annual production of 550 million tonnes of animal feed presently use (or will soon use) phytase as a supplementary enzyme in diet formulations. The commercial potential for supplementation with phytase preparations, therefore, is quite obvious. However, a major drawback to the widespread use of phytases in animal feed is the constraint of thermal stability (65–95 °C) required for these enzymes to withstand inactivation during the feed-pelleting or expansion processes. As such, a commercially successful phytase must be able to withstand brief heating at elevated temperatures prior to being administered to monogastric target species. In addition, for industrial applications in animal feed, a phytase of interest must be optimally active in the pH range prevalent in the digestive tract.

7.6 Enzyme Production Strategies

Fungal enzyme production is mainly through the use of submerged fermentation strategies, although a second method known as solid-state fermentation or the Koji process is used extensively in Asia. We will concentrate on submerged fermentation strategies, of which a typical flow diagram of the process is depicted in Figure 7.1.

Broadly speaking, the fermentation stages involved in the production of fungal enzymes are relatively similar. Large batch fermentations using inexpensive culture media are scaled up in a similar manner to antibiotic production. The downstream processing steps, though, can vary widely, and this is dependent on the ultimate process in which the enzyme preparation will be used. Similarly, those that are produced intracellularly, such as lactase, will require additional processing steps to those that are produced on an extracellular basis. Generally, though, most industrial enzymes are produced extracellularly in large batch fermentations and require little in the way of downstream processing. In fact, fewer processing steps are preferable, as this will lead to a reduction in enzyme losses. Only enzymes and material likely to interfere with the catalytic process for which the enzyme is required will be removed. Unnecessary purification is to be avoided, as each step is costly in terms of equipment, manpower and enzyme loss. As a consequence, some industrial enzyme preparations are a simple mix of concentrated fermentation broth and additives to stabilize the preparation. If required, though, a powder preparation can be prepared from a fermentation broth. Treatments such as salt or ethanol precipitation will result in the generation of a protein suspension from the spent culture media, which can then be filtered and dried to recover an extremely concentrated enzyme preparation.

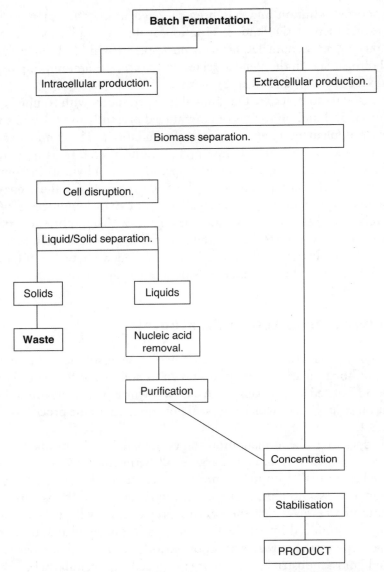

Figure 7.1   Flow diagram of fungal enzyme production.

7.7 Conclusions

Enzymes have long played a central role in mankind's attempts to utilize biological systems for a variety of purposes. Examples of harnessing their power include cheese making, brewing, baking, the production of antibiotics and in the manufacture of commodities such as leather, indigo and linen. They also find applications in areas such as detergent and paper production, the textile industry and in the food and drinks industry in products ranging from fruit juice, to coffee, tea and wine.

Revision Questions

Q 7.1   Name five uses for enzymes.

Q 7.2   Name three fungal genera which produce enzymes.

Q 7.3   What do protease enzymes do?

Q 7.4   Name the two main classes of protease.

Q 7.5   Name the three main enzymes involved in cellulose breakdown.

Q 7.6   Name the two fungal organisms which are mainly responsible for xylanase production.

Q 7.7   What does xylanase do?

Q 7.8   Name the three major types of amylase enzyme.

Q 7.9   What is the principal function of lipase?

Q 7.10   What is the principal benefit of phytase inclusion into monogastric diets?

References

AQ1

Bhat, M.K. and Bhat, S. (1997) Cellulose degrading enzymes and their potential industrial applications. *Biotechnology Advances*, 15, 583–620.

Meyer, V. (2008) Genetic engineering of filamentous fungi – progress, obstacles and future trends. *Biotechnology Advances*, 26, 177–185.

Rao, M.B., Tanksale, A.M., Ghatge, M.S. and Desphande, V.V. (1998) Molecular and biotechnological aspects of microbial proteases. *Microbiology and Molecular Biology Reviews*, 62, 597–635.

Walsh, G. (1998) *Biopharmaceuticals: Biochemistry and Biotechnology*, John Wiley and Sons, Inc., New York.

Further Reading

Journal Articles

Bhat, M.K. (2000) Cellulases and related enzymes in biotechnology. *Biotechnology Advances*, 18(5), 355–383.

Collins, T., Gerday, C. and Feller, G. (2005) Xylanases, xylanase families and extremophilic xylanases. *FEMS Microbiology Reviews*, **29**(1), 3–23.

Haltrich, D., Nidetzky, B., Kulbe, K.D. *et al.* (1996) Production of fungal xylanases. *Bioresource Technology*, **58**(2), 137–161.

Hasan, F., Shah, A.A. and Hameed, A. (2006) Industrial applications of microbial lipases. *Enzyme and Microbial Technology*, **39**, 235–251.

Liu, B.L., Rafiq, A., Tzeng, Y. and Rob, A. (1998) The induction and characterization of phytase and beyond. *Enzyme and Microbial Technology*, **22**(5), 415–424.

Nigam, P. and Singh, D. (1995) Enzyme and microbial systems involved in starch processing. *Enzyme and Microbial Technology*, **17**(9) 770–778.

Books

Anke, T. (ed.) (1991) *Fungal Biotechnology*, Chapman and Hall, London.

Beynon, R.J. and Bond, J.S. (eds) (1989) *Proteolytic Enzymes, A Practical Approach*, IRL Press, Oxford.

Bu'lock, J. and Kristiansen, B. (eds) (1987) *Basic Biotechnology*, Academic, London.

Coughlan, M.P. and Hazlewood, G.P. (eds) (1993) *Hemicellulose and Hemicellulases*, Portland Press, Colchester.

Fogarty, W. (ed.) (1983) *Microbial Enzymes and Biotechnology*, Applied Science.

Price, N.C. and Stevens, L. (1982) *Fundamentals of Enzymology*, Oxford University Press.

Walsh, G. and Headon, D. (1994) *Protein Biotechnology*, John Wiley and Sons, Inc., New York.

8
The Biotechnological Exploitation of Heterologous Protein Production in Fungi

Brendan Curran and Virginia Bugeja

8.1 Introduction

Heterologous, or recombinant, proteins are produced by using recombinant DNA technology to express a gene product in an organism in which it would not normally be made. Unlike the synthesis of recombinant DNA, a relatively simple procedure involving the cutting and joining together of DNA sequences from different organisms, recombinant protein production is fraught with difficulty because of the need to transcribe and then translate heterologous DNA into a correctly folded protein with the appropriate biological activity

Initially developed in the simple prokaryote *Escherichia coli*, heterologous protein expression technology is now possible in scores of prokaryotic and eukaryotic host systems, including Gram-positive and -negative bacteria, yeasts, filamentous fungi, insect cells, plants, mammalian cells and transgenic animals.

Heterologous gene expression systems are available for an extremely wide range of fungi, including yeasts such as *Saccharomyces cerevisiae* and *Pichia pastoris*, and many of their filamentous cousins, including *Neurospora*, *Aspergillus* and *Penicillium* species. Some heterologous proteins have been produced for use in basic research; others for commercial exploitation. Here, we restrict ourselves to a very specific brief: the production of biotechnologically relevant heterologous proteins in fungi.

Fungi: Biology and Applications, Second Edition. Edited by Kevin Kavanagh.
© 2011 John Wiley & Sons, Ltd. Published 2011 by John Wiley & Sons, Ltd.

8.2 Heterologous Protein Expression in Fungi

Regardless of their intended use, heterologous protein production requires the following steps: the insertion of the desired heterologous DNA coding sequence into appropriate regulatory sequences in specialized expression vectors; a transformation procedure for the introduction of the construct into the desired host species. This has then to be followed by the transcription and translation of this sequence into biologically active protein molecules – a process which can require post-translational modifications, such as glycosylation.

8.2.1 Heterologous DNA, Vectors, Transformation and Host Systems

8.2.1.1 Heterologous DNA

The DNA sequence due to be expressed into heterologous protein can be genomic in origin, thereby possessing introns and/or regulatory sequences from the original organism from which it has been cloned. Alternatively, it can be a cDNA sequence derived by reverse transcription of the heterologous mRNA, in which case it will lack introns and regulatory sequences. Filamentous fungi and many yeast species are capable of excising introns accurately from the mRNA transcripts of heterologous genes; indeed, some filamentous species can recognize heterologous regulatory signals from human DNA. However, efficient expression almost always requires the heterologous mRNA to be driven from the promoter of a strongly expressed host cell gene. Therefore, although heterologous genomic DNA has been successfully expressed in a small number of fungi, commercially important heterologous protein expression is normally initiated in these organisms by inserting the appropriate cDNA sequence into an expression vector that already encodes appropriate promoter and terminator sequences.

8.2.1.2 Vectors

In addition to the backbone of bacterial plasmid DNA, which is common to expression vectors in all systems and facilitates DNA manipulation/large-scale plasmid purification in *E. coli*, fungal expression vectors carry:

- a selectable marker for the intended fungal host;
- a strong promoter to drive the production of the heterologous mRNA;
- appropriate DNA sequences to ensure efficient termination of transcription and polyadenylation of the mRNA;
- appropriate sequences to ensure the correct initiation and termination of translation.

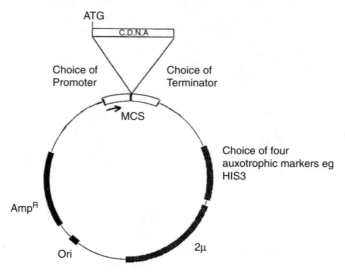

Figure 8.1 A generalized expression vector for use in *S. cerevisiae*.

The most commonly used shuttle vectors for biotechnological applications in the yeast *S. cerevisiae* are autonomously replicating because they carry appropriate sequences (Figure 8.1) from yeast 2μ DNA – a native yeast plasmid. On the other hand, many of the vectors designed for use in the methylotrophic yeast *P. pastoris* (see below), and in filamentous fungi, for example, *Penicillium* and *Aspergillus* species, are integrative vectors (Figure 8.2a) that require the heterologous DNA to be incorporated into the host cell chromosomal DNA. (Figure 8.2b).

8.2.1.3 Transformation and Selection

Although the DNA manipulation can be achieved easily in *E. coli*, for protein expression to occur the heterologous DNA construct has to be transferred into the fungal host. The constructs, therefore, are transformed into the fungal host and colonies of transformants identified using appropriate selective agar plates. Although there are a number of dominant selectable markers conferring antibiotic resistance on the transformed cells (hygromycin B resistance is a particularly versatile one), many yeast and filamentous systems exploit auxotrophic marker complementation for selection. In these cases the plasmids carry the appropriate wild-type information to complement auxotrophic alleles (e.g. *Leu2*⁻, *His3*⁻, *Trp1*⁻, etc.) in the host cells. The types of transformation process used include: enzymatically removing the cell walls and exposing the resulting sphaeroplasts to the DNA in the presence of calcium ions and polyethylene glycol; electroporation of yeast cells and fungal sphaeroplasts; and transformation of yeast cells by treating them with alkali cations (usually lithium) in a procedure analogous to *E. coli* transformation.

(a)

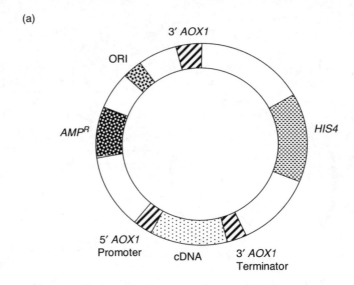

(b)

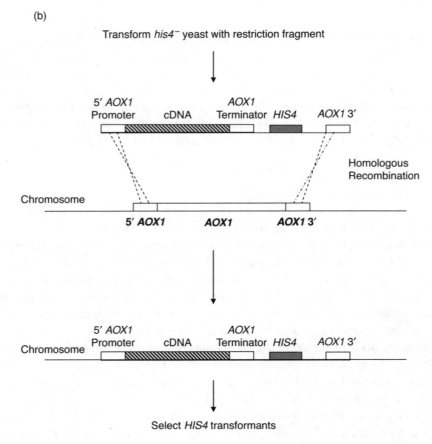

Figure 8.2 The structure (a) and integration (b) of a *P. pastoris* expression vector.

8.2.1.4 Host Systems

The type of protein product required determines the fungal host that is used. There are two basic types of biotechnologically relevant protein: enzymes, used in a wide variety of industrial applications, which have a low unit value and are required in bulk; and proteins of therapeutic value, which have a high unit value and are required in more limited amounts.

Filamentous fungi (especially *Aspergillus* species) are the host systems of choice for the high-level production of a wide variety of enzymes for industrial applications). The enzymes produced in this way include amylases (used to convert corn starch for ethanol fermentation), proteases and lipases in detergents, cellulases used in paper production and many others. This invariably involves the expression of heterologous enzymes from other fungal species, and these products can be recovered at levels up to grams/litre (Table 8.1). However, for a variety of reasons, including protease contamination and inefficient secretion of these products (shortcomings which are currently being addressed), filamentous fungi are relatively inefficient as hosts for the expression of heterologous proteins of animal/human origin.

Therefore, yeasts, and in particular *S. cerevisiae*, are preferred as hosts for the production of therapeutically important proteins (Table 8.2). Here, we focus on the biotechnological development of the two most important members of this group: *S. cerevisiae*, the oldest and most studied star, and *P. pastoris*, 'the new kid on the block', but increasingly the host of choice. These yeast hosts are much more tractable in terms of genetic manipulation and product recovery than

Table 8.1 A representative selection of heterologous proteins produced in filamentous fungi.

Host	Product	Origin	Industrial use
A. niger	Glucoamylase	*Aspergillus* sp.	Clarifying fruit juices
A. oryzae	Lipase	*Candida* sp., *Fusarium* sp., *Rhizomucor* sp., *Thermomyces* sp.	Food, textile, detergent, leather, pulp and paper
A. oryzae, T. reesei, T. longibrachiatum	Cellulase	*Trichoderma* sp.	Textiles and fruit processing
A. niger, A. oryzae	Protease	*Aspergillus* sp., *Rhizomucor* sp. or calf stomach,	Food and leather
A. niger	Glucose oxidase	*Aspergillus* sp.	Stabilizing food and beverages

Table 8.2 A representative selection of heterologous proteins produced in yeast.

Host	Product	Origin	Therapeutic Use
S. cerevisiae	Human papillomavirus vaccine	Human papillomavirus	Vaccine to protect against papillomavirus infection
S. cerevisiae	Glucagon	Human cDNA	Hormone involved in blood glucose level regulation
S. cerevisiae	Hepatitis B vaccine	Hepatitis virus	Vaccine to protect against hepatitis B infection
P. pastoris	Hepatitis B vaccine	Hepatitis virus	Vaccine to protect against hepatitis B infection
P. pastoris	Ecallantide	Human cDNA	A 60 amino-acid protein inhibitor of plasma kallikrein

their filamentous cousins; therefore, most of the high-value heterologous protein production occurs in these fungi.

8.2.2 *Saccharomyces cerevisiae*

S. cerevisiae was the first eukaryotic cell engineered to express heterologous proteins because it shared with *E. coli* many of the characteristics that make the latter such a useful host for recombinant DNA technology. *S. cerevisiae* grows rapidly by cell division, has its own autonomously replicating plasmid, can be transformed as intact cells, and forms discrete colonies on simple defined media. In addition, *S. cerevisiae* can carry out post-translational modifications of expressed proteins – essential features of many heterologous proteins that *E. coli* is unable to provide. Furthermore, it secretes a small number of proteins into the growth medium, which, as we shall see, can be exploited to simplify the purification of heterologous proteins. Finally, it has got a long, safe history of use in commercial fermentation processes and, unlike *E. coli*, *S. cerevisiae* does not produce pyrogens or endotoxins. These parameters taken together make it particularly suitable for approval by regulatory bodies charged with the responsibility of ensuring the safe production of medically important heterologous proteins.

Despite the versatility of yeast expression systems, the production of high levels of biologically active heterologous proteins is still largely a matter of trial and error. The recovery of satisfactory levels of authentic heterologous protein depends on a number of factors, including the type of expression vector used, the site of protein expression and the type of protein being expressed.

8.2.2.1 Expression Vectors

There is a wide range of yeast cloning vectors available for use in *S. cerevisiae*, but here we restrict ourselves to the self-replicating variety most commonly used for biotechnological applications (Figure 8.1). These YEp (yeast episomal plasmid) vectors are based on the autonomously replicating sequence from the endogenous yeast 2μ (so called because of its unique length) plasmid. They are present at 20–200 copies per cell and under selective conditions are found in 60–95 % of the cell population. Integrative vectors similar to those used in *Pichia* (Figure 8.2a and b) can also be used in *S. cerevisiae*; indeed, one variant of this uses homologous recombination to target multiple copies of the gene construct into middle-repetitive δ DNA sequences generated by the activity of the Ty transposable element. However, in the vast majority of cases, plasmid-borne expression vectors are used commercially.

8.2.2.2 Regulating the Level of Heterologous mRNA in Host Cells

The overall level of heterologous mRNA in the cell is determined by the copy number of the expression vector, the strength of the promoter used to drive transcription and the stability of the specific mRNA sequence.

Expression vectors based on YEp technology have a high copy number but require selective conditions to ensure their stable inheritance. High-level mRNA production is also dependent upon the type of promoter chosen to drive expression (Table 8.3). The most frequently encountered are based on promoters from the highly expressed genes that encode glycolytic enzymes. These include phosphoglycerate kinase (*PGK*), alcohol dehydrogenase 1 (*ADH1*) and glyceraldehyde-3-phosphate dehydrogenase (*GAPDH*), all of which facilitate high-level constitutive mRNA production. Constitutive expression can be disadvantageous when the foreign protein has a toxic effect on the cells. This can be circumvented by using a regulatable promoter to induce heterologous gene expression after cells have grown to maximum biomass. There are a number of regulatable promoters available. One of the most useful ones is based on the promoter of the galactokinase gene (*GAL1*), which undergoes a 1000-fold induction when glucose is replaced by galactose in the medium. Regardless of the choice of promoter, it is important that transcription of the heterologous mRNA is terminated properly, otherwise abnormally long mRNA molecules, which are often unstable, can be generated by read through along the plasmid DNA. It is for this reason that expression vectors frequently contain the 3′ terminator

Table 8.3 Promoters used to direct heterologous gene expression in *S. cerevisiae.*

Promoter	Strength[a]	Regulation	Example heterologous gene expressed using promoter
PGK, 3-phosphoglycerate kinase	++++	Constitutive	Human β-interferon
ADH1, alcohol dehydrogenase 1	+++	Constitutive	Human β-interferon
GAPDH, glyceraldehyde-3-phosphate dehydrogenase	++++	Constitutive	Human epidermal growth factor
GAL1, galactokinase	+++	1000× induction	Calf chymosin by galactose
PHO5, alkaline phosphatase	++	500× repression	Hepatits B surface antigen by phosphate
CUP1, copperthionein	+	20× induction by copper	Mouse IG Kappa chain

[a] Relative levels of mRNA expression when the promoter is active.

region from a yeast gene (e.g. *CYC1*, *PGK* or *ADH1*) to ensure efficient mRNA termination (Figure 8.1).

8.2.2.3 *Ensuring High-Level Protein Production*

The level of heterologous protein produced by the host depends upon the efficiency with which the mRNA is translated, and the stability of the protein after it has been produced.

It is vitally important to address control of protein translation and subsequent translocation when choosing the expression vector. The sophisticated translation initiation mechanism found in mammalian cells is absent from yeast. Therefore, in order to ensure efficient initiation of translation, it is important to genetically engineer mammalian cDNAs to remove regions of dyad symmetry and non-coding AUG triplets in the leader sequence of heterologous mRNAs upstream of the AUG encoding the first methionine in the protein.

Once expressed, some proteins form insoluble complexes in *S. cerevisiae* but many others do not. Other proteins can be produced as denatured, intracellular complexes which can be disaggregated and renatured after harvesting. The first recombinant DNA product to reach the market was a hepatitis B vaccine produced in this way. Some proteins are rapidly turned over by the ubiquitin degradative pathway in the cell, while others are degraded by vacuolar proteases. This can be especially true during cell breakage and subsequent purification. The powerful tools provided by a detailed knowledge of yeast genetics and biochemistry can be used to minimize this problem in *S. cerevisiae*.

The use of protease-deficient host strains can improve both the yield and the quality of heterologous proteins. One mutant (*PEP4-3*) is widely used because it is responsible for the activation of inactive vacuolar zymogen proteases; in its absence, a wide range of proteinase activities is therefore prevented. Even more impressively, our detailed knowledge of the yeast secretory pathway can be exploited to genetically engineer the heterologous protein so that it is smuggled out of the cell before it can be degraded by either vacuolar proteases or the ubiquitin degradative pathway. Indeed, secretion not only minimizes the exposure of heterologous proteins to protease activity, but, because *S. cerevisiae* only secretes a handful of proteins, it also facilitates the recovery and purification of heterologous proteins. Two of these (invertase and acid phosphatase) are targeted to the periplasmic space, which lies between the cell membrane and cell wall, the other two (α factor and killer toxin) are secreted out beyond the wall into the culture medium (Table 8.4). Entry into the secretory pathway is determined by the presence of short hydrophobic 'signal' sequences on the N-terminal end of secreted proteins. The DNA sequence for these signal peptides can be genetically engineered onto the DNA sequence for the heterologous protein of choice – thereby ensuring that it is targeted for export after being synthesized. The 'signal' sequences from all four of *S. cerevisiae*'s secretion proteins have been used in this way with varying degrees of success. A secretion vector that encodes the α-factor signal peptide is shown in Figure 8.3. A number of medically important proteins, including insulin, interferon and interleukin 2, have been successfully secreted using this type of signal peptide.

Table 8.4 Signal sequences used to direct secretion of heterologous proteins from *S. cerevisiae*.

Signal sequence	Cellular location of gene product	Example secreted heterologous protein
Invertase	Periplasm	α-1-Antitrypsin
Acid phosphatase	Periplasm	β-Interferon
α-Factor mating pheromone	Culture medium	Epidermal growth factor
Killer toxin	Culture medium	Cellulase

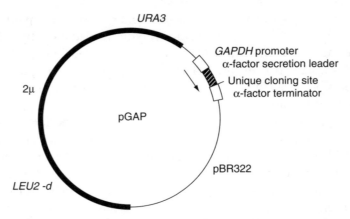

Figure 8.3 A secretion vector for use in *S. cerevisiae*.

8.2.2.4 *Ensuring Authentic Protein Structure and Function*

The objective of heterologous gene expression for commercial purposes is not just the high-level transcription and translation of the appropriate DNA. Proteins are produced because of their structure; therefore, what are required are appropriately folded, biologically active, authentic protein molecules. Many proteins of therapeutic importance undergo sophisticated post-translational modifications in mammalian cells. These vary from the removal and/or addition of small chemical moieties such as the removal of the N-terminal methionine, or the addition of an acetyl group to the N-terminal amino acid (acetylation), through the addition of large lipid molecules to generate lipo-proteins, to the complex addition of countless sugar moieties to proteins as they are synthesized and passed through the cell's endoplasmic reticulum (ER) and Golgi apparatus to produce glycosylated proteins. Simple prokaryotic expression systems like *E. coli* are unable to carry out many of these processes, and it was for that reason that eukaryotic expression systems were developed. Although not ideal in all respects, *S. cerevisiae* offers solutions to at least some of these problems. When *E. coli* failed to produce properly acetylated human superoxide dismutase, *S. cerevisiae* obliged by intracellularly expressing a soluble active protein identical to that found in human tissue – complete with acetylated N-terminal alanine.

Moreover, the yeast secretion system affords options with respect to ensuring that proteins are folded correctly, that N-terminal methionines are removed and sugar residues are added to glycoproteins. A direct comparison between the intracellular production and extracellular secretion of prochymosin and human serum albumin resulted in the recovery of small quantities of mostly insoluble, inactive protein when they were produced intracellularly, but the recovery of soluble, correctly folded, fully active protein when they were secreted.

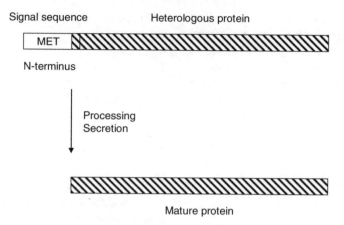

Figure 8.4 Cleavage of a secretory signal sequence from a heterologous protein.

Secretion can also be used to produce proteins that have an amino acid other than methionine at their N-terminus. If a secretory signal is spliced onto the heterologous gene at the appropriate amino acid (normally the penultimate one), then the N-terminal methionine, which is obligatory for translation initiation, will be on the secretory signal. Proteolytic cleavage of this signal from the heterologous protein in the ER will generate an authentic N-terminal amino acid (Figure 8.4). Glycosylation in yeast is of both the N-linked (via an asparagine amide) and O-linked (via a serine or threonine hydroxyl) types, occurring at the sequences Asn-X-Ser/Thr and Thr/Ser respectively. However, it is important to note that the number and type of outer core carbohydrates attached to glycosylated proteins in yeast are different to those found on mammalian proteins. Therefore, if the protein is being produced for therapeutic purposes they may cause unacceptable immunogenicity problems. One approach to overcoming this problem is to prevent glycosylation of the protein by using site-directed mutagenesis of the DNA in *E. coli* to alter one amino acid, thereby removing the glycosylation recognition site before expressing the protein in *S. cerevisiae*. This strategy was successfully used to produce urokinase-type plasminogen activator.

8.2.2.5 Limitations

Despite its ability to express a wide variety of proteins, *S. cerevisiae* has limitations. Its very primitive glycosylation system frequently hyperglycosylates heterologous proteins. Also, the production of alcohol during glucose metabolism limits the generation of biomass and, therefore, heterologous protein product. However, another yeast species, *P. pastoris*, which can grow to much higher cell densities and does not hyperglycosylate its proteins, has been developed as an alternative host system for the production of heterologous proteins. It has yet to outscore *S. cerevisiae* in terms of the number of therapeutic proteins produced

Table 8.5 Heterologous proteins of therapeutic use produced in yeasts.

Period	Fungal hosts	Number of products	Types of product
1985–1990	*S. cerevisiae*	1 (1986)	First recombinant Hepatitis B vaccine
1991–1995	*S. cerevisiae*	1	Growth factor for stimulating bone marrow
1996–2000	*S. cerevisiae*	10	5 vaccines, an anticoagulant, a tissue growth factor, 3 hormones
	P. pastoris	1	Hepatitis B vaccine
2001–2006	*S. cerevisiae*	8	5 vaccines, 2 hormones, 1 therapeutic enzyme
	P. pastoris	2	Insulin and interferon
2006–2010	*S. cerevisiae*	4	3 hormones, 1 vaccine
	P. pastoris	1	Ecallantide, a protein inhibitor of plasma kallikrein

(see Table 8.5), but new technological developments in glycobiology, which are explained below, mean that it is set to eclipse its cousin in the very near future.

8.2.3 *Pichia pastoris*

P. pastoris is one of the methylotrophs, a small number of yeast species belonging to the genera *Candida*, *Torulopsis*, *Pichia* and *Hansenula* which share a specific biochemical pathway that allows them to utilize methanol as a sole carbon source. The promoters of the genes that encode the enzymes for this pathway are extremely strong and exquisitely sensitive to the presence or absence of methanol, making them ideal for the regulation of heterologous gene expression. Offering the ease of genetic manipulation associated with *S. cerevisiae*, these species have a number of advantages over their ethanol-producing cousin:

- They grow to much higher cell densities in fermenters owing to the absence of toxic levels of ethanol.

- They use integrative vectors, which removes the need for selective media to be used in fermenters whilst at the same time offering greater mitotic stability of recombinant strains.

- They have a more authentic type of glycosylation pattern for heterologous products.

Of these species, *P. pastoris* is by far the most popular choice when it comes to producing high-value heterologous proteins of therapeutic value. With a popular commercial kit (Invitrogen, San Diego) widely availability, *P. pastoris* has been used to express over 100 heterologous proteins for both research and commercial purposes. Selection of transformants for heterologous gene expression commonly relies on complementation of an auxotrophic *his4* marker in the host cells, although a number of dominant selectable markers are also currently available. Unlike *S. cerevisiae*, integrative vectors (as opposed to autonomously replicating plasmids) are normally used in this yeast species. The gene of interest is spliced in between the promoter and terminator sequences of the *AOX1* gene in an *E. coli* vector, which also carries the *His4+* gene and further downstream of this the 3' end of the *AOX1* gene (Figure 8.2). A linear fragment bounded by AOX1 sequences is then transformed into a *His4⁻* host. This DNA construct can then undergo homologous recombination targeting the gene of interest into the chromosomal locus of the *AOX1* gene (Figure 8.2b). Such cells can grow either on methanol using an alternative alcohol oxidase locus, in which case the heterologous protein is continuously expressed, or on glucose, in which case the heterologous gene is repressed until induced by methanol. The tight level of regulation allows for extremely precise control of the expression of the heterologous gene. Integrative vectors are also available that target the constructs to the *His4* locus. Quite apart from the fact that this easily regulated promoter has practical advantages over the more cumbersome galactose-inducible ones used to regulate heterologous expression in *S. cerevisiae*, *P. pastoris* is also regarded as a more efficient and more faithful glycosylator of secreted proteins. The most widely used secretion signal sequences include the *S. cerevisiae* α-factor pre–pro sequence and the signal sequence from *Pichia*'s own acid phosphatase gene.

A number of proteins of therapeutic importance (Table 8.5) have been successfully made in *Pichia*. However, this modest number is set to increase sharply because *Pichia* has just recently been subjected to extensive genetic manipulation to engineer strains of yeast that can produce human glycoproteins which are identical in every detail to those found in the human body. This genetic manipulation entailed knocking out four genes (to prevent yeast-specific glycosylation) and introducing 14 additional glycosylation genes to generate strains of *P. pastoris* capable of producing uniformly glycosylated, sialic acid-capped proteins. Many human proteins of therapeutic value are glycoproteins that must possess the correct sugar structures in order to have authentic biological activity and to prevent unwanted immunological complications. Heretofore, yeast cells have not been able to reproduce the required pattern of sugar moieties on heterologous proteins, but now *P. pastoris* can do so, and indeed their ability to produce essentially uniform N-glycosylation is superior to that even of mammalian cells, which are the usual hosts of choice when glycosylation is a key issue of concern in heterologous protein production.

Heterologous protein production by yeast cells has come a long way since the tentative steps that culminated in the production of the first recombinant hepatitis B vaccine in 1986 – the heterologous product that is the subject of the following case study.

8.3 Case Study: Hepatitis B Vaccine: A Billion Dollar Heterologous Protein from Yeast

With sales figures in excess of US$2 billion per year, recombinant hepatitis B sub-unit vaccines produced in yeast are one of the major success stories of molecular biotechnology. Produced as a heterologous protein originally in the yeast *S. cerevisiae* and more recently in both *P. pastoris* and *Hansenula polymorpha*, the phenomenal success of the hepatitis B vaccine makes it an ideal candidate with which to illustrate the research and development of a biotechnologically important heterologous protein.

8.3.1 Hepatitis: A Killer Disease and a Huge Market Opportunity

Hepatitis B, a double-shelled virus in the class Hepadnaviridae (Figure 8.5), is responsible for the death of more than 250 000 people worldwide every year. The liver infections caused by this organism can manifest in one of two different modes: either acute or chronic viral hepatitis. In acute hepatitis, the virus is completely cleared from the body when the symptoms disappear. In chronic hepatitis, the virus persists in the liver after infection and the patient becomes a carrier of the disease. The chronic mode of the infection is regularly associated with progressive cirrhosis and primary hepatocellular carcinoma. Worldwide, a staggering 200 million men, women and children, carry the disease.

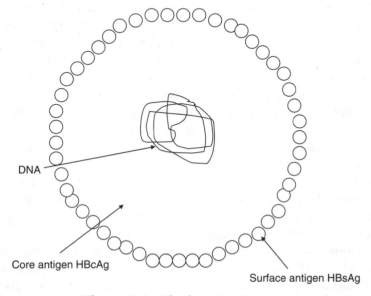

Figure 8.5 The hepatitis B virus.

In the vast majority of cases antiviral interferon therapy is either unavailable or, in the cases where symptoms have already developed, ineffective. Therefore, vaccination offers the only useful medical intervention with respect to this disease. In brutal biotechnological terms: the production of an effective hepatitis B vaccine constitutes a huge market opportunity. That is why so many companies develop and sell these products.

8.3.2 A Vaccine from Infected Carriers of the Disease

The first commercially available hepatitis B vaccine was prepared from protein particles of viral origin isolated from the plasma of chronic carriers of the disease. These extremely immunogenic 22 nm sub-viral particles, noninfective by-products of viral replication, consist of multiple molecules of a glycoprotein, called the hepatitis B surface antigen (HBsAg), embedded in a phospholipid membrane of cellular origin. These noninfective particles appear as if they are complete 42 nm viral particles to the immune system. Therefore, they elicit a strong immune response against the coat of the virus and confer resistance to subsequent viral infection.

First licensed in 1981 and used for over 10 years in the USA, this vaccine was safe, effective and well tolerated. It was less than ideal, however, because it depended upon a continuous supply of plasma and required extensive processing and safety testing. Although expensive to produce, the vaccine would have been more widely acceptable except for unbased fears that, despite elaborate safety precautions to prevent contamination, it had the potential to contain infective hepatitis B viruses or other blood-borne diseases (such as HIV), originating from the donor plasma.

Biotechnologists, therefore, turned to recombinant DNA technology to circumvent the problems associated with the human-derived product. However, given that genetic manipulation was in its infancy in the early 1980s and that, at that time, heroic efforts were needed to express even a simple human protein in *E. coli*, the expression of an effective heterologous hepatitis B vaccine was a tall order indeed. Such a challenging objective required genetic engineers to separate the DNA sequence encoding the HbsAg protein from the rest of the viral genome and then to arrange to have the information transcribed and translated into an immunogenic protein in a suitable host cell.

8.3.3 Genetically Engineering a Recombinant Vaccine

Using the previously sequenced 3200-base hepatitis B genome, the short DNA sequence encoding the 226 amino acids that comprise the major surface protein was isolated, spliced in frame with a strong promoter in an *E. coli* expression vector and transformed into the appropriate cells. Although the viral protein was

expressed, it was not glycosylated. Furthermore, the host cells failed to produce the 22 nm phospholipid–protein particles. The unassembled human HBsAg
protein was known to be 1000 times less immunogenic than the 22 nm plasma-
derived particles; scientists were not surprised, therefore, when the E. coli recombinant protein failed to elicit an appropriate immune response in animals.

However, undaunted by this failure, scientists then attempted to exploit recombinant DNA technology that had just been developed to facilitate basic gene
cloning in the simple eukaryotic organism S. cerevisiae. It was hoped that this
yeast, a eukaryote capable of glycosylating and secreting proteins, would be
able to produce immunogenic particles of glycosylated proteins where E. coli, a
prokaryote, had failed. In a proof of principle experiment, the DNA encoding
the HBsAg was spliced downstream of the yeast alcohol dehydrogenase (ADH1)
promoter in an E. coli-based shuttle vector carrying a 2μ replication origin and
the TRP1 gene (Figure 8.6a). After transformation and selection in a trp1- host,
the resulting transformants were found not only to express substantial levels
of HBsAg protein, but the proteins also aggregated into phospholipid particles
similar to those found in the plasma from human carriers of the disease. Unlike
the majority of HBsAg proteins synthesized in humans, however, the yeast-
expressed protein lacked glycosylation and, rather than being secreted, it accumulated inside the cells. Despite the lack of appropriate glycosylation, these
particles elicited the appropriate immunological response when tested in animals,
indicating that glycosylation was not needed for assembly of the particles or for
immunogenicity.

8.3.4 From Proof of Principle to Industrial Scale-Up

Having used a basic expression vector to demonstrate that yeast could produce
immunogenic 22 nm phospholipid–protein particles, thereby circumventing the
problems associated with the blood-borne source of the vaccine, an improved
expression vector (Figure 8.6b) was developed as a prelude to industrial scale-up
of vaccine production. A comparative analysis of the industrial vector and the
vector used in the proof of principle experiments illustrates many of the molecular
subtleties associated with ensuring high-level heterologous protein production
in yeast.

As can be seen in Figure 8.6, both plasmids are shuttle vectors carrying segments of plasmid DNA from E. coli and both are based on the yeast 2μ plasmid.
However, whereas the proof of principle vector has a TRP1 marker, the industrial one has a LEU-2d gene. This gene has a truncated promoter and, as it is
ineffectively transcribed, the cell requires a higher copy number of the plasmid
encoding it in order to be able to grow in the absence of leucine. On average,
each cell has 150–300 copies of a LEU-2d-carrying plasmid per cell, as against
approximately 30 copies of the TRP1-carrying plasmid. The industrial vector
also has a terminator sequence to ensure efficient termination, whereas the proof

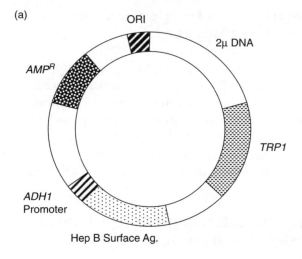

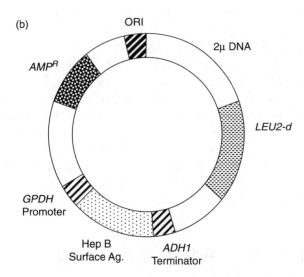

Figure 8.6 Proof of principle (a) and scale-up (b) plasmids.

of principle vector lacks such a sequence. Finally, the reasonably strong *ADH1* promoter of the proof of principle vector has been replaced in the industrial one by an extensively characterized, highly efficient *GPDH* promoter.

In short, in addition to the basic expression modules found in the proof of principle vector, the industrial one has DNA sequences to ensure a higher plasmid copy number, more efficient transcription initiation and more efficient termination of transcription. All of these lead to high-level mRNA production, thereby ensuring that the level of protein expressed is significantly higher in the industrial strain than in the proof of principle host.

8.3.5 Much More than a Development in Expression Technology

The development of high-level hepatitis B vaccine production in *S. cerevisiae* illustrated some of the subtleties of gene regulation and heterologous protein expression – aspects of which can still confound biotechnologists today. It also marked the first commercial exploitation of gene expression in a eukaryotic host, and a new era in vaccine development. However, above and beyond expression technology, the greatest impact of this billion-dollar protein has been the protection it affords millions of individuals against a debilitating, often deadly, disease.

8.4 Further Biotechnological Applications of Expression Technology

Quite apart from producing proteins of commercial value, protein expression can be manipulated in *S. cerevisiae* to provide *in vivo* tools with which to probe molecular interactions. These sophisticated heterologous protein-expression systems re-engineer promoter elements, transcription factors and signal cascade proteins to transduce heterologous molecular interactions into easily scorable phenotypes. Such assays enable biotechnologists to screen for molecules that interfere with/enhance these interactions – so-called lead molecules in drug development. Reporter genes and growth on selective media have been used to examine the molecular biology of expressed heterologous steroid and peptide receptor proteins respectively, whereas the elegant two-hybrid technique, which exploits the modular nature of transcription factors, provides a window into intracellular interactions between proteins.

8.4.1 Expression and Analysis of Heterologous Receptor Proteins

Oestrogen is an important human hormone that has been linked to breast cancer. As with many steroid hormones, oestrogen affects gene expression by binding to a cytoplasmically sequestered receptor protein and this complex then enters the cell's nucleus where it binds to promoters containing a specific receptor recognition sequence referred to as the oestrogen receptor element (ERE). Normal yeast cells do not contain either the receptor or ERE sequences. However, yeast cells have been re-engineered so that a β-galactosidase gene fused to a disabled *CYC1* promoter carrying the ERE resides at the *URA3* chromosomal locus (Figure 8.7a). The same cells also carry a plasmid constitutively expressing the receptor protein intracellularly (Figure 8.7b). When these cells are now treated with oestrogen, the hormone binds to the receptor and enters the nucleus, where it then binds to the ERE cloned into the promoter in front of the β-galactosidase

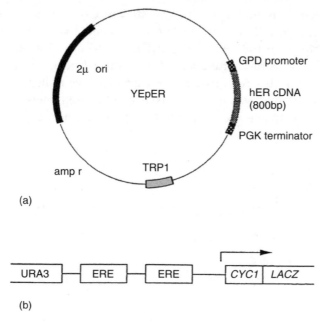

(a)

(b)

Figure 8.7 Analysis of oestrogen receptor proteins.

gene, thereby inducing expression. This enzyme can be easily assayed. The level of enzyme detected in the cells is then an index of oestrogen-induced receptor activity, allowing antagonists to be tested on the cells to identify which ones are best at inhibiting the hormones activity. This reporter system was sufficiently sensitive to analyse the effect of site-specific mutations on hormone binding efficiency and to measure the effectiveness of agonists and antagonists on hormone action.

Unlike steroid hormone receptors, which when activated by hormone bind directly to promoter elements in the DNA, peptide hormones bind to receptor proteins that are embedded in the cell membrane. These molecular interactions are then communicated to the nucleus by kinase cascades (enzymes that add phosphate groups to proteins), ultimately resulting in the phosphorylation of a transcription factor which induces gene expression via specific promoter elements. This type of signal transduction is controlled by so-called G-proteins (a heterotrimeric protein complex activated when a constitutively bound GDP molecule is replaced by a GTP molecule) that reside in the membrane next to the receptors. These protein complexes consist of three sub-units: α, which is the sub-unit in contact with the receptor protein and; β and γ, which initiate the kinase cascade. When the receptor is activated by binding of the appropriate peptide, the α sub-unit, which has a GDP molecule bound to it, undergoes a conformational change during which the bound GDP molecule is replaced by a GTP molecule with concomitant dissociation of the β and γ sub-units. The latter translocate to yet another membrane protein that then initiates the kinase cascade – ending in the phosphorylation of a specific transcription factor and activation of gene expression. The precise structure of the receptors and G proteins, and the transcription factor/promoter elements that they influence, varies from one organism to the next. Nevertheless, this overall cellular strategy for the transduction of a membrane signal into altered gene expression is highly conserved. It is for this reason that the extremely well characterized mating-signal-transduction pathway of *S. cerevisiae* can be re-engineered to analyse the interactions of human peptides and their target membrane receptors (Figure 8.8).

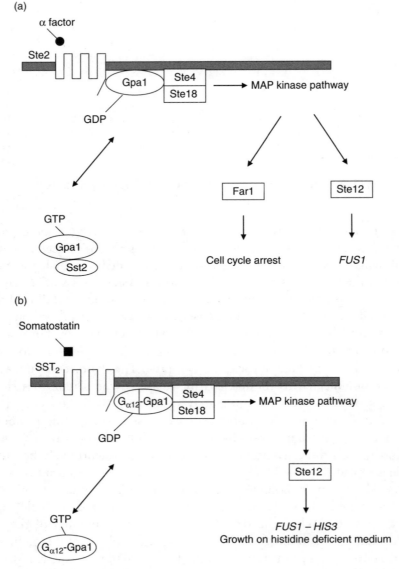

Figure 8.8 Heterologous receptor analysis using a re-engineered yeast pathway.

Haploid cells secrete small peptides, which bind to receptors in the membranes of cells of the opposite mating type in order to synchronize their cell cycles for mating. The binding of α-factor from α cells to the receptor protein of a normal haploid cell of the 'a' mating type causes the activation of a G-protein in close contact with the receptor. The resulting kinase cascade causes a number of alterations in the cell's gene expression pattern. These include the activation of genes to arrest the cell in G1 of the cell cycle and the expression of gene products ready for cell and then nuclear fusion as the synchronized cells mate together. With a view to using yeast cells in a high-throughput screening format, yeast cells have been manipulated to provide a 'readout' of cell growth in a selective medium. The success of this extremely elegant approach to developing tools for screening drugs hinges on the fact that the mating pathway of yeast cells of the 'a' mating type could be re-engineered such that they:

- expressed a human receptor protein instead of the α-factor receptor;

- expressed an α-sub-unit of the G-protein re-engineered so it could interact with the human receptor protein whilst retaining the segment that interacted with the yeast β and γ sub-units;

- carried an HIS3 wild-type gene driven by a promoter sensitive to activation by the transcription factor normally activated by the mating pathway kinase cascade;

- a deleted FAR1 gene to prevent cell-cycle arrest when the kinase cascade is activated.

These cells no longer respond to the addition of α-factor because they lack the α-factor receptor protein. However, when the appropriate human peptide is added to the cells, it binds to the heterologous human receptor protein in the membrane (Figure 8.8). The conformational change this causes is detected by the 'humanized' α sub-unit of the G-protein. When GTP replaces GDP in this sub-unit the β and γ sub-units dissociate, thereby initiating the kinase signal cascade to the nucleus. The activated transcription factor binds to the promoters of the genes normally induced by α-factor and to the promoter driving the HIS3 gene product, thereby conferring an HIS+ phenotype on the cell. Far1 activation would normally ensure that the cells arrest in G1 of the cell cycle, but as it has been deleted in this strain and so the cells progress through G1, allowing them to divide in His-selective medium.

In short, this re-engineered yeast cell transduces the heterologous human receptor–agonist interaction into a scoreable HIS+ phenotype. As G-protein-coupled receptors represent the targets for the majority of presently prescribed pharmaceutical drugs, this system has exciting potential for the development of high-throughput screening technology.

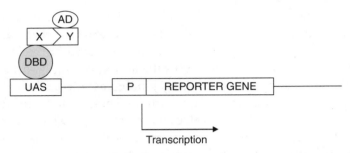

Figure 8.9 The two-hybrid system – only when 'X' interacts with 'Y' can the reporter gene get activated.

8.4.2 2-Hybrid Analysis: An Expression System that Turns Protein–Protein Interactions into a Scorable Phenotype

This extremely clever expression system hinges on the fact that transcription factors consist of two separate domains: the DNA binding and activation domains (DBD and AD). When expressed separately from different expression vectors, the two domains cannot interact and, therefore, a functional transcription factor cannot be constructed. However, if the domains are expressed as fusion proteins with two proteins (X and Y) and the two proteins interact with one another in the cell, then the two transcription factor domains are brought into contact with one another and a functional transcription factor is regenerated, which can then drive gene expression (Figure 8.9).

Yeast vectors are available in which the DNA encoding the DBD of the yeast *GAL4* transcription factor and the DNA sequences encoding the AD of the same protein are on separate 2μ-based expression plasmids carrying different selectable markers. The genes for the proteins under analysis for potential interaction with one another are inserted as gene fusions with the sequence for the DBD and AD respectively in these vectors. The plasmids are transformed into separate haploid cells carrying complementary genetic markers, defective in *HIS4*, and one of which has integrated into its chromosomal DNA a gene construct consisting of a GAL4-inducible promoter fused to the coding sequence for HIS4+. The two strains are then mated. If the two proteins interact, then this will bring the DBD and AD together in the cell, thereby reassembling the dissociated transcription factor. Such cells will express the His4 gene product and be able to grow on selective His-medium. In cases where the expressed fusion proteins do not interact, the two transcription factor domains remain apart and, although the DBD can bind to the His4 promoter, it cannot activate transcription because it lacks the AD (Figure 8.9). Such a diploid strain will be unable to express *HIS4* and will, therefore, not grow on the selective medium. Not only has this so-called yeast two-hybrid system been successfully applied to study a broad spectrum of protein--protein interactions from many different species, it can also be used to detect small molecules that interfere with the targeted protein–protein interaction, thereby providing yet another tool for drug discovery.

8.5 Conclusions

The production of heterologous proteins in fungi includes the production by filamentous fungi of heterologous proteins (mainly of fungal) origin for industrial applications, the production of high-value therapeutic proteins mainly by yeast species and the use of yeast cells to probe the protein–protein interactions of heterologous proteins from a wide range of organisms. Although heterologous proteins production was initially an empirical science, with hosts being tested to identify which one(s) provided the most authentic and highest yielding protein production, the next-generation technologies promise to be much more considered in their approach. The most impressive example of this to date is the successful re-engineering of the yeast *P. pastoris* to produce human proteins with glycosylation patterns that are more reproducible than those produced by mammalian host cells. However, this is just the beginning: global transcriptomic analysis of fungal cells is identifying the rate-limiting factors in the protein folding and secretion pathways, which are two rate-limiting aspects of heterologous expression. Guided by this much more rational approach, fungi in general, and yeast in particular, key players in this technology over the past two decades, are set to assume an even more central role in the production of heterologous proteins in the years to come.

Revision Questions

Q 8.1 Why are recombinant proteins much more difficult to genetically engineer than recombinant DNA molecules?

Q 8.2 Why is heterologous cDNA, rather than genomic DNA, normally used in expression vectors?

Q 8.3 List the important features of an expression vector.

Q 8.4 List three different types of DNA transformation used in fungi.

Q 8.5 Why was *S. cerevisiae* the first eukaryotic heterologous expression system?

Q 8.6 Compare and contrast heterologous protein expression vector systems in *P. pastoris* and *S. cerevisiae*.

Q 8.7 Outline the important parameters that impact on successful heterologous protein production.

Q 8.8 Why was it not possible to use *E. coli* as the expression system for the hepatitis B vaccine?

Q 8.9 Compare and contrast the plasmids used as proof of principle and scale-up of the hepatitis B vaccine.

Q 8.10 Outline the cloning strategy used to use *S. cerevisiae* as a surrogate oestrogen receptor protein testing system.

Q 8.11 Outline the cloning strategy used to use *S. cerevisiae* as a surrogate G-protein signal transduction testing system.

Q 8.12 Explain the basis of the yeast two2-hybrid system.

Further Reading

Journal Articles

Bonander, N., Darby, R.A.J., Grgic, L. *et al.* (2009) Altering the ribosomal subunit ratio in yeast maximizes recombinant protein yield. *Microbial Cell Factories*, 8, 10. Available from http://www.microbialcellfactories.com/content/8/1/10.

Gasser, B., Saloheimo, M., Rinas, U. *et al.* (2008) Protein folding and conformational stress in microbial cells producing recombinant proteins: a host comparative overview. *Microbial Cell Factories*, 7, 11. Available from http://www.microbialcellfactories. com/content/7/1/11.

Hamilton, S.R., Davidson, R.C., Sethuraman, N. *et al.* (2006) Humanization of yeast to produce complex terminally sialylated glycoproteins. *Science*, 313, 1441–1443.

Lubertozzi, D. and Keasling, J.D. (2009) Developing *Aspergillus* as a host for heterologous expression. *Biotechnology Advances*, 27, 53–75.

Nevalainen, K.M.H., Te'o, V.S.J. and Bergquist, P.L. (2005) Heterologous protein expression in filamentous fungi. *Trends in Biotechnology*, 23, 468–474.

Pausch, M.H. (1997) G-protein-coupled receptors in *Saccharomyces cerevisiae*: high-throughput screening assays for drug discovery. *Trends in Biotechnology*, 15, 487–494.

Porro, D. and Branduardi, P. (2009). Yeast cell factory: fishing for the best one or engineering it? *Microbial Cell Factories*, 8, 51. Aavailable from http://www. microbialcellfactories.com/content/8/1/51.

Rai, M. and Padh, H. (2001). Expression systems for production of heterologous proteins. *Current Science*, 80, 1121–1128.

Uetz, P., Giot, L., Cagney, G. *et al.* (2000) A comprehensive analysis of protein–protein interactions in *Saccharomyces cerevisiae*. *Nature*, 403, 623–627.

Valenzuela, P., Medina, A., Rutter, W.J. *et al.* (1982) Synthesis and assembly of hepatitis B virus surface antigen particles in yeast. *Nature*, 298, 347–350.

Wrenn, C.K. and Katzenellenbogen, B.S. (1993) Structure–function analysis of the hormone binding domain of the human estrogen receptor by region-specific mutagenesis and phenotypic screening in yeast. *Journal of Biological Chemistry*, 268, 24089–24098.

Books

Curran, B.P.G. and Bugeja, V. (2009). The biotechnology and molecular biology of yeast, in *Molecular Biology and Biotechnology* (ed. J.M. Walker and R. Rapley), Royal Society of Chemistry, Cambridge, pp. 159–191.

Gamelkoorn, G.J. (1991). Case study: hepatitis B vaccines a product of rDNA techniques, in *Biotechnological Innovations in Health Care* (ed. G. Turnock), Butterworth-Heinemann, Oxford, pp. 141–164.

Glick, B.R. and Paternak, J.J. (2003) Molecular biotechnology principles and applications of recombinant DNA, in *Heterologous Protein Production in Eukaryotic Cells*, 3rd edn, ASM Press, Washington, DC, pp. 163–189.

Price, V.L., Taylor, W.E., Clevenger, W. *et al.* (1991) Expression of heterologous proteins in *Saccharomyces cerevisiae* using the *ADH2* promoter, in *Gene Expression Technology* (ed. D.V. Goeddel), Methods in Enzymology, vol. **185**, Academic Press, San Diego, CA, pp. 308–318.

9
Fungal Proteomics

Sean Doyle

9.1 Introduction

The term 'fungal proteomics' can be defined as the study of the intracellular and extracellular protein complement of fungi. Historically, individual proteins were isolated by a combination of chromatographic techniques, enzymatically characterized and subjected to N-terminal or partial amino acid sequence analysis for identification. However, the advent of genome sequencing and protein mass spectrometry (MS), allied to high-resolution separation techniques for proteins (e.g. molecular mass-based separation by electrophoresis) has meant that hundreds, if not thousands, of proteins can be simultaneously isolated, separated and identified from an individual fungus. Although this large-scale approach has drawbacks, such as the generation of huge amounts of data, it represents the current situation with respect to the study of fungal proteomics.

The strategies for undertaking fungal proteomic investigations are continually evolving; however ,all share the following themes:

1 Availability of a full or partial genome sequence (or cDNA) for the fungus of interest, or establishment of the extent of fungal genomic information on related fungi in publicly available databases (e.g. Pubmed; http://www.ncbi.nlm.nih.gov/pubmed).

2 Optimization of protocols for both fungal culture and intracellular and/or extracellular protein isolation.

3 Separation of proteins by techniques such as chromatography, or preferably electrophoresis (e.g. sodium dodecyl sulfate-polyacrylamide gel electrophoresis (SDS-PAGE), Figure 9.1).

Fungi: Biology and Applications, Second Edition. Edited by Kevin Kavanagh.
© 2011 John Wiley & Sons, Ltd. Published 2011 by John Wiley & Sons, Ltd.

(a)

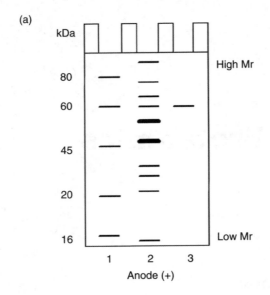

(b)

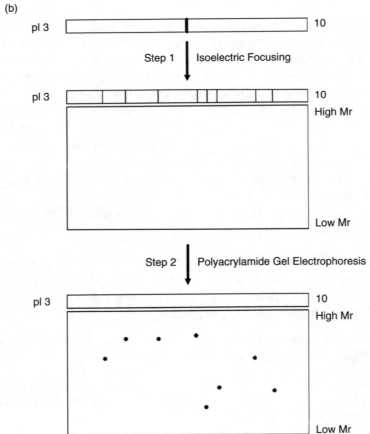

Figure 9.1   Schematic representation of SDS-PAGE and two-dimensional poly-acrylamide gel electrophoresis (2D-PAGE). (a) SDS-PAGE separation of proteins by molecular mass. Lane 1 illustrates proteins of known molecular mass ('marker proteins'), lane 2 contains a protein extract containing proteins of different molecular mass and lane 3 contains a purified protein preparation (single band). (b) 2D-PAGE consists of two distinct steps, isoelectric focusing (IEF) followed by PAGE. In step 1, native proteins are separated according to charge on an IEF strip. This strip is then placed on top of a PAGE gel and step 2 involves the additional separation of proteins by molecular mass to yield protein 'spots'. These can be excised and digested with trypsin for mass spectrometric analysis.

4   Fragmentation of individual proteins into peptides by digestion with the proteolytic enzyme trypsin.

5   Analysis of each peptide mixture by liquid chromatography–mass spectrometry (LC–MS). This technique separates the peptides from one another, determines the mass and sequence of each and then compares these data to DNA or cDNA databases.

6   Subsequent identification of a peptide or set of peptides coded by a specific gene – following *in silico* translation, which allows one to conclude that these peptides must have originated from the protein encoded by this gene.

This chapter will focus primarily on proteomics of filamentous fungi such as *Aspergillus* spp., primarily because these organisms are reservoirs of valuable protein products and are effectively 'black boxes' with respect to our knowledge of the detailed biochemical mechanisms used by these fungi to survive in the environment. Although the biotechnological potential of filamentous fungi in such areas as antibiotic and hydrolytic enzyme production has been exploited for a long time, research into fungal proteomics has lagged significantly behind that of bacteria and viruses for a number of reasons:

1   No filamentous fungal genome sequences were available until the 2000s, which meant that protein MS was of limited use in facilitating protein identification.

2   Filamentous fungal genes were known to contain multiple introns, unlike bacterial, yeast and viral genes. This meant that interrogation of available genomic DNA databases (following *in silico* translation) with peptide mass data was often of limited value for protein identification, since inadequate bioinformatic tools were available for intron–exon splice site identification in genes.

3   Intracellular protein isolation from fungi is particularly difficult owing to the rigid nature of the fungal cell wall, allied to the presence of large amounts of interfering carbohydrate polymers.

4 The level of proteins secreted by fungi is low in many wild-type organisms
   and is dependent upon the culture medium and conditions used for growth.

Consequently, fungal proteomics is an emerging and important area, where
modern proteomic techniques allow rapid identification of fungal proteins
of biomedical or biotechnological importance. In addition, proteomics can
be further defined in terms of either the study of protein modifications or
protein–protein interactions. This chapter will describe methods used for protein
extraction and isolation, electrophoretic techniques used for protein purification
prior to identification, protein MS and present selected examples of the uses of
proteomics in the study of fungal virulence and commercial potential.

9.2 Protein Isolation and Purification

9.2.1 Cell Lysis Strategies

The extraction or isolation of proteins from fungi is a prerequisite for subsequent analysis. Unlike animal cells, which do not possess a cell wall and where
the cell membrane can be readily lysed using detergents, the rigid cell wall represents a significant barrier to the efficient extraction of intracellular proteins from
fungi. Consequently, a number of vigorous techniques have been devised to enable the release of proteins from fungal mycelia, including hyphal maceration
in liquid N_2 using mortar and pestle, bead-beating, sonication and rapid pressure changes (French Press technology). One or a combination of these physical
techniques must be used to disrupt fungal mycelia, which when allied to strict
temperature control (2–8 °C), use of protease inhibitors, pH control and optimal
mycelia : extraction buffer ratio will yield efficient release of intracellular contents. Mycelial lysis is generally followed by high-speed centrifugation to remove
insoluble material such as intact mycelia, cell-wall fragments or cell debris, to
yield a clarified, protein-rich supernatant. This supernatant can then be further
processed by ammonium sulfate fractionation, dialysis or volume reduction by
ultrafiltration, prior to protein purification by chromatographic techniques.

9.2.2 Chromatography

A detailed description of chromatographic techniques is beyond the scope of this
chapter; however, ion-exchange chromatography (cation- or anion-exchange)
facilitates protein separation by charge, gel filtration enables protein separation
based on molecular mass and affinity chromatography allows protein isolation
by virtue of the specific affinity between the protein of interest and an immobilized ligand. Table 9.1 provides information of the relative merits and de-
merits of each chromatographic approach. Once the protein of interest has been

Table 9.1 Advantages and disadvantages of alternative chromatographic techniques for fungal protein isolation.

Chromatographic technique	Advantages	Disadvantages
Ion exchange	High resolution, speed, high protein-capacity resins, crude protein preparations suitable for use	Salt interference, total protein purity rarely achieved
Gel filtration	Separation based on molecular mass, high purity achieveable, no salt interference	Low resolution, pre-fractionation required, time consuming
Affinity	Excellent protein purity achievable, rapid, no pre-fractionation required	Requires affinity ligand, harsh protein release conditions
Immunoaffinity	Excellent protein purity achieveable, rapid[a]	Requires protein-specific purified IgG, immunoaffinity-purified protein must be IgG free.

[a] Chromatography is rapid; however, antibody generation may be time consuming.

isolated, it can then be characterized in terms of activity (if an enzyme), sequence, immunologically (i.e., antibody reactivity) or protein–protein interaction. Chromatography is generally used for the preparative isolation of proteins for further use, rather than as an analytical technique which facilitates global protein isolation for sequence characterization and identification.

9.2.3 Protein Extraction Prior to 2D-PAGE

Protein separation by 2D-PAGE is an important separation technique that exploits both protein charge and molecular mass to rapidly (1–2 days) yield individual, purified proteins, in sufficient amount, which can be analysed and identified by MS following enzymatic digestion by trypsin. This represents a significant advantage over chromatographic purification, which can take days or weeks to complete and often will yield only one or a few pure proteins. Fortunately, mycelial proteins can be extracted directly into reagents which are compatible with subsequent separation by 2D-PAGE. Here, mycelia are ground in liquid N_2 and sonicated in 'solubilization buffers' that contain high concentrations of urea, thiourea and detergents, along with ampholytes. Protein extracts prepared in this way can be applied directly to isoelectric focusing strips (pH range 4–7 or 3–10) to facilitate separation by isoelectric point prior to subsequent separation by molecular mass via protein electrophoresis (Figure 9.1).

9.2.4 Protein Extraction and Enzymatic Fragmentation without Prior Purification

Protein fractionation in not always required prior to analysis, and the term 'shotgun proteomics' has been introduced to describe the trypsin-mediated digestion of a complex mixture of proteins, followed by LC–tandem mass spectrometry (MS/MS) identification of most constituent proteins following comparison of individual peptide sequences with *in silico* databases. Such analyses require significant computing power and represent the limit of current proteomic approaches for global protein identification. In practice, this technique is carried out as follows. Prior to enzyme digestion, total protein can be extracted from mycelia ground in liquid N_2 in the presence of high concentrations of protein denaturing agents (e.g. 6 M guanidine-HCl) and reducing agents (β-mercaptoethanol or dithiothreitol (DTT)). This serves to solubilize and release the vast majority of cellular proteins and, if carried out in the presence of protease inhibitors, minimizes most nonspecific protein degradation. Solubilized protein can then be chemically modified with alkylation agents (e.g. iodoacetamide) to prevent disulfide bridge re-formation and stabilize C residues in the carboxyamidomethyl form. This step is followed by dialysis into ammonium bicarbonate, which results in removal of the denaturing, reducing and alkylation agents, and is almost always accompanied by protein precipitation. The protein precipitate is then treated with trypsin, which enzymatically fragments all proteins present into constituent peptides. This peptide mixture can then be subjected to LC–MS/MS analysis to facilitate global protein identification.

9.2.5 Protein Recovery from Culture Supernatants

In addition to the production and presence of mycelial (intracellular) proteins, fungi also secrete a wide range of enzymes into the extracellular environment; indeed, the pattern of secreted enzymes is often dependent on the available carbon source. Fungal culture can take place on either solid matrices (solid-state fermentation (SSF)) or in liquid culture, sometimes referred to as submerged fermentation. In both scenarios, the concentration of secreted enzymes is generally low (nanograms/milliliter to micrograms/millilitre), and so a concentration step is often required prior to subsequent analysis and characterization. Initial enzyme recovery from SSF is generally via resuspension in aqueous buffered solutions (occasionally containing low concentrations of detergents). Once resuspended SSF material or submerged culture supernatants are available, enzyme concentration is effected by (i) ammonium sulfate precipitation, (ii) protein ultrafiltration, (iii) lyophilization or (iv) trichloroacetic acid (TCA) precipitation. Once one or more of these concentration steps has been performed, and the volume has been reduced by up to a factor of 50–100, analytical tests such as protein estimation or protein electrophoresis are performed to estimate and visualize all proteins present. It should be noted that many fungi, especially basidiomycete species, produce large amounts of extracellular carbohydrate polymers that can interfere with the isolation of secreted enzymes. Consequently, a high-speed centrifugation step is often used to remove this material prior to protein concentration, when investigating the extracellular proteome (secretome) of many fungal species.

9.3 Electrophoretic Techniques

9.3.1 SDS-PAGE

SDS-PAGE is one of the most useful and straightforward techniques for protein fractionation or separation. SDS is a powerful anionic detergent which binds to, and denatures, proteins following co-incubation at 95 °C for 3–5 min. Moreover, complete protein denaturation is enabled by inclusion of reducing agents during this denaturation step, which serve to cleave all intra- and inter-molecular disulfide bridges. SDS-PAGE is based on the principle that if all protein molecules present in a porous polyacrylamide matrix are fully denatured and reduced, and possess equivalent (negative) charge, then they will migrate towards the anode (+) when subjected to an electric field. Proteins of low molecular mass will migrate farthest, while those with the highest molecular mass will migrate least. In this way, proteins are separated from one another by size to yield a 'ladder' of individual proteins in the polyacrylamide gel (Figure 9.1). Following electrophoresis it is necessary to fix and stain the proteins in the gel using specific staining reagents, such as Coomassie brilliant blue (CBB), silver staining or Amido Black. Staining is essential to allow protein visualization in the gel; fortunately, the most commonly used stain, CBB, does not interfere with subsequent trypsin digestion of excised protein bands and mass spectrometric identification of fractionated proteins. Silver stains can interfere with trypsinization and subsequent LC–MS identification, but good results can occasionally be achieved.

The immunological identification of proteins separated by SDS-PAGE, using polyclonal or monoclonal antibodies, is known as western blotting. Here, an SDS-PAGE gel containing separated but unstained proteins is sandwiched against either nitrocellulose or polyvinylidene fluoride (PVDF) membrane and subjected to electrotransfer to move the proteins from the gel onto the membrane. Once all proteins are replica-transferred to the membrane, a 'blocking' step is carried out to minimize nonspecific antibody binding to the membrane following by visualization of immunoreactive proteins by a combination of antibody, antibody–enzyme conjugate and substrate addition (Figure 9.2). Immunoproteomics is a term used to describe the combined used of immunological and proteomic techniques for protein identification. It will be appreciated that duplicate PAGE analysis, where one gel is CBB stained (gel-1) and the other subjected to western (immuno) blot analysis (gel-2) will allow mass spectrometric

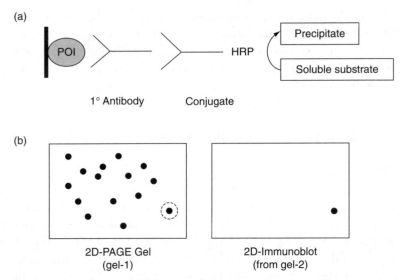

Figure 9.2 Immunoproteomics. (a) The protein of interest (POI) is recognized by an antibody present in serum or by purified IgG. An anti-species IgG-enzyme (e.g. horseradish peroxidase, HRP) conjugate reveals this interaction, with the aid of a precipitating substrate. (b) Duplicate 2D-PAGE analysis is performed and gel-1 is stained with CBB dye while gel-2 is subjected to electrotransfer onto a suitable membrane. This membrane is subsequently probed with antisera using an immunoblot procedure as shown in (a) to detect immunoreactive proteins (2D-immunoblot). The corresponding protein (circled) in gel-1 can then be excised and identified by protein MS. Note: SDS-PAGE can also be used for immunoproteomic investigations; however, the resolution is lower than 2D-PAGE.

identification of a protein from gel-1 and co-analysis of the immunoreactivity of the same protein in gel-2. This approach is widely used to identify immunoreactive fungal antigens associated with disease states, such as allergy, in humans.

Although SDS-PAGE is an extremely powerful and widely used technique in fungal proteomics, it does not possess sufficient resolving power to completely separate every protein in a complex mixture prior to further mass spectrometric analysis. In addition, because MS detection is such a high-sensitivity technique, it can detect protein presence even if a protein band is not visualized by CBB staining! Consequently, 2D-PAGE is the method of choice for fungal protein separation prior to analytical MS.

9.3.2 2D-PAGE

2D-PAGE is actually two techniques, IEF and SDS-PAGE, combined into a single process. It is also referred to as 2-DE (two-dimensional electrophoresis). Here, either total mycelial or secretome protein extracts are obtained from fungi under

4 ← ------------------------- p/Range ------------------------- → 7

High Mr

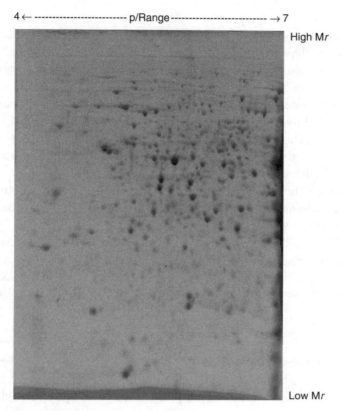

Low Mr

Figure 9.3 2D-PAGE of a mycelial protein extract from an ascomycete species stained with CBB protein stain. IEF was carried out over a pH range 4–7 and 10% PAGE was employed for molecular mass separation. Each 'spot' corresponds to an individual protein monomer from the organism. In general, different protein spot intensity corresponds to different relative amounts.

native conditions and without significant salt contamination. Following addition of IEF sample buffer, protein extracts (250–400 μg) are applied to isoelectric focusing strips followed by high-voltage conditions (up to 8000 V) for 24 h. Under these conditions, the charged proteins migrate in the electric field until they reach their isoelectric point pI, at which point they cease movement as a result of loss of charge – the proteins become focused. Once the IEF stage is completed, the entire strip is removed and placed adjacent to an SDS-PAGE gel, electrophoresis is commenced and the proteins leave the IEF strip, enter the SDS-PAGE gel and migrate according to molecular size. After electrophoresis, the gel is stained using CBB, or the more sensitive colloidal Coomassie dye (detection limit: 0.1 μg protein/spot), to detect protein spots (Figure 9.3); fluorescent stains including Sypro Ruby or Deep Purple™ dyes also allow protein visualization; however, these require fluorescent scanners for detection. Once protein visualization is complete, individual protein spots can be either manually or automatically excised from

2D-PAGE gels for trypsin digestion and mass spectrometric analysis. It should be noted that proteins present in mycelial or secretome extracts can also be labelled with fluorescent dyes (termed Cy2, Cy3 and Cy5) prior to IEF and SDS-PAGE separation; this approach not only removes the requirement for post-2D-PAGE gel staining, but also facilitates the quantitative assessment of differential protein expression under different experimental conditions on a single PAGE gel. This process is known as 2D-DIGE (2D-difference gel electrophoresis) and requires an extensive instrumentation environment, including fluorescent scanners and automatic protein 'spot-picking' devices. In fungal proteomics, 2D-DIGE has been primarily employed for the quantitative assessment of differential protein expression either when the organism has been exposed to environmental stress (e.g. oxidative stress) to identify protective proteins, or to different carbon sources to identify hydrolytic enzyme expression.

Like SDS-PAGE, 2D-PAGE gels can be electrotransferred and subjected to immunological interrogation using antisera (Figure 9.2). Thus, immunoproteomic approaches are possible for proteins fractionated by both SDS-PAGE and 2D-PAGE, although the resolution and sensitivity of 2D-PAGE far exceeds that of SDS-PAGE.

9.3.3 Offgel Electrophoresis

A solution-based fractionation technique, known as 'offgel electrophoresis', can also be used for protein or peptide separation prior to mass spectrometric analysis. Indeed, peptide fractions obtained following offgel fractionation can be directly analysed by LC–MS to aid protein identification. However, this technique has not yet gained widespread application for fungal proteomic investigation and will not be discussed further.

9.4 Protein Mass Spectrometry

9.4.1 Genome Databases

Before detailing protein MS techniques, and in particular the further consideration of fungal proteomics, it is necessary to be aware of the availability of computer databases which store large amounts of DNA and cDNA sequences from fungi, and indeed other organisms. These databases contain huge amounts of publicly available nucleic acid sequence data, including that from approximately 100 sequenced fungal genomes, which has been deposited by researchers from all over the world. The internet addresses of, and portals to, many of these sites are given in Table 9.2. The pace at which fungal genomic and cDNA data are deposited at these locations is rapidly accelerating owing to the advent of

Table 9.2 Genomic, proteomic and general fungal web sites.

Aspergillus genomes	http://www.cadre-genomes.org.uk/
Multi-fungal genome database	http://mips.gsf.de/genre/proj/fungi/fungal_overview.html
Aspergillus nidulans genome	http://www.broad.mit.edu/annotation/genome/aspergillus _nidulans/Home.html
Aspergillus niger	http://genome.jgi-psf.org/Aspni5/Aspni5.home.html
US Government Genome Sequencing Initiative	http://www.jgi.doe.gov/
	http://genome.jgi-psf.org/programs/fungi/index.jsf
Fungal portal	http://fungalgenomes.org/blog/
Fungal portal	http://www.fgsc.net/
Swissprot	http://www.ebi.ac.uk/uniprot/
Expasy	http://expasy.org/
MASCOT	http://www.matrixscience.com/

high-throughput next-generation DNA sequencing and the increased interest in fungi as a source of enzymes for carbohydrate degradation with a view to biofuel production. Most importantly, these databases represent the source of raw data, often referred to as *in silico* information, which is interrogated by, or compared against, fungal peptide information (e.g. peptide masses or sequences) to yield definitive fungal protein identification. Thus, at the heart of fungal proteomics and protein identification lies the ability to compare the mass or sequence of tryptic peptides isolated from a particular protein against the theoretical tryptic peptides present in a predicted protein sequence encoded by a gene in the aforementioned *in silico* databases (Figure 9.4).

The data present in these databases can either be downloaded to local servers or accessed or interrogated over the internet. In addition, it is important to note that, although extensive DNA sequence information is available, there can often be minimal information with respect to gene structure (intron/exon presence or splicing), whether or not a particular gene is expressed and what, if any, post-translational protein modifications occur to the encoded proteins. A specific issue with respect to fungal genomic data is that many predicted genes are only classified as encoding 'hypothetical' or 'predicted' proteins, as either the proteins have never been identified or, even if they have previously been detected, no function has been assigned to them. This 'functional proteomics' challenge represents one of the major issues in fungal biology today.

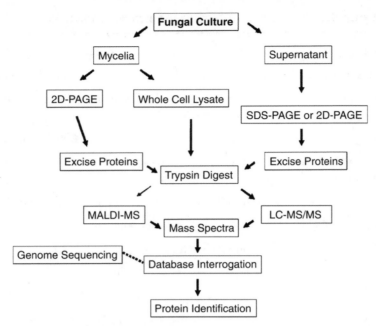

Figure 9.4 Schematic diagram of the overall process involved in fungal proteomics. Fungal cultures are initially separated into mycelia and culture supernatants. Protein fractionation can then be carried out using electrophoresis
(SDS-PAGE or 2D-PAGE) or chromatography (not shown). Alternatively, whole-
cell lysates can be trypsin digested without prior fractionation. Ultimately,
peptide mixtures derived either from individual proteins (following purification) or protein mixtures (shotgun proteomics) are analysed by LC–MS/MS, and
to a lesser extent by MALDI-ToF MS, to generate mass or sequence information. These data are then used to interrogate publicly or local genomic/cDNA
databases to facilitate protein identification.

9.4.2 Protein Fragmentation

Prior to MS analysis of either a gel-purified protein or protein mixture, it must be
digested into constituent peptides using trypsin. Trypsin is the enzyme of choice
for protein fragmentation, as it cleaves specifically at the C-terminal side of either lysine or arginine to yield a population of peptides, each terminating in a
positively charged amino acid residue. The presence of this positive charge is key
to peptide flight during matrix-assisted laser desorption ionization–time-of-flight
mass spectrometry (MALDI-ToF MS or MALDI MS). Moreover, many of the
peptides generated fall within the mass range 500–2500 Da, which is compatible with the optimal detection range of most MS instrumentation. High-grade
trypsin is almost always used, which means that no contaminating protease activities will be present and trypsin is chemically treated to prevent auto-digestion
and inhibit endogenous chymotryptic activity. Any of these interferences could

reduce the efficiency and sensitivity of subsequent peptide detection by MS. A detailed description of the methodology of protein digestion will not be addressed here; however, target protein amounts can range from nanogram/microgram to milligram amounts and the substrate protein : trypsin is generally 20 : 1. It should be noted that, in theory, digestion of a protein should result in release and detection of all tryptic peptides. However, in practice, this rarely if ever occurs. From a chemical perspective, this inefficient detection of all constituent tryptic peptides occurs for the following reasons: (1) inhibition of peptide bond cleavage, (2) loss of large hydrophobic peptide by adherence to plastic or insolubility, (3) release of multiple short peptides of low molecular mass (<300 Da), (4) nonspecific binding of peptides to filters used for sample preparation prior to MS analysis and (5) modified lysine or arginine residues. From an instrumentation viewpoint, peptides may not efficiently ionize during MALDI-ToF MS analysis or can be irreversibly bound to LC columns during LC–MS, thereby resulting in absence of detection.

9.4.3 Mass Spectrometry

It should be clear at this point that when reference is made to protein MS one is generally inferring identification of proteins by comparison of peptide data with genomic or cDNA databases. The fidelity of protein identification, therefore, is significantly dependent upon the accuracy of peptide mass, and sequence, determination by MS. Modern mass spectrometers can accurately determine the masses of either peptides or peptide fragments for subsequent interrogation against the corresponding theoretical tryptic peptide masses or sequences present in predicted protein sequences encoded by genes or cDNA in *in silico* databases (Figure 9.4). Peptides must generally be ionized to facilitate detection in a mass spectrometer, and this is achieved either by laser or electrospray techniques.

9.4.4 MALDI-ToF MS

In this type of protein MS, a peptide digest, derived from a purified protein, is mixed with an energy-absorbing matrix material (e.g. α-cyano-4-hydroxycinnamic acid; HCCA) and spotted onto a metal target plate. After drying, this sample is placed into the vacuum chamber of a MALDI-ToF mass spectrometer where it is subjected to ionization by laser light (337 nm), in a vacuum and at high voltage. The HCCA facilitates laser energy transfer to constituent peptides which vaporize and, because they are positively charged, electrostatically 'fly' in the time-of-flight detector. Peptide ions are thereby separated according to mass : charge, or *m/z*, ratio to yield a peptide mass fingerprint or MS spectrum (Figure 9.5). It is worthwhile recalling that not all peptides present in the protein will be detected, and that MALDI-ToF MS is suitable for analysing

(a)

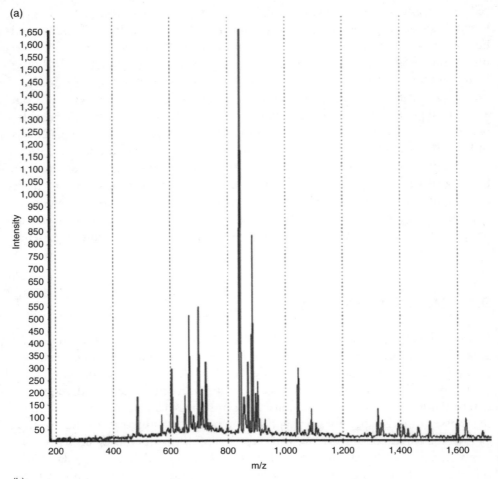

(b)

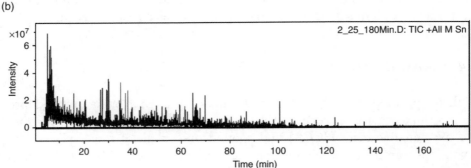

Figure 9.5 MALDI-ToF MS spectrum and LC–MS/MS total-ion chromatogram. (a) The MALDI-ToF MS spectrum is a plot of signal intensity versus m/z value of the positively charged peptide ions, and optimal data acquisition is in the m/z range 500–2500. The m/z values correspond to the precise mass of the peptide ions derived from the protein of interest and are automatically used to interrogate *in silico* databases which contain all theoretical tryptic peptide

other biomolecules (e.g. nucleic acids and carbohydrates) and intact proteins also. However, alternative matrix materials, such as 2′,5′-dihydroxybenzoic acid or sinapinic acid, and target deposition strategies need to be employed to enable such analyses. While MALDI-ToF MS is a rapid and automated technique, whereby multiple specimens can be analysed sequentially and individual specimen analysis can take as little as 5 min, it tends to be less sensitive than other types of protein MS and requires relatively pure protein samples to identify proteins with high confidence.

9.4.5 Electrospray Mass Spectrometry

Unlike MALDI-ToF MS, electrospray or nanoelectrospray ionization (ESI or nanoESI) techniques generate gaseous ionized molecular species (e.g. peptides) directly from a liquid phase. This type of MS for analysis of peptide mass and sequence has grown in popularity over the past 10 years owing to the ability to link the initial LC separation of peptides, via ESI, to high-mass-accuracy analysers. In practice, a mixture of peptides from a pure or relatively pure protein, or a peptide mixture from a relatively complex mixed-protein digest is initially fractionated by reversed phase high-performance liquid chromatography. As peptides elute from the LC column, they are ionized by nanoESI or ESI and subject to highly accurate mass determination prior to interrogation against *in silico* databases to enable protein identification. This so-called 'LC–MS approach' is superior to MALDI-ToF MS insofar as highly confident protein identification is possible using significantly smaller amounts of starting material. Moreover, since peptide purification occurs during the LC step, sample purity is less of an issue than when using MALDI. Optimal LC run times are, on average, 30 min duration, and so throughput is less than with MALDI-ToF MS.

Peptides detected by both MALDI-ToF and LC–MS can also be subjected to repeat ionization to yield smaller peptide fragments. This is sometimes referred

---

Figure 9.5 (Continued) masses predicted from DNA or cDNA sequences. Protein hits are generally expressed as (i) percentage sequence coverage or (ii) peptide number (observed)/peptide number (theoretical). Higher scores imply greater confidence in the identification. (b) Shotgun proteomics. This total-ion chromatogram (ion intensity versus time (min)) shows all peptides detected by LC–MS/MS (via ESI) following trypsin digestion of a total fungal protein lysate (Figure 9.4). Sample amount used for analysis is generally 1.5 µg total protein and an LC run time of 180 min is necessary to enable optimal peptide fractionation. Instrument software (primarily) and operator expertise then match peptide mass/sequence data to translated DNA/cDNA sequences in *in silico* databases to enable multiple protein identification. Hundreds of individual proteins in a single sample can be confidently identified using this approach.

to as either MS/MS or MS^n. Here, the observed sub-fragmentation pattern for any peptide can be compared with the theoretical sub-fragmentation pattern to yield an improved identification of the peptide sequence, and thereby *in silico* protein 'hit', with even higher confidence or specificity. This type of analysis is not performed manually, but carried out using commercially available software that comes as standard with most commercial instrumentation.

9.4.6 Shotgun Proteomics

So far in this chapter the emphasis has been placed on the requirement to trypsin-digest a purified or semi-purified protein prior to mass spectrometric analysis. However, owing to improvements in LC fractionation for the separation of peptides and, more importantly, in the algorithms used to assign detected peptides to a corresponding protein in a database, it is now possible to simultaneously identify many of the proteins present in a complex mixture by protein MS. This approach is known as 'shotgun proteomics' and bypasses the need to purify the protein of interest prior to trypsinization and MS analysis. Figure 9.4 describes how shotgun proteomic analyses can be carried out. The resultant mixture, after trypsin digestion, contains peptides derived from all proteins present in the initial sample, where hundreds of proteins may yield many thousands of distinct peptides. Following fractionation and detection of this mixture by LC–MS/MS, a list of proteins is obtained, which have been identified by comparison against the appropriate *in silico* database. Some proteins may be identified even from as little as a single peptide, while in most cases the detection of two peptides per protein, or sequence coverage of 5–10 %, is necessary for confident identification of constituent proteins. Although shotgun proteomics yields large amounts of data from a single experiment, it can be time consuming both in terms of LC run time (2–3 h) and subsequent data analysis (Figure 9.5).

9.4.7 Quantitative Proteomics

Protein identification using the proteomics strategies described so far is primarily a qualitative event. In others words, we can detect the presence of a protein but generally have no quantitative estimate as to how much is actually present, or if the amount of a particular protein changes under different conditions. Quantitative proteomics is in part enabled by comparative 2D-PAGE analysis of two experimental conditions followed by image analysis to detect differential protein expression; however, gel-free strategies are also available. These generally involve chemically labelling proteins using either cysteine-specific (isotope-coded affinity tag (ICAT)) or lysine-specific, commercially available reagents (isobaric tag for relative and absolute quantification (iTRAQ)), followed by trypsin digestion and LC–MS/MS analysis. Protein quantification can then be achieved by computation of ion signal intensities of differentially labelled peptide pairs in comparative mass spectra. The importance of this technique in fungal proteomics cannot be overestimated, as it provides a means to (i) investigate the effect of culture conditions on the extent of protein expression, (ii) study altered global or specific protein expression following gene deletion and (iii) explore altered virulence factor expression during infection.

9.5 Fungal Proteomics

9.5.1 *Trichoderma* Proteomics

Trichoderma harzianum is a mycoparasitic fungus and can protect plants against the deleterious effects of plant pathogenic fungi like *Botrytis cinerea* and *Rhizoctonia solani*. This means that *T. harzianum* can be considered as a biocontrol agent. Moreover, since *Trichoderma* spp., in general, secrete valuable carbohydrate-degrading enzymes, they have attracted much interest from the fungal proteomic community. Prior to 2004, little information was available on *T. harzianum* proteomics. Thus, initial proteomic investigation of *T. harzianum* involved 2D-PAGE fractionation and MALDI and LC–MS/MS identification of 25 intracellular proteins, including heat-shock proteins and glycolytic enzymes. These findings were followed by those which examined the types of protein produced by *Trichoderma atroviride* (formerly *T. harzianum* P1) in response to exposure to cell-wall extracts of the plant pathogenic fungus *R. solani*. Here, increased expression of eight *T. atroviride* enzymes was detected by protein mass spectrometric techniques and the proteins identified included N-acetyl-β-D-glucosaminidase, endochitinase, vacuolar protease A, superoxide dismutase, trypsin-like protease, a serine protease and a hypothetical protein. It will be immediately apparent that the majority of proteins identified represent hydrolytic or protective enzymes which were most likely produced to degrade *R. solani* biomolecules, including cell-wall material and constituent proteins. Thus, it is clear that cell-wall extracts from one fungus can generate a biological response in another fungal species and that proteomics can reveal these types of interaction. Interestingly, many of the *T. atroviride* enzymes were present in multiple 'spots' on 2D-PAGE gels. This is a characteristic of this technique and arises because of the high resolution associated with IEF, which enables separation of different charged forms of the same protein. It will be noted that a single 'hypothetical protein' was also detected, which means a protein with no known function or homology to a previously identified protein. Detection of this type of protein during fungal proteomic studies is common, and, as mentioned previously, represents a significant functional genomics/proteomics challenge for the future. Proteomics can also provide an insight into the complex pattern of protein expression in mixed cultures involving *T. atroviride*, *B. cinerea* and plant leaves. These types of mixed-proteomic study use the high specificity of detection of protein MS (mainly LC–MS/MS) to identify the types of protein produced by each of three species under conditions which mimic infection and biocontrol conditions. For instance, expression of cyclophilin A, a protein involved in protein folding, and superoxide dismutase, a free radical scavenger, was up-regulated in *B. cinerea* upon co-incubation with *T. atroviride* and bean plant leaves.

Subcellular organelles (e.g. mitochondria, nuclei or microsomes) or specific protein fractions (e.g. high-molecular-mass proteins) can be specifically enriched from cells or cell lysates prior to proteomic analysis. This pre-enrichment strategy is often termed 'sub-proteomics'. Sub-proteomic investigations of *Trichoderma* spp. have also been enabled by protein MS. In particular, the mitochondrial proteome of *T. harzianum* and that of the 26S proteosome of the industrially important species *Trichoderma reesei* have been dissected by combined electrophoretic and MS analyses. The relevance of these studies lies in the fact that organelle purification *preceded* fractionation and identification, and thus served as an enrichment step to improve the detection of low-abundance and highly localized proteins within the cell. These studies identified a number of unique proteins, many of which appear to serve regulatory, as opposed to catalytic, functions in *Trichoderma* spp. Importantly, since *T. reesei* is used for the commercial production of enzymes, it is imperative that the protein degradation system (which includes the 26S proteosome) in this microorganism is intensively and comprehensively studied. This will ultimately ensure that any barriers to protein stability and/or release can be circumvented, and protein production optimized to improve yields of secreted enzymes.

9.5.2 *Aspergillus* Proteomics

Aspergillus fumigatus is an opportunistic human pathogen and causes significant mortality in immunocompromised patients. *Aspergillus niger* and *Aspergillus oryzae* respectively produce enzymes and metabolites which have applications in the food industry. Finally, many genome sequences from *Aspergillus* spp. are now available (13 genomes from seven species; genome size range: 28.6–36.7 Mb) and it is clear that many genes encode proteins of unknown function, thus highlighting the requirement for both functional proteomic and genomic studies of these species to illuminate the biological roles of these proteins. The difficulty associated with investigating *Aspergillus* spp. proteomics is exemplified by the fact that only about 10 reports of 2D-PAGE and protein MS in these species were published between 2002 and 2007. The increased interest in these important species for both biomedical (need for improved diagnostics and new drug targets) and industrial reasons (valuable source of enzymes) has intensified the research and, consequently, the number of publications now emerging.

Proteomic and immunoproteomic studies of *A fumigatus* have investigated the intracellular proteome, the types of protein secreted by the organism under different culture conditions, conidial- and biofilm-associated proteins and immunoreactive or antigenic proteins present in the organism. Moreover, method development for optimal protein extraction from *A. fumigatus* and investigation of the intracellular proteome of *A. fumigatus* have received much attention. Combined, the work of many groups has resulted in the identification of hundreds of intracellular proteins in *A. fumigatus*, many of which are of unknown function. Predominant amongst the proteins that have been identified are heat-shock proteins, glycolytic enzymes, mitochondrial enzymes and catalases. Sub-proteomic pre-enrichment of proteins present in *A. fumigatus* cell lysates, by affinity chromatography, has revealed the unexpected identification of glutathione *S*-transferase activity in a putative translation elongation factor protein; gel-permeation chromatographic enrichment, prior to 2D-PAGE and MALDI-ToF MS, led to the detection of low-abundance, but high-molecular-mass, siderophore synthetases produced in response to iron-depleted conditions.

In an effort to identify mechanisms by which *A. fumigatus* may become tolerant or resistant to antifungal drugs such as voriconazole, caspofungin or amphotericin B, a number of research groups have investigated the fungal proteomic response subsequent to drug exposure. The rationale here is that alteration (elevation or decrease) of intracellular or extracellular protein expression will improve our understanding of fungal drug resistance and, more importantly, enable researchers to develop strategies to combat resistance mediated by these proteins, and thereby enhance the potency of the aforementioned antifungal drugs. In amphotericin B exposure studies, the differential expression (at least a twofold difference in expression) of 85 proteins (76 up-regulated and 9 down-regulated) was detected, compared with normal growth conditions. These included cell-stress proteins, transport proteins and enzymes involved in ergosterol biosynthesis (which is targeted by amphotericin B). Moreover, the unexpected alteration in the levels of enzymes involved in protein secretion was evident; the significance of this finding remains unclear.

Although possible, ethanol (i.e. biofuel) production from lignocellulose biomass via enzymatic hydrolysis and fermentation is quite constrained owing to the relatively high cost, and associated minimal efficiency, of the enzymes required to break down cellulose to fermentable sugars. Proteomic studies on *Aspergillus* spp. have focused on identification of carbohydrate-degrading enzymes for use in substrate generation with potential for biofuel production. One approach has led to the identification of two thermostable β-glucosidases, secreted by *A. fumigatus*, which could degrade cellulose in a superior fashion to preexisting enzymes from other *Aspergillus* spp. In general, fungi represent a rich source of commercially relevant enzymes for use in a range of biotechnological processes – as described in detail elsewhere in this book.

2D-PAGE and MALDI-ToF MS enabled identification of 57 proteins in *A. fumigatus* conidia that were overrepresented compared with those present in fungal mycelia. In particular, enzymes associated with anaerobic fermentation (e.g. alcohol dehydrogenase and pyruvate decarboxylase) were evident, suggesting a nonaerobic-type metabolism in resting conidia. Also, stress-resistance

enzymes, such as those responsible for pigment biosynthesis, inactivation of reactive oxygen species (catalase A, thioredoxin reductase and peroxiredoxins) and
conidial surface formation, were found, and it is thought that these may contribute to the protection of A. *fumigatus* conidia against the innate or adaptive
animal immune system. Conidial germination in *Aspergillus nidulans* has also
been studied using proteomics and expression of 241 proteins was altered ($P <$
0.05) at the early phase of germination. In fact, 40 of 57 proteins which were
identified by MALDI-ToF MS were associated with detoxification of reactive
oxygen species (as also noted for A. *fumigatus* above), energy metabolism, protein synthesis and protein folding process. Simultaneous analysis of gene expression using molecular techniques such as northern blot and reverse transcriptase
(RT)-PCR analyses confirmed the altered protein levels found in A. *nidulans*
conidia. However, it should be noted that coincidence of gene and protein
expression is not seen on all occasions during combined fungal genomic and
proteomic analyses.

Proteomics has a significant role to play in confirming the accuracy of gene
prediction by high-throughput bioinformatics tools used for gene identification
(sometimes called 'gene or gene model calling'), because tryptic peptide detection can only result from transcription and translation of actual genes. As noted
earlier, fungal genes contain multiple introns, which can make the absolute delin-
eation of the start and stop codon of a gene difficult to define, even using the most
sophisticated bioinformatics tools. Moreover, although highly reliable software
tools are used to predict intron : exon splice sites (>95 % confidence), they may
not always correctly predict (*in silico*) the corresponding cDNA sequence of a
fungal gene. In this regard, it will be recalled that a protein sequence corresponds
to the cDNA sequence; thus, the protein sequence will contain tryptic peptides
which derive from both the corresponding internal regions of an exon and from
exon : exon splice regions (Figure 9.6). Moreover, tryptic peptide sequences are
often identified that do not correspond to the internal region of an exon and,
therefore, must result from an exon : exon fusion, thereby confirming the existence of an intron : exon splice site, and the accuracy of the splice-site prediction
software (Figure 9.6). These important concepts have been clearly elucidated
using LC–MS/MS analysis of tryptic peptides derived from A. *niger* proteins separated only by SDS-PAGE. In this study, tryptic peptides were mapped against
two A. *niger* genome sequences, one of which (ATCC1015) was sequenced by
the Joint Genomes Initiative (Table 9.2) and the other (CBS 513.88) by the Dutch
biotechnology company DSM. Data from 19 628 mass spectra yielded 405 peptide sequences which were mapped to 214 different A. *niger* genomic locations,
of which 6 % were not found to be the best predicted gene model. Consequently,
the peptide data were used to modify, or correct, the *in silico* genome annotation process and, in addition, confirm the prediction of 54 intron–exon splice
sites. This experimental approach shows how proteomic data can be used to
validate fungal genome annotation, intron : exon splice site identification and
mRNA translation.

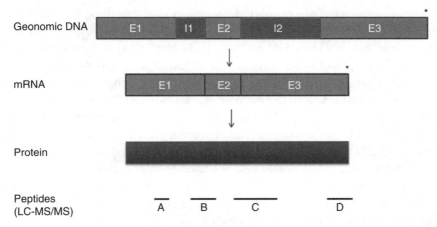

Figure 9.6 The role of protein MS in confirming *in silico* splice site identification in fungal genes. Following fungal genome sequencing, *in silico* analysis had predicted the presence of two introns (I1 and I2) and a stop codon in exon 3 (E3) in the gene of interest. LC–MS/MS analysis of tryptic peptides from the protein revealed the presence of four peptides (A–D). Peptide A was originally encoded by within exon 1. Peptides B and C could only have been detected if I1 and I2 respectively had been present in DNA and subsequently excised during mRNA synthesis. Peptide D did not contain a C-terminal K or R residue, and so its detection by LC–MS/MS strongly indicates correct stop codon (*) annotation in genomic DNA.

The secretome of *A. niger*, following culture in different carbon sources, has also been explored using protein fractionation by SDS-PAGE, subsequent LC–MS/MS shotgun proteomics and interrogation of fungal genomic databases. This work resulted in the identification of 200 secreted proteins which were encoded by genes containing *in silico* predicted signal peptides. Secretome composition was also observed to respond to changes in culture conditions, where protease secretion was up-regulated under carbon starvation conditions and pectinolytic enzyme secretion increased when galacturonic acid was used as the carbon source. Interestingly, one protein which was quite similar to Asp-hemolysin from *A. fumigatus*, and lacked a signal peptide, was detected in the secretome.

A. niger is used by the biotechnology industry because of its extensive protein secretion ability, via a process involving intracellular secretory organelles, which greatly facilitates downstream processing and enzyme recovery. Indeed, both transcriptomic and proteomic investigation of the factors affecting protein secretion by this filamentous fungus have been undertaken; however, only discussion of proteomic approaches will be detailed here. Following D-xylose induction of cellulase and hemicellulase secretion, the protein composition of secretory organelles in *A. niger*, compared with sorbitol presence only (i.e. no induction), was investigated. Microsomal membrane fractions, enriched in ER and Golgi component, were isolated by subcellular fractionation followed by shotgun proteomics involving LC–MS/MS. This led to the identification of over 1000 proteins which were predicted to partake in protein secretion. Some of the most abundant types of protein identified included protein disulfide isomerases, mannosyltransferases, chaperones and GTPases. Interestingly, recruitment of the 20S proteosomal complex to the microsomal fraction was noted only under conditions of D-xylose induction of protein secretion, which suggests involvement of this complex under conditions of induced protein secretion. Since no altered gene expression of 20S proteosomal components was observed, quantitative PCR or microarray analysis could not have been used to reveal this interaction. Thus, this work not only demonstrates a role for the 20S complex in protein secretion, but clearly shows the essential, and complementary, role of proteomics for elucidating protein–protein interactions, *in vivo*.

Many hypothetical proteins, more correctly called proteins of unknown function, were also detected by LC–MS/MS. Detection of such unknown-function proteins not only confirms expression of cognate genes, but also leads to hypotheses that they may play a role in protein secretion. This type of supposition can be further investigated by gene knockout technologies, where genes of interest can be specifically deleted and mutant phenotypes, in this case with respect to protein secretion, assessed by comparison with wild-type *A. niger*. Any alteration in mutant ability to secrete proteins would contribute to assignment of gene function. Specifically, the quantitative proteomic investigation of protein secretion from an *A niger* wild-type and mutant strain has also been achieved using iTRAQ allied to LC–MS/MS. Although full details of the mutant *A. niger* strain used were unavailable, there was a statistically significant increase ($P < 0.001$) in secretion of a range of glycosidases, cellulases, hemicellulases and a pectinase compared with wild-type cultures. This approach clearly highlights the potential of gene knockout strategies to positively influence global protein secretion, allied to highly sensitive quantitative proteomic analysis, to maximize evaluation of the effect of gene loss on protein secretion.

9.6 Specialized Proteomics Applications in Fungal Research

Three specific proteomics applications, (i) detection of post-translational modifications (PTMs), (ii) whole-protein or 'top-down' proteomics and (iii) intact conidial proteomics, merit some additional commentary.

PTMs of fungal proteins include, but are not limited to, glycosylation, phosphorylation, acetylation, phosphopantetheinylation, ubiquitination and methylation. While DNA sequence analysis may predict PTM sites or motifs within proteins, direct biochemical analysis of the protein is necessary to confirm that the modification actually occurs *in vivo*. Glycosylation aids protein stability or helps target localization, while phosphorylation is generally a regulatory switch to activate protein–protein interaction and ubiquitination directs proteins for degradation within the cell. Methodologies for detection of these PTMs are well described in the scientific literature.

However, protein acetylation and phosphopantetheinylation have emerged as significant PTMs of fungal proteins, as the former appears to regulate chromatin activation and, hence, expression of fungal gene clusters, while the latter is essential for natural product or secondary metabolite biosynthesis (e.g. antibiotics, siderophores or toxins) via non-ribosomal peptide synthesis. Modern mass spectrometers can readily detect acetyl mass (42 Da) within peptides and thereby confirm protein acetylation. Non-ribosomal peptide synthetases must be post-translationally activated by a 4′-phosphopantetheinyl transferase (4′-PPTase) activity, by attachment of 4′-phosphopantetheine (from coenzyme A) to specific S residues within the synthetase. Non-ribosomal peptide-synthetase-derived peptides, containing 4′-phosphopantetheine (358.09 Da), can also be detected by MS, and this approach has been successfully used in A. *fumigatus* proteomics to both confirm synthetase activation and 4′-PPTase activity.

Protein PTMs can also be detected using whole-protein MS. This technique is known as Fourier-transform ion cyclotron resonance (FT-ICR) MS and is the optimal approach for the analysis of intact proteins because it can reveal subtle differences in mass, associated with PTM, without prior trypsin fragmentation. FT-ICR MS has been used in fungal proteomics to determine the extent of histone acetylation and deacetylation, and the nature of the amino acids bound to the 4′-phosphopantetheine arms of non-ribosomal peptide synthetases for subsequent non-ribosomal peptide formation.

Finally, MALDI-ToF MS strategies have been developed using intact conidial proteomic analysis for strain typing, based on analysis of the species- or strain-specific mass spectrum patterns observed. Here, fungal conidia are mixed with MALDI-compatible matrices and spotted onto target plates for ionization. The technique, also called MALDI-ToF intact cell mass spectrometry (MALDI-ToF ICMS) uses matrix material such as 2′,5′-dihydroxybenzoic acid and sinapinic acid to enable specimen ionization. The spectra obtained appear to be specific for particular fungal species (e.g. *Trichoderma*, *Fusarium*, *Penicillium* or *Aspergillus* spp.) and may also facilitate strain identification (e.g. *Fusarium* spp.). Although many species-specific ions, in the *m/z* range 1500–15 000, can be detected by MALDI-ToF ICMS, the nature of the actual compounds is often unknown.

9.7 Conclusion

Fungal proteomics has come of age. Advances in proteomic technologies for individual and large-scale protein identification, allied to the plethora of fungal genome sequences becoming available, have created ideal conditions for the exploration of fungal proteomes. It is clear that fungal proteomes are highly dynamic and demonstrate exquisite responsiveness to the environment in which the organism finds itself. Proteomics provides the tools to explore these changes in detail. Thus, fungal proteomics research will accelerate our understanding of both fungal biology and disease-causing mechanisms, and yield identification of many new enzymes for biotechnological exploitation.

Revision Questions

Q 9.1   Give an overview of the strategies employed to investigate fungal proteomes.

Q 9.2   What factors have delayed the investigation of fungal proteomes?

Q 9.3   Why are proteins treated with iodoacetamide, or other alkylation agents, prior to MS analysis?

Q 9.4   What are the advantages and disadvantages of 2D-PAGE for protein isolation prior to MS analysis?

Q 9.5   What is 'shotgun proteomics'? Describe, in detail, this approach to protein identification.

Q 9.6   Describe the key applications of proteomics for the study of fungi.

Q 9.7   Why is proteomics a complementary, possibly even a superior, technology to gene expression for assessing fungal responses to altered environmental conditions?

Q 9.8   List the types of PTM which may be associated with fungal proteins. How can proteomics reveal the nature of selected modifications?

Q 9.9   What type of MS is used to study whole-conidia proteomics? Why?

Q 9.10   What matrix materials are used for peptide or protein ionization during MALDI ToF MS?

Further Reading

Journal Articles

Bhadauria, V., Zhao, W.S., Wang, L.X. *et al.* (2007) Advances in fungal proteomics. *Microbiological Research*, **162**, 193–200.
Bouws, H., Wattenberg, A. and Zorn, H. (2008) Fungal secretomes – nature's toolbox for white biotechnology. *Applied Microbiology and Biotechnology*, **80**, 381–388.
Braaksma, M., Martens-Uzunova, E.S., Punt, P.J. and Schaap, P.J. (2010) An inventory of the *Aspergillus niger* secretome by combining *in silico* predictions with shotgun proteomics data. *BMC Genomics*, **11**, 584.

Carberry, S., Neville, C.M., Kavanagh, K.A. and Doyle, S. (2006) Analysis of major intra-cellular proteins of *Aspergillus fumigatus* by MALDI mass spectrometry: identification and characterisation of an elongation factor 1B protein with glutathione transferase activity. *Biochemical and Biophysical Research Communications*, **314**, 1096–1104.

Kniemeyer, O., Lessing, F. and Brakhage, A.A. (2009) Proteome analysis for pathogenicity and new diagnostic markers for *Aspergillus fumigatus*. *Medical Mycology*, 47(Suppl 1), S248–S254.

Marra, R., Ambrosino, P., Carbone, V. *et al.* (2006) Study of the three-way interaction between *Trichoderma atroviride*, plant and fungal pathogens by using a proteomic approach. *Current Genetics*, **50**, 307–321.

Stack, D., Frizzell, A., Tomkins, K. and Doyle, S. (2009) Solid phase 4′-phosphopantetheinylation: fungal thiolation domains are targets for chemoenzymatic modification. *Bioconjugate Chemistry*, **20**, 1514–1522.

Suárez, M.B., Sanz, L., Chamorro, M.I. *et al.* (2005) Proteomic analysis of secreted proteins from *Trichoderma harzianum*. Identification of a fungal cell wall-induced aspartic protease. *Fungal Genetics and Biology*, **42**, 924–934.

Tan, K.C., Ipcho, S.V., Trengove, R.D. *et al.* (2009) Assessing the impact of transcriptomics, proteomics and metabolomics on fungal phytopathology. *Molecular Plant Pathology*, **10**, 703–715.

Wiese, S., Reidegeld, K.A., Meyer, H.E. and Warscheid, B. (2007) Protein labeling by iTRAQ: a new tool for quantitative mass spectrometry in proteome research. *Proteomics*, **7**, 340–350.

Book

Siuzdak, G. (2003) *The Expanding role of Mass Spectrometry in Biotechnology*, MCC Press, San Diego, CA.

10
Fungal Infections of Humans

Derek J. Sullivan, Gary P. Moran and David C. Coleman

10.1 Introduction

The ubiquity of fungi in the environment has already been alluded to in earlier chapters of this book. Fungal species have evolved and adapted to live in a wide variety of environments and ecological niches and, consequently, constitute a very diverse group of organisms. There are fewer than 100 000 species of fungi that have been identified to date; however, this is likely to be just the tip of the iceberg and it has been conservatively estimated that there are probably at least 1.5 million fungal species inhabiting our planet. Given this vast number of species and their prevalence in the environment, it is hardly surprising that humans unwittingly come into contact with many different types of microscopic fungus every day and that humans offer a potential source of nutrients for some of these fungal species. For the most part, transient exposure to fungi or fungal colonization occurs without the knowledge of the affected individual. This is primarily due to the inherent low virulence of most fungi, especially when confronted with the full arsenal of the human immune system. However, some species of fungi are long-established members of the human microflora (i.e. commensals in the oral cavity and the gastrointestinal tract) and under certain conditions some fungi can cause disease and even death. As many as 200 fungal species have been associated with human infections (known as mycoses); however, only a handful of these species are responsible for the vast majority of infections. Fortunately for us, fungi are far more important pathogens of plants (see Chapter 12) and insects.

Some of the most common microbial infections in humans are caused by microscopic fungi (e.g. thrush, dandruff and 'athlete's foot'). These are superficial

Fungi: Biology and Applications, Second Edition. Edited by Kevin Kavanagh.
© 2011 John Wiley & Sons, Ltd. Published 2011 by John Wiley & Sons, Ltd.

and relatively innocuous; however, fungi can also cause far more devastating diseases, such as invasive aspergillosis and systemic candidosis, both of which have very high associated mortality rates. The incidence of these latter infections has been increasing in recent decades, and this has fuelled a heightened interest in mycoses and the fungal species responsible for them amongst the clinical and scientific communities.

The types of infection caused by fungi can be classified in a number of ways. One division is based on whether the infection occurs in an otherwise healthy host (i.e. primary mycoses) or whether the host has an underlying medical condition causing impaired immune function (i.e. opportunistic mycoses). These groups of diseases can be further subdivided depending on whether the infection is confined to the outer layers of the epithelia (i.e. superficial mycoses) or whether the infecting organisms penetrate through this barrier into the bloodstream and disseminate throughout the body (i.e. systemic or disseminated mycoses).

10.2 Superficial Mycoses

The human body is covered by skin, hair and nails which, given their location in the body, are continuously exposed to the environment and, consequently, a wide variety of environmental microbes. For the most part, the keratinised epithelia which comprise the outer layers of the skin constitute an effective barrier that excludes microorganisms from gaining entry to deeper tissues. In addition, the skin also produces secretions, including sweat, sebum, transferrin and antimicrobial peptides known as defensins, that have the ability to kill many bacterial and fungal species. The skin is also equipped with intra-epithelial T and B cells, as well as a range of phagocytes. However, a small number of fungal species have evolved mechanisms of overcoming these defensive mechanisms and can actively colonize the skin surface, becoming established as members of the normal skin microbial flora. From time to time (mainly for reasons that are still unclear) these fungi can cause disease (Table 10.1). Two examples of such infections are pityriasis versicolor and tinea nigra. The former is caused by a yeast-like organism known as *Malassezia furfur*. This species thrives on the fatty acids found in sebum secreted by the skin and affects pigment-producing cells, resulting in a pink rash on pale skin and hypopigmentation in darker skin. Interestingly, this species and the related species *Malassezia globosa* have also been associated with dandruff, a common ailment characterized by increased shedding of skin cells from the scalp. Tinea nigra, a rare dermatomycosis characterized by a rash caused by the mould species *Hortaea werneckii*, results from the production of melanin by the fungus that causes the formation of brown macular patches on the palms and soles of the feet. As well as infecting the skin, fungi can also infect hair and nails. For example, fungi belonging to the genus *Trichosporon* cause a disease in hair known as white piedra (from the Spanish for stone), while *Trichophyton rubrum* causes the nail infection onychomycosis, also known as tinea unguium.

Table 10.1 Examples of superficial mycoses.

Site of infection	Disease	Examples of causative species
Skin	Pityriasis versicolor	*Malassezia furfur*
	Dandruff	*Malassezia globosa*
	Tinea nigra	*Hortaea werneckii*
	Ringworm (e.g. tinea capitis)	*Trichosporon/Microsporon* spp.
	Athlete's foot (e.g. tinea pedis)	*Trichosporon/Microsporon* spp.
Hair	White piedra	*Trichosporon beigleii*
Nail	Tinea unguium	*Trichophyton rubrum*
Subcutaneous	Chromoblastomycosis	*Fonsecaea* spp.
	Sporotrichosis	*Sporothrix schenkii*
	Mycetoma	*Pseudallescheria boydii*

Owing to the confinement of these infections to the extreme outer layers of the body there generally is no cellular immune response to the pathogens responsible for the disease. However, if the infecting fungi penetrate deep enough into the tissues to elicit an immune response, the infections are referred to as cutaneous mycoses. The most important examples of these infections are known collectively as tinea (more commonly known as ringworm). These infections can occur in various locations in the body, ranging from the feet (i.e. tinea pedis, better known as athlete's foot) to the head (i.e. tinea capitis) and are caused by keratin-degrading fungi collectively known as the dermatophytes, a group of organisms that includes species such as *Trichophyton* spp. and *Microsporum* spp. These infections are usually self-limiting and can be treated relatively easily using topical antifungal drugs, such as members of the azole family and terbinafine, although oral drugs may be administered in severe cases.

Rarely, fungi manage to penetrate deeper through the epidermis and cause infection in the underlying subcutaneous tissues (sometimes penetrating as deep as underlying muscle and bone). These infections are usually the result of the fungus gaining access to these tissues following trauma (e.g. wounds, splinters and bites). Examples of these infections are chromoblastomycosis, sporotrichosis and mycetoma.

10.3 Opportunistic Mycoses

As mentioned previously, the innate and adaptive human immune systems are remarkably adept at protecting the human body from infection by fungi. Consequently, in normal healthy individuals, systemic fungal infections are relatively uncommon. However, some fungi can capitalize on defects in the host

Table 10.2 Examples of risk factors for opportunistic fungal infection.

HIV infection and AIDS

Solid-organ transplantation

Anti-cancer chemotherapy

Granulocytopenia

Premature birth

Old age

Use of corticosteroids

Use of broad-spectrum antibiotics

Central vascular catheters

Gastrointestinal surgery

Colonization with fungus (e.g. *Candida* spp.)

(see Table 10.2 for specific risk factors) and overgrow and cause infection. These infections (Table 10.3) are known collectively as opportunistic mycoses, owing to the fact that the fungi that cause them are opportunists that exploit the imbalance between the host and the pathogen that occurs, for instance, when patients' defence systems are not functioning adequately. The two most important opportunistic fungal pathogens are yeast species belonging to the genus *Candida* (which cause candidosis) and moulds belonging to the genus *Aspergillus* (which cause aspergillosis).

Table 10.3 Examples of opportunistic mycoses.

Causative species	Disease
Candida spp.	Oropharyngeal candidosis/denture stomatitis (OPC) Vulvovaginal candidosis (VVC) Chronic mucocutaneous candidosis (CMC) Invasive candidosis (IC)
Aspergillus spp.	Invasive aspergillosis (IA) Aspergilloma Allergic bronchopulmonary aspergillosis (ABPA)
Cryptococcus neoformans	Cryptococcal meningitis
Pneumocystis jiroveci	Pneumocystic mneumonia

10.3.1 Candidosis

The genus *Candida* is comprised of approximately 200 yeast species, most of which have no known teleomorphic (i.e. sexual) reproductive phase. They are ubiquitous in the environment (often associated with plants and animals), but little more than a dozen have been associated with human commensalism or disease (see Table 10.4). Most of these *Candida* species are carried innocuously by a large proportion of humans, particularly on the epithelial surfaces of the mouth, gastrointestinal tract, vaginal tract and skin. They typically grow as ovoid blastospores; however, under specific conditions most can produce filamentous cells known as pseudohyphae, while *Candida albicans* and *Candida dubliniensis* can produce true mycelium and refractile spore-like structures known as chlamy-dospores (see Figure 10.1). Although these species are usually harmless in healthy individuals, when the host's immune defences are compromised in any way they have the potential to overgrow and cause infection which can, depending on the circumstances, be severe. Predisposing factors (see Table 10.2) to candido-sis include immunosuppression (e.g. due to HIV infection, anticancer therapy and treatment with immunosuppressive drugs used in organ transplantation),

Table 10.4 *Candida* species most commonly associated with human disease.

Species	Frequency
Candida albicans	Common
Candida glabrata	Common
Candida parapsilosis	Common
Candida tropicalis	Common
Candida dubliniensis	Infrequent
Candida krusei	Infrequent
Candida guilliermondii	Infrequent
Candida lusitaniae	Infrequent
Candida kefyr	Rare
Candida norvegensis	Rare
Candida famata	Rare
Candida inconspicua	Rare
Candida metapsilosis	Rare
Candida orthopsilosis	Rare

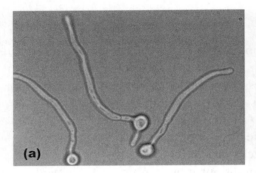
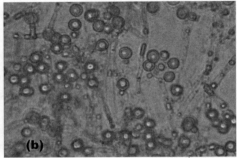

Figure 10.1 (a) *C. albicans* yeast cells (blastospores) producing hyphae and (b) chlamydospores.

catheterization (which allows the direct inoculation of the yeast cells into tissue and blood vessels), premature birth (immature immune system), extreme old age (defective immune system), use of broad-spectrum antibiotics (disruption of the normal bacterial microflora), use of corticosteroids (disruption of local immune response), gastrointestinal surgery (direct inoculation of yeast cells into the blood stream) and prior colonization with *Candida* spp. (most cases of candidosis are acquired endogenously from the patient's normal microbial flora). Once *Candida* cells have overcome the (usually impaired) immune response they can cause a wide range of infections. These range from superficial infections of the skin and the mucous membranes of the oral cavity and the vagina to cases when the cells penetrate through the epithelia and are disseminated throughout the body by the blood to infect a wide variety of organs, including the kidney, liver and brain.

One of the most common fungal infections is vulvovaginal candidosis (VVC), an infection of the vulva and vaginal area, also known as vaginal thrush. This infection can occur in apparently otherwise normal healthy women, approximately 75 % of whom will become infected at least once during their reproductive years. The symptoms of the disease include discomfort, itching, erythema and the production of a white discharge. While VVC is usually easily treated using topical antifungal agents, a small proportion of women suffer from recurrent infections which are particularly recalcitrant to conventional therapies. The predisposing factors for VVC are not clear; however, an association with pregnancy, diabetes, antibiotic use and HIV infection have been suggested.

Candida species are also associated with infections of the mucosal tissues of the mouth and oropharynx. These infections are known as oropharyngeal candidosis (OPC) and can occur following antibacterial therapy or in individuals whose immune systems have been compromised, such as those infected with HIV, neonates with immature immune systems, patients receiving steroid therapy for asthma and patients receiving head and neck radiotherapy therapy for the treatment of cancer. In addition, patients wearing dentures who practice poor oral hygiene can present with overgrowth of *Candida* spp. and inflammation of oral tissues in contact with the denture (i.e. denture stomatitis). There are several forms of OPC (Plate 10.1), the most common of which is known as pseudomembranous candidosis (more commonly known as oral thrush). This is characterized by the presence of creamy-white patches (comprised of buccal cells, host protein exudate and candidal yeasts and hyphae) overlying red patches primarily found on the palate and dorsum of the tongue. Other forms of OPC include erythematous candidosis and angular cheilitis. OPC causes a high degree of discomfort, particularly when eating and swallowing. It was originally treated using topical agents such as nystatin lozenges; however, severe cases are now treated with systemic oral azole drugs, especially fluconazole. Recurrent infections (due to acquired or intrinsic resistance to antifungals) can also occur in specific patient populations, and in these cases amphotericin B is often the most effective treatment option. OPC is most commonly observed in individuals infected with HIV. The vast majority of AIDS patients suffer from oral candidosis at some stage during their disease progression, and in the early stages of the AIDS epidemic OPC was often used as one of the diagnostic markers for HIV infection. On occasion, the infection can extend into the oesophagus, resulting in oesophageal candidosis. Fortunately, the introduction of highly active antiretroviral therapy (HAART) for the treatment of AIDS in the developed world has led to a significant reduction in the incidence of OPC in HIV-infected individuals.

Candida spp. can also cause infections of the nail bed and skin (e.g. nappy rash and intertrigo). In addition to these infections, *Candida* spp. can also cause a more severe skin disease that is known as chronic mucocutaneous candidosis (CMC). This infection can occur in HIV-infected patients, individuals with certain types of cancer (e.g. thymoma) and patients with endocrine and immune dysfunction (e.g. autoimmune polyendocrinopathy candidiasis ectodermal dystrophy (APECED)).

The infections described so far are superficial and confined to the outer layers of the skin and mucosal surfaces. However, under specific circumstances *Candida* cells can sometimes penetrate through these barriers, eventually reaching the bloodstream, thus causing transient candiaemia and ultimately resulting in disseminated infections in a wide range of organs (e.g. kidney, spleen liver) and systems (e.g. urinary tract). Invasive candidosis (IC) usually occurs in patients with severe neutropaenia (i.e. patients with severely depleted neutrophil levels, often due to treatment for cancer and haematological malignancy), patients who have undergone major abdominal surgery, patients receiving prolonged broad-spectrum antibacterial therapy and in patients with catheters (Table 10.2). Consequently, these infections are usually only encountered in hospitalized individuals who are already very ill. How the yeast cells gain access to the bloodstream is not entirely clear. However, antineoplastic chemotherapy can damage the epithelia of the gastrointestinal tract sufficiently to allow the yeasts to translocate across the gut wall into the local blood vessels. Similarly, intestinal surgery and intravenous catheters can allow the yeasts direct access to the bloodstream. Under most circumstances the human immune system can eradicate low numbers of invading candidal cells (transient candidaemia); however, in patients with reduced numbers of neutrophils, the yeasts are allowed to overgrow and spread throughout the body. The symptoms associated with invasive candidosis are very difficult to discriminate from those of systemic infections caused by other pathogens (e.g. bacteria). One of the earliest indicators of systemic fungal infection is persistent fever that does not respond to broad-spectrum antibacterial therapy; however, in some cases skin lesions can also appear. If left untreated, candidaemia can result in candidal cells being distributed to organs such as the kidney, liver and brain, ultimately leading to death. Because of the underlying illness of most invasive candidiasis patients, it can be difficult to attribute patient death to *Candida* infection. However, attributable mortality rates have been estimated to range between 14.5 and 49%, far higher than for other common causes of systemic infection (e.g. methicillin-resistant *Staphylococcus aureus*). In addition to the personal costs of candidal infection, in the USA alone in 2002 the attributable financial costs associated with candidaemia were calculated to be as high as US$1.7 billion per year. Invasive candidosis is usually treated with azole and polyene drugs (especially expensive lipid formulations of amphotericin B). In addition, novel azoles (e.g. voriconazole) and the new class of drugs known as the echinocandins offer the chance of improved survival rates in infected patients.

10.3.1.1 Diagnosis and Epidemiology of Candida Infections

In order to effectively treat candidal infections it is imperative that the yeast is detected and identified as rapidly as possible. In the case of superficial infections, such as VVC and OPC, this is relatively straightforward. Swabs can be taken from the affected area and inoculated onto routine mycological agar plates. The chromogenic medium CHROMagar Candida has been shown to be particularly useful in enumerating and identifying the yeast species present, with different species of *Candida* yielding colonies with characteristic colours (Plate 10.2). However, the diagnosis of candidaemia is more problematic. As mentioned earlier, the symptoms of candidaemia are nonspecific, and can easily be confused with other bloodstream infections. When candidaemia is suspected, a blood sample should be taken and inoculated into culture medium and analysed using manual or automated culture methods. However, in some cases it can take up to 48 h for growth to be detected, and it has also been reported that in many cases of candidaemia the blood culture fails to detect evidence of *Candida*, mainly due to the relatively low levels of *Candida* cells usually present in the blood of patients with candidaemia. This can result in a delay in the provision of appropriate therapy and increased mortality. Other tests to detect *Candida* spp. in blood include the detection of anti-*Candida* antibodies and *Candida* antigens in blood samples. In order to improve the speed, sensitivity and specificity of fungal infection diagnosis, molecular methods based on the polymerase chain reaction (PCR) are currently being assessed. This test relies on the detection and amplification of very small amounts of candidal DNA in blood samples using the thermostable enzyme *Taq* polymerase. PCR is exquisitely sensitive (it can detect as little as 10 *Candida* cells per millilitre of blood) and results can be obtained within a few hours. However, the exquisite sensitivity of PCR can lead to problems, with false-positive results occurring due to problems associated with contamination and the detection of nonviable *Candida* cells. For these reasons, despite the potential offered by this technique, it has still to gain acceptance by the clinical community.

The epidemiology of candidosis is constantly changing and mirrors developments in other aspects of human disease. The incidence of superficial forms of the disease, particularly OPC in the developed world, increased dramatically during the 1980s and early 1990s. This increase was due primarily to the HIV pandemic, which began in the 1980s and which resulted in large numbers of individuals at risk of infection. However, the introduction in the late 1990s of combination therapies, including novel anti-HIV agents such as protease inhibitors, which lead to a reconstitution of the T helper cell count and inhibit candidal proteinases, has resulted in a marked decline in the prevalence of OPC in developed countries. Despite this, candidal infections can still occur in HIV-infected patients in whom HAART has failed or in cases where the patient does not follow the full course of therapy (i.e. noncompliance). Unfortunately, owing to their high cost, these therapies are rarely provided in developing countries around the world where HIV infection continues to be a major health problem.

While the incidence of invasive candidosis was observed to increase during the 1980s, recent studies suggest that it has now levelled off and *Candida* species are now recognized as the fourth most common cause of nosocomially acquired (hospital-based) bloodstream infections (the leading causes are Gram-positive bacteria such as *S. aureus* and *Staphylococcus epidermidis*). The most likely reasons for the persistence in the prevalence of candidal systemic infection are the development of more aggressive anticancer therapies and more powerful immunosuppressive drugs associated with organ transplantation. These have resulted in a greater number of individuals surviving life-threatening diseases; however, these patients are usually extremely ill and are immunocompromised for increasingly long periods of time. Consequently, these patients provide an ever-growing population of at-risk individuals in our hospitals who are very susceptible to fungal infection. In order to protect patients from these infections and to decrease the incidence of systemic candidal disease, many hospitals treat their 'at-risk' patients prophylactically with antifungal drugs, although this runs the risk of causing the development of resistance in colonizing strains or in the selection of species such as *C. glabrata* and *C. krusei*, both of which have reduced susceptibility to azole drugs.

The *Candida* species most frequently associated with human infection is *C. albicans*. This species is widely regarded as being the most important human yeast pathogen, and when people refer to *Candida* it is often assumed that the species in question is *C. albicans*. However, in many epidemiological studies it is clear that *Candida* species other than *C. albicans* can cause both superficial and systemic disease (see Table 10.4). The exact proportion of species responsible for cases of candidiasis can vary from country to country and even between different hospitals in the same geographical region. In the majority of studies *C. albicans* is the most commonly identified species in clinical samples. However, species such as *C. glabrata* and *C. parapsilosis* are often also identified as contributors to human disease, while in some studies in Asia the most commonly isolated *Candida* species in blood samples was found to be *C. tropicalis*. The reasons for the apparent increased prevalence of these species are not clear. However, since *C. glabrata* can rapidly develop resistance to antifungal drugs, it has been suggested that prior treatment with these drugs can select for infections caused by this species. In the case of *C. parapsilosis*, this species is often found in biofilms growing on plastic surfaces and has been primarily associated with infections resulting from procedures requiring the use of intravenous lines and catheters.

The fact that some species are inherently resistant (e.g. *C. krusei*) or can develop resistance (e.g. *C. glabrata*) to many antifungal drugs has significant implications for the choice of antifungal therapy. Therefore, in order to prescribe the optimum therapy it is important to be able to identify the species responsible for a particular infection as rapidly as possible. Unfortunately, it is quite difficult to readily discriminate between different *Candida* species. Their cell and colony morphologies are, for the most part, very similar, with the yeast cells of most species being similar in size and shape, while colonies cultured on routine diagnostic agar (such as Sabouraud's dextrose agar) are often very similar in size, shape, texture and colour. However, as mentioned earlier, the chromogenic agar medium CHROMagar Candida has been shown to be a useful aid in species identification, with the most clinically important *Candida* species being distinguishable on the basis of colony colour. In diagnostic medical mycology laboratories there are two 'gold standard' tests that have been used for decades for the identification of *C. albicans*. These rely on the fact that *C. albicans* produces germ tubes (the first stage of hyphal development as the hyphae emerge from the cell) when incubated in serum and chlamydospores (thick-walled refractile spores of unknown function) when cultured on particular nutrient-depleted media. Until recently, *C. albicans* was the only species known to produce germ tubes and chlamydospores; however, in 1995, a novel, closely related species, *C. dubliniensis*, was identified in the oral cavities of HIV-infected individuals, which can also exhibit these two morphological characteristics. Other commercially available tests used routinely to identify specific *Candida* species include the analysis of carbohydrate assimilation profiles and serological tests. Molecular tests, in particular those based on PCR and microarray technology, have also been developed to allow the rapid identification of specific species and offer great potential for future rapid and accurate diagnostic tests.

While it is certainly very important to be able to discriminate between *Candida* species, for epidemiological studies it is often also very useful to be able to distinguish between strains within a species. This is particularly important when tracking the source of an infection in an outbreak, in determining if a recurrent infection is due to reinfection with the original or a new strain or whether more than one strain is present in a clinical sample, all of which can have a significant effect on the course of therapy for an infection. A wide variety of methods of differentiating between strains have been developed. The earliest methods used were based on the comparison of phenotypic characteristics (e.g. sugar assimilation profiles, colony morphology, antifungal drug-resistance profiles); however, these tests have poor discriminatory power and have largely been superseded by molecular methods, which are far more discriminatory. Molecular strain-typing methods are mainly based on the comparisons of the genetic content of individual strains. Until recently, the commonly used method for strain discrimination was DNA fingerprinting using species-specific probes homologous to regions of DNA repeated in the candidal genome. This generates a bar code-like pattern which is strain specific and, with the aid of computer software, can be compared quantitatively with the patterns obtained for other strains. DNA fingerprint analysis allows strains to be tracked epidemiologically and has revealed that the C. albicans species is comprised of more than 17 clades. Discriminatory DNA fingerprinting probes have been used in the epidemiological analysis of C. albicans, C. glabrata, C. parapsilosis and C. dubliniensis, amongst others. The karyotype (chromosome content) and the electrophoretic mobility of certain enzymes can also be used as a means of strain comparison. Owing to problems concerning the subjectivity of interpreting and comparing fingerprint and karyotype patterns, these methods have largely been superseded by multilocus sequence typing (MLST) analysis. This technique relies on the comparison of sequences of ~500 bp segments of seven housekeeping genes, and the data, which can be generated rapidly, are reproducible and not subjective. Because the sequence data can be stored in publicly accessible databases, many hundreds of strains can be compared. MLST has been shown to be as discriminatory as DNA fingerprinting and has identified the same clade structure in C. albicans and C. dubliniensis.

These methods have been applied widely to investigate the origin of Candida infections. In the case of the majority of superficial and systemic infections it appears that the infecting strains are mainly acquired endogenously from the patients' own colonizing flora, confirming the observation that prior colonization with Candida is a risk factor for candidosis. However, there are numerous cases in the literature describing the identification of strains present in a patient and on environmental surfaces and on the hands of health-care workers, suggesting that in some cases of disease the source of infection may be acquired from an exogenous source.

10.3.1.2 Candidal Virulence Factors

The capacity of a pathogenic microorganism to cause disease is usually the result of a complex interaction between the microbe and its host. Since Candida species mainly cause disease in immunocompromised individuals, it would be

Table 10.5 Virulence factors of *C. albicans*.

Adhesins (e.g. Hwp1, Als proteins)

Yeast ↔ hyphal dimorphism

Phenotypic switching

Extracellular hydrolases (e.g. Sap proteinases and lipases)

---

easy to conclude that the most important factors contributing to the establishment of candidal infections are purely related to the host. While host factors undoubtedly play a significant role in the development of candidosis, traits associated with the yeast are also very important. The capacity of a microorganism to cause disease in a host is known as its virulence. Clearly, certain microbes are more predisposed to cause disease than others, and this is usually due to the possession of specific attributes related to the ability to grow *in vivo* and to cause damage to host tissue. These attributes are often referred to collectively as virulence (or virulence-associated) factors. In the case of some bacterial species, it is very easy to identify virulence factors (e.g. the toxins produced by *Vibrio cholerae* and *Clostridium botulinum*); however, in the case of *Candida* species it is quite difficult to establish which of its phenotypic characteristics contribute to pathogenesis. Since *C. albicans* remains the most important yeast pathogen, its virulence mechanisms have been studied far more intensively than any other fungal pathogens of humans. From these studies a number of traits expressed by *C. albicans* have been identified as being putative virulence factors (Table 10.5).

One of the most important requirements for any microbe, whether commensal or pathogenic, is the ability to adhere to host tissue. This requires that the microbial cell be able to recognize and to stick to host ligands, such as extracellular matrix proteins and cell-membrane components. This allows the organisms to establish a foothold on the host surface, which prevents it from being dislodged by blood or host secretions, such as saliva or sweat. So far, a number of such proteins have been identified in *C. albicans*. Adhesins have been identified that bind to a range of host proteins, such as fibronectin and components of complement, as well as to host-cell-surface carbohydrate moieties of membrane glycolipids and glycoproteins. In addition, a large family of related proteins, known as the Als (agglutinin-like sequence) family have been shown to be involved in yeast–host cell interactions. One of these proteins, Als3, which is only expressed by *C. albicans* hyphae, has been proposed to act as an invasin and has been shown to sequester iron, suggesting that this protein is an important multifunctional virulence factor. The surfaces of hyphal *C. albicans* cells have also been shown to express a protein called hyphal wall protein (Hwp1), which appears to act as a substrate for host enzymes (transglutaminases) that can covalently link the hyphal Hwp1 proteins directly to proteins on the surfaces of the epithelial cells.

One of the most important phenotypic characteristics of *C. albicans* (and *C. dubliniensis*) is its ability to grow as ovoid blastospores and elongated hyphal filaments; the transition between yeast and hyphal forms is known as dimorphism. Both yeast and hyphal cells have been observed in tissue samples, and it is possible that each form may contribute to different stages or different types of candidal infection, as mutants locked in either the yeast or filamentous forms have a reduced capacity to cause disease. A range of environmental conditions have been shown to induce hypha formation (e.g. incubation in serum, growth temperature greater than 35 °C, nutrient starvation and pH greater than 6.5); however, the precise nature of the genetic switch(es) involved is complex and, as yet, not fully understood. Interestingly, several virulence factors have been associated with specific morphological forms (e.g. the adhesin Hwp1 and several aspartyl proteinases (see below)). In reality, *C. albicans* is in fact polymorphic, as, in addition to yeast and hyphae, it can also grow as pseudohyphae and chlamydospores; however, the role of the latter morphological forms in pathogenesis is not clear.

As in the case of most other pathogens, *C. albicans* has been shown to produce a wide range of extracellular enzymes that can digest host proteins. In particular, *C. albicans* possesses a family of 10 closely related secreted aspartyl proteinases (Saps) that have the ability to hydrolyse a wide range of host proteins, ranging from matrix components to immune-system proteins such as antibodies. In addition, *C. albicans* also encodes a family of lipases and phospholipases that have also been implicated in virulence. Interestingly, genes encoding individual proteinases and lipases are quite stringently regulated, only being expressed in specific morphogical or switch phenotypes.

Clearly, candidal pathogenesis is a complicated and multifactorial process, with a complex array of different virulence factors interacting in specific microenvironments to contribute to the survival and proliferation of the organism (see Plate 10.3 for a summary of events). However, it is becoming clear that it is naïve to think of *Candida* cells acting in isolation. In nature, it is now believed that the majority of microbes exist in complex communities of microorganisms attached to surfaces and covered in extracellular matrix. These communities are known as biofilms, and it is now believed that the ability to form biofilm is an integral component of the candidal offensive armoury. Candidal biofilms are of particular interest because it has been demonstrated that cells in biofilm are more resistant to the activity of antifungal drugs. In addition to *C. albicans*, *C. parapsilosis* has also been demonstrated to form biofilms, particularly on the surfaces of plastic catheters, possibly explaining the high level of association between this species and catheter-related infections.

The recent completion of the genome sequences for a range of pathogenic and nonpathogenic yeast species offers great potential to further our understanding of these species and how they cause disease. Genome sequence analysis has, for instance, allowed the discovery of a parasexual mating cycle in *C. albicans* that is related to the ability of this species to undergo phenotypic switching between white and opaque cells. The availability of these genomic sequences also allows comparisons to be made between species and has confirmed that virulence is associated with expansion of specific gene families, such as the ALS and SAP families. Post-genomic studies investigating global transcriptional and proteomic responses in a range of infection models are also currently being used to identify the functions of previously uncharacterized genes and novel patterns of gene expression in disease.

10.3.2 Aspergillosis

In addition to pathogenic yeast species, some filamentous fungi and moulds, under specific circumstances, have the ability to cause serious infections in humans. The most important of these opportunist moulds belong to the genus *Aspergillus*. Approximately 20 *Aspergillus* species have been associated with human infections; however, the vast majority of cases of aspergillosis are caused by only a handful of species, especially *Aspergillus fumigatus*, *Aspergillus niger*, *Aspergillus terreus* and *Aspergillus flavus*, with *A. fumigatus* being the most commonly identified cause of infection. *A. fumigatus* is a saprophytic fungus that is ubiquitous in the environment, and is particularly associated with soil and decaying vegetable matter. Although it has been thought that this fungus is asexual, it has been shown recently to have the capacity to reproduce sexually in the laboratory. In the wild, *A. fumigatus* grows as a mass of branching hyphae; however, it also produces vast numbers of asexual spores, known as conidia, in structures known as conidiophores (Figure 10.2). Conidia are released into the environment (especially in construction and demolition sites) and can be carried great distances by air currents, thus providing the means to contaminate food and water sources. The concentration of conidia in air can range from 1 to 100 per cubic metre. Therefore, they are routinely inhaled by humans and their size (approximately 2–3 μm) allows them to penetrate deep into the lower respiratory tract. In normal healthy individuals these are easily detected and destroyed by

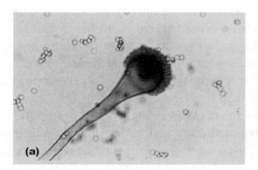
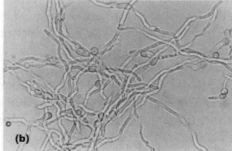

Figure 10.2 *A. fumigatus* (a) conidiophore and (b) germinating conidia.

alveolar macrophages present as part of the innate pulmonary immune system. However, in certain subsets of the population the spores can settle and cause a wide range of diseases.

The most important form of aspergillosis is known as invasive aspergillosis (IA). Patients at risk of acquiring this infection are generally profoundly neutropaenic for prolonged periods of time. Patients most at risk of acquiring IA include those receiving immunosuppressive treatment for organ transplantations (especially bone marrow transplants) and those receiving cytotoxic anti-neoplastic therapy. Reduced neutrophil counts and impaired alveolar macrophage function prevent the host from destroying *Aspergillus* conidia that are inhaled into the alveoli, allowing the conidia to germinate and the developing germlings to proliferate. If not detected and treated in time the hyphae produce hydrolytic enzymes and cause tissue damage, eventually breaching the alveolar wall, before ultimately reaching and penetrating into the circulatory system. Fragments of hyphae can then be spread to a wide range of organs throughout the body, resulting in disseminated aspergillosis.

Clearly, the human body represents a very different environment for aspergilli in comparison with their normal habitat in compost. However, these species have the ability to adapt to growth *in vivo* when the immune response is compromised. Identification of virulence factors or virulence-associated traits in opportunistic pathogens, such as *A. fumigatus*, is notoriously difficult. However, virulence-associated traits identified in these fungi include thermal tolerance, production of proteinases, production of gliotoxin and rapid response to environmental stress. Completion of its genome sequence in 2005 has facilitated the investigation of the molecular basis of *A. fumigatus* pathogenesis and has facilitated the development of *Aspergillus* microarrays. Global transcriptional profiling studies using these arrays have revealed that alterations in temperature are an important trigger for virulence, and that genes involved in iron and nitrogen assimilation, as well as alkaline stress response, also play an important role in adaptation to growth *in vivo*.

The symptoms of IA are nonspecific, but usually include fever and sometimes chest discomfort and cough (sometimes with blood). The mortality rate of IA is very high (50–100 %), especially once the fungi have spread to other organs, particularly the brain. Clearly, early diagnosis of IA is essential to allow optimum treatment; but this is very difficult to achieve, as the available diagnostic methods are not sufficiently effective. The gold standards of diagnosis include histopathological analysis and culture of biopsy and broncoalveolar lavage fluid taken from the infected area of the lung. However, improved radiographic imaging methods (e.g. CT scans) and serological tests to detect fungal galactomannan have been demonstrated to facilitate accurate and early diagnosis. Although the potenial usefulness of PCR analysis of blood samples to detect fungal DNA has been investigated for almost 20 years, this method has not achieved widespread usage for technical reasons. Despite all of the tests available, definitive diagnosis of IA is still often only obtained at post-mortem. Because of the problems associated with diagnosis, cases of suspected aspergillosis are often treated empirically (e.g. in patients with persistent fever that is unresponsive to antibacterial therapy) with antifungal agents. Liposomal amphotericin B is one of the most commonly used drugs to treat aspergillosis; however, recently developed drugs, such as itraconazole, voriconazole and caspofungin, have also been shown to be effective. Since acquired resistance to these drugs is relatively rare in *A. fumigatus*, attempts can be made to minimize the risk of IA by using prophylactic doses of drug during periods of profound neutropaenia. Other strategies to reduce infection risk in susceptible patients include pre-emptive antifungal therapy and the use of high-efficiency air filters to remove conidia from the air in the patients' rooms.

Aspergillus spp. can also cause a range of other diseases, often in non-immunocompromised individuals. One example is aspergilloma, an infection that develops due to *Aspergillus* species colonizing areas in the lung that have been damaged (e.g. by tuberculosis and sarcoidosis scars and cavities). The fungus proliferates and forms a large ball-like hyphal mass (the aspergilloma). The infection is often asymptomatic; however, fever and coughing up blood (haemoptysis) can also develop. It is usually treatable using conventional antifungal drugs; however, in extreme cases the fungal ball has to be surgically excised. Another disease caused by *Aspergillus* species is allergic bronchopulmonary aspergillosis (ABPA). This is an allergy to toxic products produced by *Aspergillus* cells that chronically colonize the upper respiratory tract, especially in asthmatic and cystic fibrosis patients.

There are, of course, many other filamentous fungi in the environment that can cause infection in severely immunocompromised individuals. These include the dematiaceous fungi (so called because of their dark colour) (e.g. *Alternaria* and *Pseudallescheria* spp.), *Fusarium* spp. and the zygomycetes, such as *Mucor* and *Rhizopus* spp. While infections caused by these species are still relatively rare, their incidence is increasing due to the growing numbers of patients with severely compromised immune systems for longer periods of time. It is particularly noteworthy that it is often difficult to identify many of these moulds in clinical samples; and in any case, many of them are not susceptible to most of the commonly used antifungal drugs. Consequently, these mycoses represent an increasing challenge for clinicians.

10.3.3 Cryptococcosis

Cryptococcus neoformans is the most pathogenic *Cryptococcus* species, although the closely related species *Cryptococcus gatii* is also an opportunistic human pathogen. Cryptococci are sexually reproducing basidiomycetous yeast-like fungi that produce a characteristic carbohydrate capsule (an important virulence factor) which is required for pathogenesis. The natural environmental reservoir of *C neoformans* is believed to be the soil, particularly in areas with high levels of pigeon guano. Infectious propagules of the fungus are thought to be carried by air currents and are regularly inhaled by humans. In the majority of cases the fungi are cleared by the immune system or a transient nonsymptomatic infection can sometimes occur. However, in susceptible individuals, especially HIV-infected and AIDS patients, the yeast cells can disseminate from the lungs into the bloodstream. Yeasts ultimately reach the central nervous system and the meninges, resulting in cryptococcal meningitis, a very serious infection with high mortality rates. The most common treatment is with a combination of amphotericin B and 5-fluorocytosine. While the levels of cryptococcosis were very high during the 1980s and 1990s, the introduction of HAART for the treatment of AIDS has resulted in a marked decrease in the incidence of the disease.

10.3.4 *Pneumocystis* Pneumonia

Pneumonia caused by the ascomycetous fungal species *Pnemocystis jiroveci* (until recently known as *Pneumocystis carinii* or *P. carinii* f. sp. *hominis*) is one of the most commonly encountered opportunistic infections associated with AIDS, although a wide range of other immunocompromised individuals can also acquire the disease, which is commonly referred to as PCP. PCP has also been associated with outbreaks of infection in malnourished children in crowded institutions. *P. jiroveci* was originally considered to be a protozoan; however, in depth phenotypic and genotypic analyses have confirmed it is more closely related to fungi. It is believed that this species is prevalent subclinically in the lungs of a large proportion of the human population (particularly young children) and is either latent or is continuously encountered by individuals through person-to-person exposure. In either case, when the human T cell count becomes depleted (e.g. by HIV infection or immunosuppressive therapy) the infection becomes activated, resulting in pneumonia-like symptoms, which if untreated can lead to death due to hypoxia. PCP is difficult to diagnose because *P. jiroveci* cannot be cultured *in vitro* and it is difficult to treat because *P. jiroveci* cell membranes do not contain ergosterol (the target of common antifungal drugs such as the azoles and amphotericin B). Consequently, the disease is treated using trimethoprim-sulfamethoxazole, which can have serious side effects. Fortunately, owing to the recent success of anti-HIV HAART therapy, the incidence of PCP is decreasing, although it still remains as the most common AIDS-defining opportunistic infection in developed countries.

10.4 Endemic Systemic Mycoses

Not all systemic fungal infections require an immunocompromised host. There are several examples of fungi that are primary pathogens (i.e. naturally virulent for humans) and which can cause symptoms of disease in individuals who

Table 10.6 Examples of endemic systemic mycoses.

Causative species	Disease
Histoplasma capsulatum	Histoplasmosis
Blastomyces dermatitidis	Blastomycosis
Coccidioides immitis	Coccidioidomycosis
Paracoccidioides brasiliensis	Paracoccidioidomycosis

are apparently otherwise healthy, as well as in patients who are immunocompromised (Table 10.6). These infections are usually most often found in very specific geographic locations and, hence, are often referred to as endemic mycoses.

10.4.1 Histoplasmosis

Histoplasmosis is caused by the dimorphic fungus *Histoplasma capsulatum*. This sexually reproducing fungal species is naturally found in the soil (often associated with avian and bat guano) and is endemic in tropical areas of the world and in the Ohio and Mississippi river basins in the USA. In the soil and at 25 °C in the laboratory the fungus exists in a hyphal/mould form; however, in the human host and at 37 °C *in vitro* it exists as small yeast-like spherules (conidia). It is interesting to note that in *C. albicans* both the yeast and hyphal forms are thought to be important in virulence, while in *H. capsulatum* (and other systemic mycoses) only the conidial form is associated with disease. Inhalation of the infectious propagules by normal healthy individuals usually results in a self-limiting subclinical infection, although in a small number of cases the patient may complain of mild flu-like symptoms. In cases where the patient happens to be immunocompromised (e.g. infected with HIV), histoplasmosis can be life threatening. An important virulence factor of *H. capsulatum* is the ability of this species to survive and proliferate within macrophages following phagocytosis. In certain circumstances, the *H. capsulatum* cells can remain latent and become reactivated in tissue years following the original exposure to the fungus, appearing once the immune system begins to deteriorate. Histoplasmosis hit the news headlines in 1997 when the singer/songwriter Bob Dylan was hospitalized due to the illness.

10.4.2 Blastomycosis, Coccidioidomycosis and Paracoccidioidomycosis

The causative agent of blastomycosis, *Blastomyces dermatitidis*, is very similar to *H. capsulatum*, in that it is also a sexually reproducing dimorphic fungus that can cause primary human infections. As with histoplasmosis, blastomycosis is acquired by inhaling conidia from contaminated damp soil and is endemic in large areas of North America. In the majority of cases the infection causes little or no symptoms; however, progressive pulmonary and/or systemic infection can rarely occur.

Coccidioides immitis and *Coccidoides posadasii* are the causative agents of coccidioidomycosis (also known as Valley fever). These grow in the hyphal form in their natural habitat (alkaline soils in dry and arid regions of the southwest USA and northern Mexico). These fungi produce spore-like structures known as arthroconidia, which when inhaled into the lungs of normal healthy individuals develop into multinucleate spherical structures, called spherules, that are filled with endospores. This results in respiratory infection, which is normally self-limiting, although infected individuals sometimes complain of cough and fever. However, in approximately 5 % of cases the disease can disseminate and progress to become a more serious disease. Another disease caused by a dimorphic fungus is paracoccidioidomycosis, which is caused by the dimorphic pathogen *Parracoccidioides brasiliensis*. This infection is endemic in Central and South America and, as occurs with the other systemic mycoses, results from the inhalation of spores originating from the soil. The primary disease is usually subclinical; however, for reasons as yet unknown, symptomatic infections are primarily diagnosed in males.

10.5 Mycotoxicoses

Like all organisms, fungi produce by-products of metabolism as they grow. The majority of these low molecular weight secondary metabolites are harmless to humans; however, some can deleteriously affect human health, and these are referred to as mycotoxins. It should also be remembered that some fungal metabolites are beneficial to humans (e.g. antibiotics produced by fungi). While there are approximately 300–400 recognized mycotoxins, only 10 or so are commonly observed in disease, which are collectively known as mycotoxicoses. The best known mycotoxins are the aflatoxin family and the ergot alkaloids.

There are four major aflatoxins (B_1, B_2, G_1 and G_2); however, B_1 is the most toxic (see Figure 10.3). In biochemical terms the aflatoxins are difuranocoumarin derivatives and are produced by a range of *Aspergillus* species, particularly

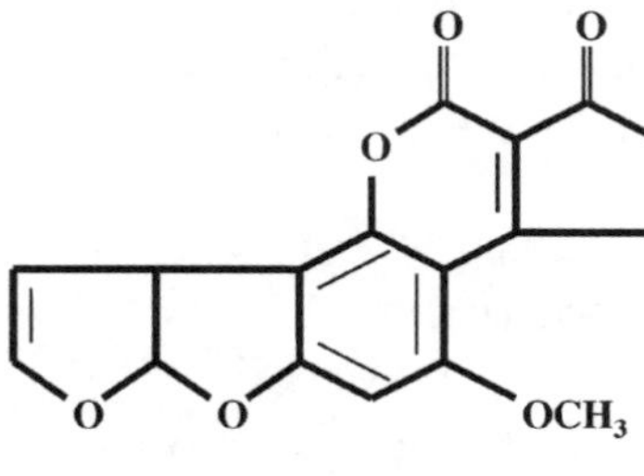

Figure 10.3 The chemical structure of aflatoxin B_1.

A. flavus. These fungi frequently contaminate and produce toxins while growing on crops such as corn and peanuts that are subsequently consumed by animals and humans. Consumption of meat and milk from cows who have been exposed to the mycotoxins can also result in exposure of humans to the toxins. While acute aflatoxicosis is a relatively rare phenomenon, ingested aflatoxins are notorious for their carcinogenicity as a result of chronic exposure. The mutagenic nature of aflatoxins is believed to be due to the DNA-damaging properties of aflatoxin metabolic derivatives. The most common form of disease associated with dietary exposure to aflatoxins is liver cancer; however, it may also be implicated in other forms of cancer. Because *Aspergillus* species are ubiquitous in the environment, it is impossible to prevent foodstuff contamination with fungi. However, this contamination can be minimized by the use of stringent production, storage and monitoring procedures.

Ergotism (also known as St Anthony's fire) is a disease associated with the ingestion of cereals contaminated with the fungus *Claviceps purpurea*. This fungus, which infects the flowers of grasses and cereals, produces a range of alkaloids, of which ergotamine is the best known. Ergotamine is related to the hallucinogen LSD, and ingestion of cereals and cereal-derived products such as rye bread that are contaminated with the fungus can result in serious symptoms of disease, such as convulsions and gangrene.

It has been suggested that mycotoxins might also exert a damaging effect through inhalation, rather than by ingestion. Most homes, office buildings and factories provide many niches suitable for the growth of a myriad of filamentous fungi, including *Aspergillus* spp., *Claviceps* spp. and *Stachybotrys* spp., and it has been suggested that mycotoxins produced by these species might contribute to the phenomenon known as 'sick building syndrome'. This syndrome has been associated with an ill-defined group of nonspecific symptoms (usually including fatigue, minor respiratory problems and headache) that are only experienced within a particular building. The aetiology of sick building syndrome is not known; however, poor air quality and ventilation, cleaning chemicals and microbial contamination have all been suggested as contributory causes. However, the role of fungi in the syndrome is a source of considerable conjecture.

Spores and volatile by-products from a wide range of filamentous fungal species can also act as allergens. Since most fungi come into contact with humans by inhalation into the respiratory tract, the symptoms of allergy usually occur in the sinuses and lungs. Symptoms are similar to other allergies, such as hay fever and asthma, and are usually caused by IgE-mediated type I hypersensitivity reactions.

10.6 Concluding Remarks

Considering the vast number of fungi that most of us come into contact with everyday, it is surprising that fungal diseases are less prevalent than perhaps might be expected. The fact that fungal infections are relatively rare is testament to the amazing efficiency of the human immune system; and from the descriptions of the mycoses described above, it is clear that when the immune system fails the fallout can be catastrophic for the patient. As great improvements continue to be made in medical science, especially in organ transplantation and cancer treatment, there will be an ever-increasing number of severely ill patients with profound immunosuppression for longer time periods in our hospitals. Therefore, the number of individuals at risk of acquiring life-threatening fungal infections is growing all the time. Mycologists are becoming increasingly able to prevent, diagnose and treat these diseases, thus improving the prognosis for a wide range of patients.

Over the past 30 years the prevalence of many fungal infections has changed dramatically for a wide variety of reasons. For instance, in the early 1980s it would have been impossible to forecast the huge increase (and subsequent decrease during the late 1990s) in the incidence of oral candidosis associated with AIDS. Therefore, it is essential that clinicians, epidemiologists and mycologists remain vigilant and ready for future changes in the epidemiology of fungal diseases. Future medical advances, viral epidemics and bioterrorism could all alter the spectrum of fungal pathogens plaguing mankind in the future. The armamentarium of antifungal agents available to clinicians continues to grow (e.g. the echinocandins, voriconazole); however, the greatest challenge facing mycologists is the development of improved, rapid diagnostic techniques to allow antifungal drugs to be administered in sufficient time to maximize the chances of survival of infected patients.

Revision Questions

Q 10.1 What are the most common superficial fungal infections and what are the causative agents?

Q 10.2 Name the four *Candida* species most commonly associated with human infections.

Q 10.3 What are the most important risk factors for opportunistic *Candida* infections?

Q 10.4 What are the most important virulence factors associated with the ability of C. *albicans* to cause disease?

Q 10.5 What types of infection can be caused by *Aspergillus* species?

Q 10.6 By what name is the fungal species formerly known as *Pneumocystis carinii* now known?

Q 10.7 In histoplasmosis, what is the pathogenic form of the infecting fungus?

Q 10.8   What is the most important form of aflatoxin and what are its deleterious effects on humans?

Further Reading

Journal Articles

Butler, G., Rasmussen, M.D., Lin, M.F. *et al.* (2009) Evolution of pathogenicity and sexual reproduction in eight *Candida* genomes. *Nature*, **459**, 657–662.

Cairns, T., Minuzzi, F. and Bignell, E. (2010) The host-infecting transcriptome. *FEMS Microbiology Letters*, **307**, 1–11.

Calderone, R.A. and Fonzi, W.A. (2001) Virulence factors of *Candida albicans*. *Trends in Microbiology*, **9**, 327–335.

Dagenais, T.R.T. and Keller, N.P. (2009) Pathogenesis of *Aspergillus fumigatus* in invasive aspergillosis. *Clinical Microbiology Reviews*, **22**, 447–465.

Niermann, W.C. (2005) Genomic sequence of the pathogenic and allergenic filamentous fungus *Aspergillus fumigatus*. *Nature*, **438**, 1151–1156.

Pfaller, M.A. and Diekema, D.J. (2010) Epidemiology of invasive mycoses in North America. *Critical Reviews in Microbiology*, **36**, 1–53.

Book

Calderone, R.A. (ed.) (2002) *Candida and Candidiasis*, ASM Press, Washington, DC.

Web Sites

http://www.doctorfungus.org/.
http://www.mycology.adelaide.edu.au/.
http://www.aspergillus.man.ac.uk.

11

Antifungal Agents for Use in Human Therapy

Khaled H. Abu-Elteen and Mawieh Hamad

11.1 Introduction

Treating human fungal infections by chemicals and medicinal plant extracts is a century-old tradition. Potassium iodide was first introduced as an antifungal agent in 1903. This was followed by Whitefield's ointment in 1907 and undecylenic acid in the 1940s. Garlic extracts (allicin), seed and leaf oils (tea tree, orange, olive), probiotic therapy (yogurt), benzoic acid, zinc and selenium were also used during the early days of antifungal therapy. In fact, some of these agents are still in use as folk medicines. In the last 60 years, however, systematic antifungal chemotherapy has advanced significantly through the introduction of drugs capable of treating most forms of human mycosis. Antifungal chemotherapy, the mainstay treatment of most fungal infections, relies on the capacity of various drugs to inhibit the synthesis or disrupt the integrity of cell walls, plasma membranes, cellular metabolism and mitotic activity (Table 11.1). Antifungal agents currently available for clinical use belong to four major classes: polyenes, azoles, 5-fluorocytosine and echinocandin. Allylamines, thiocarbamates, nikkomycin, aureobasidin, sordarin and others are also available to treat specific forms of mycosis. Antifungals differ in structure, solubility, mode of action, pharmacokinetic profile, spectrum of activity, therapeutic effect (fungistatic or fungicidal), degree of toxicity and ability to induce resistance. They are either topical (treat local infections) or systemic (treat systemic and disseminated infections) and they can be naturally derived (antibiotic) or chemically synthesized (synthetic).

Fungi: Biology and Applications, Second Edition. Edited by Kevin Kavanagh.
© 2011 John Wiley & Sons, Ltd. Published 2011 by John Wiley & Sons, Ltd.

Table 11.1   Commonly used antifungal drugs.

Class and agents	General mechanism of action	Susceptible fungi
I. Active against plasma membrane integrity or synthesis		
Polyenes (AMB and nystatin)	Autooxidation of ergosterol and formation of free radicals which compromise plasma membrane integrity and increase permeability	Different species of *Candida, Aspergillus, Histoplasma, Coccidioides, Cryptococcus* and *Saccharomyces*
Azoles (imidazoles and triazoles) *Echinocandins* (cilofungin)	Inhibit ergosterol synthesis by blocking the activity of the P450-dependent enzyme 14 α-demethylase	Different species of *Candida, Aspergillus* and *Cryptococcus*
Allylamines (terbinafine and naftifine), *Thiocarbamates* (tolnaftate)	Inhibit the activity of squalene epoxidase, which is responsible for the cyclization of squalene to lanosterol	Different species of *Aspegillus, Fusarium, Penicillium, Trichoderma, Acremonuim* and *Arthrographis*
Octenidine and *Pirtenidine*	Affect ergosterol biosynthesis by inhibition of 14α – demethylase	*C. albicans, S. cerevisiae*
Sphingofungin	Interrupts sphingolipid synthesis by inhibiting serine palmitoyltrasferase activity	Different fungal species
Floimycin (concanamycin A) Hydroxypyridones	Inhibitor of V-type Proton-ATPase Inhibits ATP-synthesis and cellular uptake of essential components	Different fungal species
II. Active against cell-wall components		
Echinocandins (caspofungin, micafungin anidulafungin)	Inhibit cell-wall glucan synthesis by blocking (1,3)-β-D glucan synthase activity	*Candida* spp.
Nikkomycin, Aureobasidin, Polyoxins	Inhibit chitin synthesis and assembly	*Candida* and *Cryptococcus* spp.
Pradimicin A *Benanomycin A*	Calcium-dependent complexing with saccharides of manno protein and disruption of membrane causing leakage of intracellular potassium	*Candida* and *Aspergillus* spp. and *C. neoformans*

Table 11.1 (*Continued*)

Class and agents	General mechanism of action	Susceptible fungi
III. Active against cellular anabolism		
Sordarin (sordaricin methyl ester)	Disrupt displacement of tRNA from A site to P site and movement of ribosomes along the mRNA thread by blocking the activity of elongation factor EF2 and the large ribosomal subunit stalk rpPO	*C. albicans* and *S. cerevisiae*
5-Fluorocytosine	Inhibits pyrimidine metabolism by disrupting nucleic acid and protein synthesis	Different species of *Candida*, *Aspergillus* and *Cryptococcus*

11.2 Drugs Targeting the Plasma Membrane

Agents that disrupt plasma membrane integrity and/or synthesis represent the vast majority of Food and Drug Administration (FDA)-approved drugs prescribed to treat pre-systemic and systemic fungal infections. They include polyenes, azoles, allylamines and thiocarbamates, octenidine and pirtenidine, and morpholines amongst others.

11.2.1 Polyenes

The discovery of nystatin (fungicidin; $C_{47}H_{75}NO_{17}$) and amphotericin B (AMB or fungizone; $C_{47}H_{73}NO_{17}$) in the 1950s has led to the isolation and characterization of numerous antibiotics. AMB, first isolated in the 1957 from *Streptomyces nodosus*, was the first systemic antifungal drug to be commercially available. More than 200 polyenes have been described so far; SPK-843 (*N*-dimethylaminoacetyl-partricin A 2-dimethylaminoethylamide diascorbate), a water-soluble polyene with a heptaene structure, is a recent promising addition. However, concerns regarding stability, solubility, toxicity and absorption profiles of most such compounds greatly limit the number of polyenes that gain approval as drugs.

11.2.1.1 General Properties of Polyenes

Polyenes have a large macrolide ring consisting of 12–37 carbon atoms closed by an internal ester of lactone and 6–14 hydroxyl groups are distributed at alternate carbon atoms along the ring (Figure 11.1). AMB has a free carboxyl

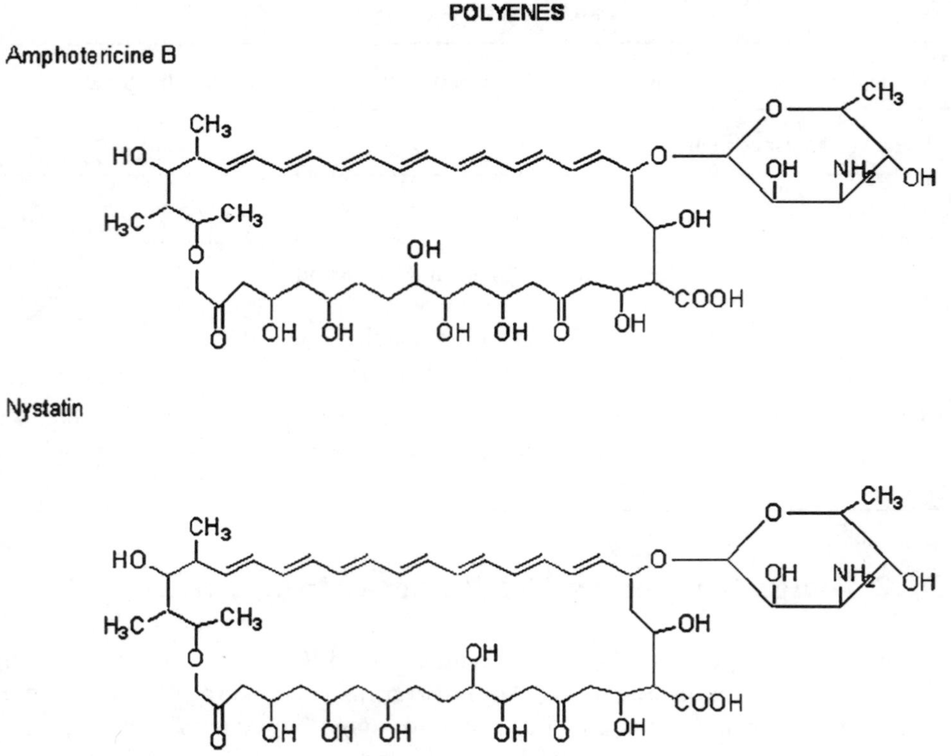

POLYENES

Amphotericine B

Nystatin

Figure 11.1   Chemical structure of polyene antifungal agents.

group and a primary amine group, which makes it amphoteric and capable of forming channels across cell membranes. The amine group present in some polyenes is associated with an amine sugar linked to the macrolide ring through a glycosidic bond. The carbohydrate moiety in AMB, candicidin, trichomycin, pimaricin, candidin and nystatin is mycosamine ($C_6H_{13}O_4N$). Polyenes show limited solubility in water and nonpolar organic solvents, but they dissolve easily in polar solvents like dimethyl sulfoxide and dimethyl formamide.

Although AMB remains the mainstay treatment of systemic mycoses, low water solubility, high toxicity and ineffectiveness against mould diseases in immunocompromised hosts limit its therapeutic potential. Three lipid formulations (AMB lipid complex, AMB cholesteryl sulfate and liposomal AMB) have been developed and approved for clinical use. These delivery systems offer several advantages over conventional AMB. Usage of higher doses (up to 10-fold) of the parent drug, higher tissue concentration of the parent drug in primary reticuloendothelial organs and reduced toxicity are some of the advantages. All three formulations are indicated for patients with systemic mycoses who are either intolerant to conventional AMB or have pre-existing renal dysfunction. There is no consensus regarding the use of lipid AMB formulations in immunocompromised patients with invasive mould infections (aspergillosis and zygomycoses).

11.2.1.2 *Mechanism(s) of Action*

The interaction between AMB (large polyenes) and membrane sterols generates aqueous pores, each consisting of an annulus of eight AMBs linked hydrophobically to sterols (Figure 11.2). The hydroxyl residues in AMB face inwards to give an effective pore diameter of 0.4–1.0 nm. This leads to leakage of cytoplasmic components, which amongst other consequences results in inhibition of aerobic and anaerobic respiration, cell lysis and death. Like AMB, nystatin, when available in sufficient amounts, tends to form pores in the membrane that lead to K^+ leakage and cell death.

Pore formation is only part of a complex and multifaceted process that produces the full extent of polyene activity against target cells. For example, inhibition of fungal growth depends on the ability of appreciable amounts of polyene to directly bind target cells. Polyenes also induce oxidative damage to fungal plasmalemma, which may contribute to their fungicidal activity. Chemotherapeutic effects can also result due to the ability of AMB to induce oxidation-dependent stimulation of macrophages. The selective mode of action of polyenes is related to their differential affinity to membrane sterols on target cells: ergosterol in fungi and cholesterol in mammals. In other words, specificity to ergosterol rather than cholesterol greatly minimizes potential adverse effects of

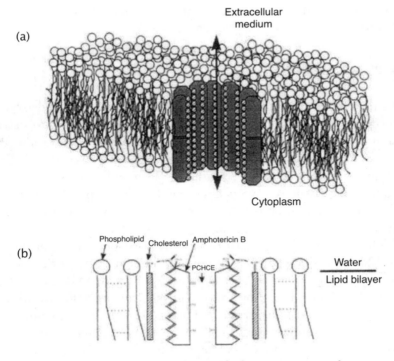

Figure 11.2 Schematic representation of the interaction between amphotericin B and cholesterol in a phospholipid bilayer.

polyenes in mammalian hosts. Resistance of sterol-lacking *Mycoplasma laidlawii* to polyenes is evidence of this type of association. In sterol-containing media, cells tend to gradually show signs of sensitivity to polyenes as sterols start to incorporate into the plasmalemma. Furthermore, sensitive fungi can be protected from the inhibitory effects of polyenes by adding sterols into the growth media to compete with membrane sterols. Polyene toxicity is also dependent on the fatty acyl composition of membrane phospholipids; changes in the ratio of various phospholipids may affect the internal viscosity and molecular motion of lipids within the membrane.

11.2.1.3 Spectrum of Activity

Significant activity of polyenes against most *Candida* spp., *Cryptococcus neoformans*, dimorphic fungi (*Sporothrix schenkii*) and some dematiaceous fungi is well established. Compared with AMB, SPK-843 has superior *in vitro* inhibitory activity against *Candida* spp., *C. neoformans* and *Aspergillus* spp. Synergistic effects have been reported in mice infected with *C. neoformans*, *C. albicans*, *Aspergillus* spp. or *Histoplasma capsulatum* following treatment with AMB plus 5-fluorocytosine and in mice infected with *Coccidioides immitis* following treatment with AMB plus tetracycline or rifampicin. Nystatin is used against cutaneous, vaginal and oesophageal candidiasis as well as cryptococcosis. It is often used as a prophylactic agent in patients at risk of developing fungal infections (e.g. AIDS patients with low CD4$^+$ counts).

11.2.1.4 Pharmacokinetics

Following intravenous (IV) administration, AMB dissociates from deoxycholate; >90 % of freed AMB becomes protein bound and redistributes to tissues. An IV infusion of 0.6 mg/kg in adults yields a peak serum concentration (PSC) of 1–3 mg/L, which rapidly falls to a prolonged plateau phase of 0.2–0.5 mg/L. AMB is found primarily in the liver and the spleen and to a lesser extent in the kidneys and lung. Tissue reservoirs elute the drug back to blood as its plasma levels fall. The concentration of AMB in peritoneal, pleural, and synovial fluids is usually less than half that in serum.

AMB follows a biphasic pattern of elimination from serum; an initial half life $t_{1/2}$ of 24–48 h followed by a prolonged elimination $t_{1/2}$ of up to 15 days owing to slow release from tissues. Detectable levels of AMB can stay in bile for up to 12 days and in urine for up to 35 days following administration. AMB can remain in liver and kidneys for as long as 12 months following cessation of therapy. At therapeutic doses, 2–5 % of AMB is excreted in urine and bile. Elimination rates are unchanged in anephric patients or those on haemodialysis; in hyperlipidaemic patients on haemodialysis, however, drug concentration may drop due to AMB–lipoprotein complex binding to dialysis membranes. In children, the volume of distribution is smaller (<4 L/kg) than in adults, rate of clearance is larger (>0.026 L/(h kg)) and PSC is approximately half that reached in adults receiving an equivalent dose.

The pharmacokinetic profile of liposomal AMB significantly differs from that of the conventional micellar form. Liposomal AMB is selectively taken up by the reticulo-endothelial system to be concentrated in the liver, spleen and lungs. Lipid-rich particles are also ingested by monocytes, which helps to channel the drug to sites of infection. Lipid formulations of AMB have significantly less nephrotoxicity than deoxycholate AMB even at higher doses. The pharmacokinetic profile of SPK-843 seems to be compatible with the drug being used in clinical settings.

11.2.1.5 Administration and Dosage

The deoxycholate AMB comes as a lyophilized powder containing 50 mg AMB and 41 mg deoxycholate and an infusion solution of 5 % dextrose in water. An infusion of ~1 mg is often used to test for immediate reactions (fever, hypotension, chills and dyspnoea), especially in patients with cardiac or pulmonary problems. Depending on illness severity, two to three escalating doses can be given every 8 h or once daily until a full therapeutic dose is reached. Doses can then be given once daily or once every other day. For most indications, the therapeutic dose is 0.4–0.6 mg/kg daily or 1 mg/kg every other day. Tolerance is assessed by starting the treatment with a low dose (0.05–0.1 mg), which is then gradually increased as tolerated. The addition of 10 to 15 mg hydrocortisone hemisuccinate to the infusion can decrease some of the acute reactions. Injectable formulations of nystatin have been discontinued owing to high toxicity; however, the drug is safe to use in oral (0.1–1.0 million *units*/dose) and topical forms, as absorption through gut and skin mucocutaneous membranes is minimal.

11.2.1.6 Adverse Effects

Nephrotoxicity is the most significant adverse effect of AMB treatment. Manifestations may include decreased glomerular filtration rate, decreased renal blood flow, casts in urine, hypokalaemia, hypomagnesaemia, renal tubular acidosis, and nephrocalcinosis. The drug can cause changes in tubular cell permeability to ions; tubuloglomerular feedback (increased delivery and reabsorption of chloride ions in the distal tubule) may also reduce glomerular filtration rates. Tubuloglomerular feedback is amplified by sodium deprivation and is suppressed by previous sodium loading. This may explain renal arteriolar spasm and calcium deposition during periods of ischaemia and direct tubular or renal cellular toxicity following intake of AMB. Sodium loading is effective in attenuating nephrotoxicity, especially reduced glomerular filtration rates. Cancer patients concomitantly receiving AMB with high sodium content antibiotics (e.g. carbenacilin) experience less severe nephrotoxicity than those receiving AMB alone or in combination with low-sodium-content antibiotics. However, sodium loading requires close monitoring to avoid hypernatraemia, hyperchloraemia, metabolic acidosis and pulmonary oedema. Prostaglandin and tumour necrosis factor (TNF-α) may play a role in AMB-induced azotaemia. Renal azotaemia is reversible following cessation of therapy; however, return to pretreatment levels could take months. Anaemia (normocytic normochromic) commonly develops due to suppressed erythropoietin synthesis and is exacerbated by reduced renal function and red blood cell production. Anaemia is reversible within several months after treatment cessation. Renal toxicity caused by aminoglycosides and cyclosporin is often worsened by AMB therapy.

AMB-induced hypokalaemia may result from increased renal tubular cell membrane permeability to potassium or from enhanced excretion via activation of Na^+/K^+ exchange. It requires parenteral administration of 5–15 mmol/L supplemental potassium per hour. Magnesium wasting may also be associated with AMB therapy, but it is more profound in cancer patients with divalent cation losing nephropathy associated with antineoplastic (cisplatin) therapy. Fever, chills and rigors are frequently associated with AMB therapy and are thought to be mediated by TNF and interleukin-l (IL-1) produced by AMB-induced peripheral monocytes. Acute reactions to AMB can be minimized with corticosteroids, paracetamol (acetaminophen), aspirin or pethidine (meperidine). Aspirin should be avoided in AMB infusion-associated thrombocytopenia. Slow infusion of the drug, rotation of infusion site, small dose of heparin in the infusion, hot packs, inline filters, avoidance of AMB concentrations >0.1 g/L or infusion of AMB through a central venous line can help to minimize thrombocytopenia. Nausea, vomiting, anorexia, headaches, myalgia and arthralgia are infusion-related reactions that associate with the initiation of AMB therapy. Clinical trials with SPK-843 indicate that it results in lower renal toxicity than AMB or liposomal AMB.

11.2.1.7 Resistance to Polyenes

Resistance to polyenes remains insignificant despite decades of clinical use. Decreased binding capacity of polyenes to mutant strains of *C. albicans* occurs due to one of several mechanisms: (i) decreased total ergosterol content without concomitant changes in sterol composition; (ii) presence of low-affinity polyene-binding sterols (3-hydoxysterol or 3-oxosterol) instead of ergosterol; (iii) reorientation or masking of existing ergosterol so that binding with polyene is thermodynamically less favoured. The majority of polyene-resistant *Candida* isolates belong to species other than *C. albicans* (*Candida tropicalis*, *Candida lusitaniae*, *Candida glabrata* and *Candida guilliermondii*) and tend to have low membrane ergosterol content. *Aspergillus fennellial* mutants resistant to polyenes have sterols other than ergosterol. Nystatin-resistant strains of *Saccharomyces cerevisiae* have 5,6-dihydroergosterol instead of ergosterol as the main membrane sterol.

11.2.2 Azoles

Inhibition of fungal growth by azoles was first described in the 1940s and the fungicidal properties of *N*-substituted imidazoles were described in the 1960s. Since then, several azoles have been developed to treat various forms of mycosis. More than 40 β-substituted 1-phenethylimidazoles have potent antifungal activity. Imidazoles in clinical use include clotrimazole, miconazole, econazole and ketoconazole. Triazole drugs include itraconazole, fluconazole, voriconazole, lanconazole, ravuconazole, posaconazole and isavuconazole.

11.2.2.1 *General Properties of Azoles*

Imidazoles are five-membered ring structures that contain two nitrogen atoms ($C_3H_4N_2$) with a complex side chain attached to one of the nitrogen atoms. Triazoles have a similar structure, but they contain three nitrogen atoms in the ring ($C_2H_3N_3$). The ring structure in both imidazoles and triazoles contains a short aliphatic chain in which the second carbon is linked to a halogenated phenyl group (Figure 11.3). The molecular weight of azoles is 279–700 kDa, depending on the specific compound. The unsubstituted imidazole and the N-C covalent linkage between the imidazole and the rest of the molecule are two features required for antifungal activity. Imidazoles come as white crystalline or microcrystalline powders; they are soluble in most organic solvents; clotrimazole and itraconazole are insoluble in water.

11.2.2.2 *Mechanism of Action*

Antifungal activity of azoles is mediated through inhibition of cytochrome P450 (CYP)-dependent membrane ergosterol synthesis. CYPs are integral components of the smooth endoplasmic reticulum and the inner mitochondrial membrane of eukaryotic cells. An iron protoporphyrin moiety located at the active site plays a key role in metabolic and detoxification reactions. The principle molecular target of azoles is CYP-Erg 11P, which catalyses the oxidative removal of the 14 α-methyl group in lanosterol and/or eburicol. The drug binds to iron in protoporphyrin via a nitrogen atom in the imidazole or triazole ring. Inhibition of

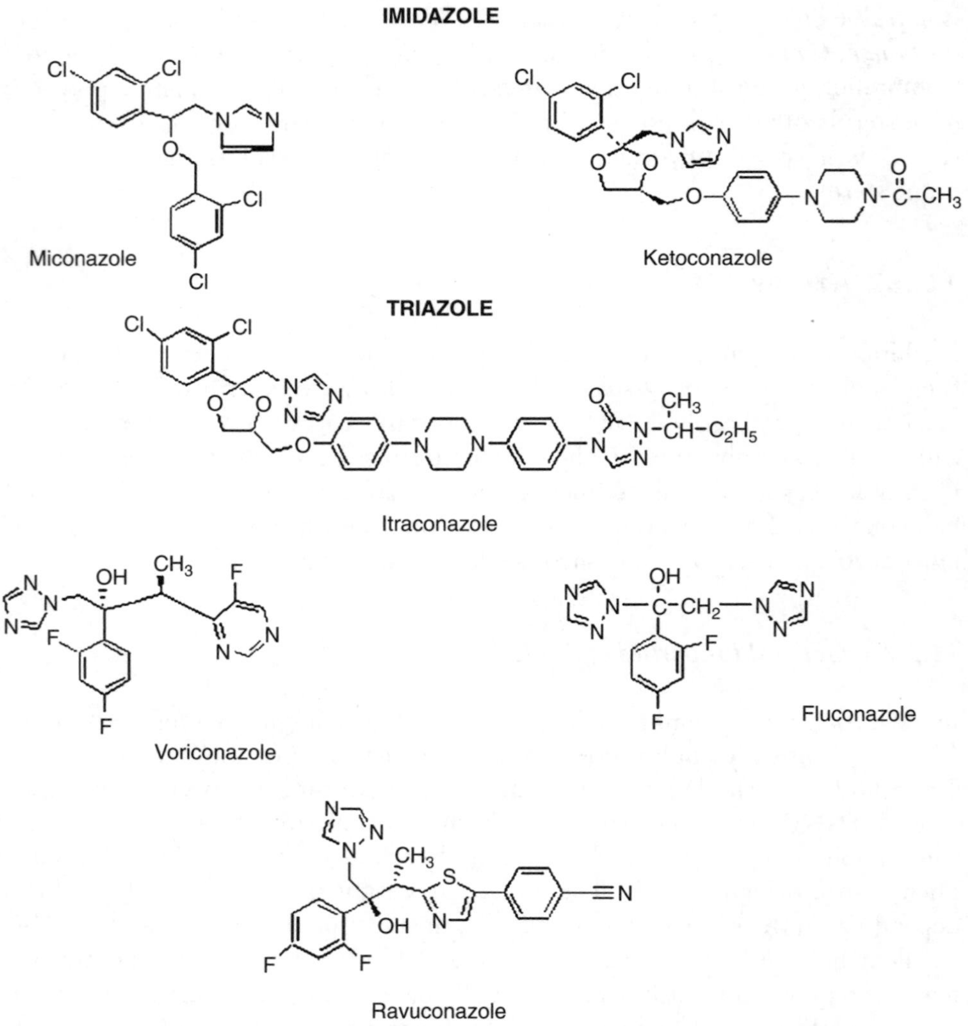

IMIDAZOLE

Miconazole

Ketoconazole

TRIAZOLE

Itraconazole

Voriconazole

Fluconazole

Ravuconazole

Figure 11.3 Chemical structure of imidazole and triazole antifungal agents.

14 α-demethylase leads to depletion of ergosterol and accumulation of sterol precursors (Figure 11.4). Depletion and replacement of ergosterol with unusual sterols alters membrane permeability and fluidity. The anti-dermatophyte abafungin (abasol; $C_{21}H_{22}N_4OS$) can also disrupt ergosterol synthesis in much the same way as azoles. Fluconazole and itraconazole can affect the reduction of obtusifolione to obtusifoliol, which results in the accumulation of methylated sterol precursors. Mammalian cholesterol synthesis is also blocked by azoles at the 14 α-demethylation stage (Figure 11.5). However, the dose required to reach inhibition in mammalian cells is much higher than that in fungi; the IC_{50} of voriconazole against CYP-dependent 14 α-demethylase in rat liver cells is 7.4 μM compared with 0.03 μM in fungi.

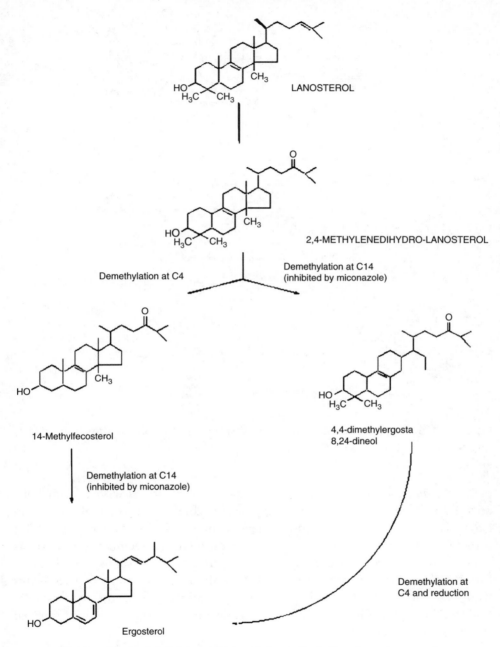

LANOSTEROL

2,4-METHYLENEDIHYDRO-LANOSTEROL

Demethylation at C4

Demethylation at C14
(inhibited by miconazole)

14-Methylfecosterol

4,4-dimethylergosta
8,24-dineol

Demethylation at C14
(inhibited by miconazole)

Demethylation at
C4 and reduction

HO
Ergosterol

Figure 11.4 Inhibition of sterol demethylation by miconazole.

Miconazole and ketoconazole can inhibit the membrane ATPase in *C. albicans* and other yeasts; hence, the rapid collapse of the electrochemical gradient and drop of intracellular ATP levels. At growth inhibitory concentrations (ICs), miconazole and ketoconazole can block the activity of *C. albicans* membrane glucan synthase, chitin synthase, adenylcyclase and 5-nucleotidase. Incubation

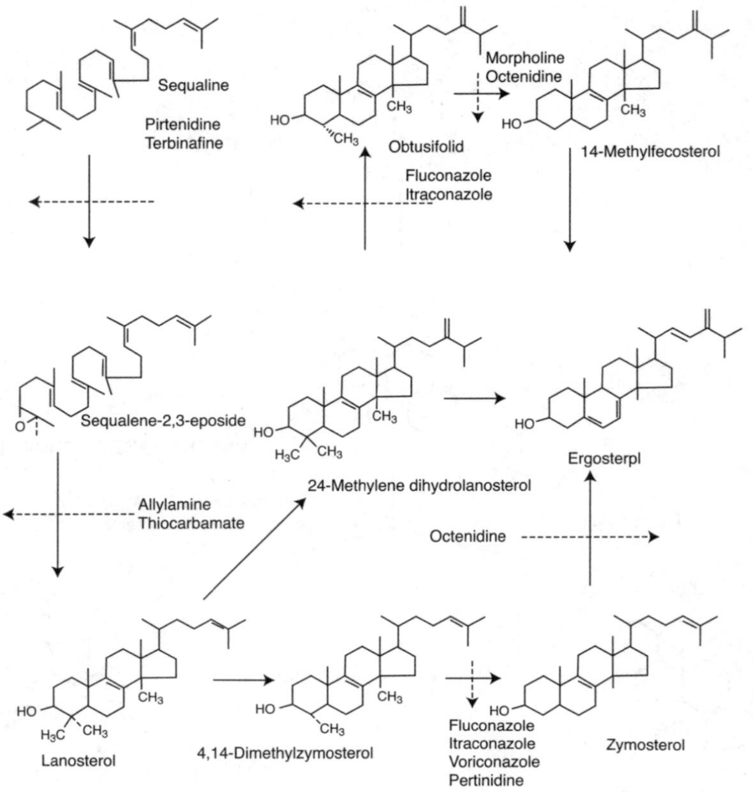

Figure 11.5 Ergosterol biosynthetic pathways and inhibition sites of clinically important antifungal agents.

of *C. albicans* and other yeasts at fungistatic concentrations with clotrimazole, miconazole, econazole, voriconazole, posaconazole or ketoconazole results in extensive changes in plasma and nuclear membranes. Azoles can also inhibit cytochrome C oxidative/peroxidative enzymes, alter cell membrane fatty acids (leakage of proteins and amino acids), inhibit catalase systems and decrease fungal adherence. Ketoconazole can disrupt the transformation of *C. albicans* from budding to pseudomycelial form, the prevailing type in infected hosts. At fungicidal concentrations, azole treatment disintegrates mitochondrial internal structures and nuclear membranes.

11.2.2.3 Spectrum of Activity

Azoles exhibit significant activity against candidiasis, cryptococcosis, coccidiodomycosis, blastomycosis and histoplasmosis. In experimental animals with aspergillosis or sporotrichosis, a partial therapeutic response can be achieved with itraconazole. Voriconazole and fluconazole synergize with neutrophils and monocytes to enhance killing of *Aspergillus fumigatus* and *C. albicans*. Azoles are also active against fusariosis, *Scedosporium prolificans* infections and clinical isolates of filamentous species. The new triazole isavuconazole (BAL4815) is active against several species of *Candida*, *Cryptococcus*, *Aspergillus*, *Absidia*, *Rhizopus* and *Rhizomucor*, and dimorphic fungi. Its activity against *Candida* spp. is superior to that of itraconazole or voriconazole. Dermatophytosis is reasonably responsive to some benzimidazole derivatives (e.g. chlormidazole).

11.2.2.4 *Pharmacokinetics*

Azoles in use differ substantially in their pharmacokinetic properties. Clotrimazole, one of the earliest azoles to be considered for systemic use, is unstable to sustain adequate blood concentrations. Miconazole, the first azole to be approved for treatment of systemic fungal infections, is administered IV to establish adequate blood concentrations. This leads to local and systemic toxicity and restricted ability to reach the cerebrospinal fluid (CSF), urine and joints.

Ketoconazole produces adequate serum concentration following oral administration, but levels of absorption are dependent on gastric acidity; antacids or H_2-blockers greatly reduce absorption. Administration of 200 mg ketoconazole in combination with 400 mg cimetidine in an acid solution increases PSC from baseline levels (1.3 µg/mL) to 5.6 µg/mL. Following the administration of 200 mg, mean PSC in young adults is 3.9 µg/mL, compared with 2.6 µg/mL in older patients. In bone marrow transplant recipients on prophylactic ketoconazole therapy, mean serum concentration drops to <0.5 µg/mL after 3 weeks of daily intake. Absorption in patients on peritoneal dialysis is greatly reduced and distribution into peritoneal fluid is negligible. About 85 % of ketoconazole binds plasma proteins and small percentages associate with haemoglobulin. More than 70 % of the normal dose is excreted within 4 days; ~80 % is excreted in faeces and 20 % in urine.

Fluconazole is available in oral and IV forms. It is effective in children, but appropriate milligram/kilogram dosage adjustments are required depending on the patient's clinical status. The drug is rapidly and completely absorbed; serum concentrations using the oral route are comparable to those of the IV route. Steady-state serum concentration is attained within 5–10 days, but an initial loading dose is recommended (twice the daily dose). Fluconazole is evenly distributed in body tissues, crosses the blood–brain barrier and penetrates into the vitreous and aqueous humours of the eye. It is minimally metabolized in the liver and excreted largely unchanged in the urine. Terminal elimination $t_{1/2}$ is ~60 h if glomerular filtration rates are 20–70 mL/min. Three hours of haemodialysis generally reduces drug serum concentration by 50 %, but peritoneal dialysis clears the drug.

Itraconazole is highly lipophilic and almost insoluble in water or diluted acids. It is only ionized at low pH (gastric juice); high concentrations can be reached in polar organic solvents, cyclodextrins and acidified polyethylene glycol. Average bioavailability after a single oral dose is ~55 %, depending on whether the drug is administered in capsule or solution form. Absorption is dependent on stomach acidity; it is enhanced by the presence of food in the stomach, but reduced in the presence of antacids. Serum elimination $t_{1/2}$ is dose dependent; it is about 15–25 h after the first dose, which increases to 34–42 h once a steady state is reached (2 weeks). Acute leukaemia and AIDS patients show reduced drug absorption. Itraconazole plasma concentration is low, but tissue concentration is two to three times greater, and adipose tissue concentration may be up to 20 times that of plasma levels. Drug concentration in the eye fluid, CSF and saliva is negligible, but it significantly increases when sputum, bronchial exudates and pus are present. The drug persists in tissues for long periods; it is highly metabolized in the liver, with 54 % of it excreted in faeces and 35 % in urine. The pharmacokinetics of itraconazole are not affected by renal impairment and the drug is not removed by haemodialysis. Drug metabolism is somewhat reduced in patients with hepatic impairment.

11.2.2.5 Adverse Effects

Gastrointestinal, hepatic, metabolic, haematological and endocrinological toxicities may associate with azole use (Table 11.2). Carcinogenic side effects also occur due to drug–drug interactions. At conventional doses, azoles are well tolerated even if administered for prolonged periods of time. However, nausea and vomiting may occur in up to 10 % of patients receiving ketoconazole at conventional doses. At higher doses ($\geq$1600 mg/dL), however, such effects occur in >50 % of patients. Other adverse effects include headache, fever, fatigue, abdominal pain, diarrhoea, nonfatal uriticaria, exfoliative dermatitis and anaphylaxis. Hypersensitivity reactions like pruritis, rash and eosinophilia are rare.

Transient nonfatal elevations of hepatic transaminase and alkaline phosphate enzymes during the first 2 weeks occasionally occur in ~10 % of patients receiving ketoconazole. Transient abnormalities of liver function occur in ~3 % of patients receiving fluconazole. Endocrinological toxicity is associated with ketoconazole in a dose- and time-dependent manner. Serum testosterone levels are reduced and may cause decreased libido and potency.

Decreased serum cortisol in response to adrenocorticotropic hormone stimulation occurs in patients receiving ketoconazole at doses >400 mg/day. At doses >600 mg/day, itraconazole causes endocrinological toxicity due to accumulation of steroid precursors with aldosterone-like effects. These adverse effects are manifested as hypokalaemia and hypertension. At doses >1200 mg/day, ketoconazole reduces mean serum LDL cholesterol levels by 27 % with no effect on triglycerides; lanosterol concentration increases by ~45 %. Miconazole causes normocytic or microcytic anaemia in 45 % of patients and thrombocytosis in

Table 11.2 Tolerability and drug interactions of membrane targeting antifungals.

Drug	Adverse effects	Drug–drug interactions
AMB deoxycholate	Nephrotoxicity, fever, chills, phlebitis, hypokalem, anaemia, GI disturbance	Azotaemia with aminoglycosides, cyclosporin, pulmonary toxicity with granulocyte transfusion
Lipid AMB formulations	Reduced azotaemia	
5-FC	Bone marrow suppression, hepatotoxicity, GI disturbance	Minimal
Imidzoles Miconazole	Headache, pruritus, thrombophlebitis, hepatotoxicity, autoinduction of hepatic degrading enzymes	
Ketoconazole	GI disturbance, hepatotoxicity	Drugs that induce hepatic microsomal enzymes, e.g. rifampicin, cyclosporine, antacids, H_2-receptor blockers
Itraconazole	GI disturbance, rare hepatotoxicity	Rifampicin, phenytoin, cyclosporine
Fluconazole	GI disturbance, rare hepatotoxicity, rare Stevens–Johnson syndrome	Phenytoin, warfarin, cyclosporine
Saperconazole	GI disturbance, rare hepatotoxicity	As itraconazole

30 % of patients; thrombocytopenia due to fluconazole therapy is rare. A daily dose of 100 mg ketoconazole increases serum concentration of sulfonylurea; therefore, close monitoring of serum concentration is indicated.

11.2.2.6 Resistance to Azoles

In clinical practice, C. albicans resistance to azoles appears only with protracted exposure. Resistance may involve reduced permeability to azoles and possible alterations in the cell membrane rather than perturbations in cytoplasmic enzymes (Table 11.3). Pathogenic strains resistant to both polyenes and azoles have higher lipid content and lower polar/neutral membrane lipid ratios than with doubly

Table 11.3 Biochemical basis of azole resistance (Ghannoum and Rice, 1999).

Mechanism	Cause	Comments
Alteration in drug target (14α-demethylase)	Mutations which alter drug binding but not binding of the endogenous substrate	Target is active (i.e. can catalyse demethylation) but has a reduced affinity towards azoles
Alteration in sterol biosynthesis	Lesions in the $\Delta^{5(6)}$ – desaturase	Results in accumulation of 14α-methyl fecosterol instead of ergosterol
Reduction in the intercellular concentration of target enzyme	Change in membrane lipid and sterols, overexpression of specific drug efflux pumps (CDR1, PDR5 and BEN^r)	Poor penetration across the fungal membrane, active drug efflux
Overexpression of antifungal drug target	Increased copy number of the target enzyme	Results in increased ergosterol synthesis, contributes to cross-resistance between fluconazole and itraconazole

Reproduced with permission, Ghannoum, M.A. and Rice, L.B., Clinical Microbiol. Rev., 12(4), 501–517, 1999, © 1999 American Society for Microbiology.

sensitive or only azole-resistant strains. In other words, altered membrane sterols that result in reduced binding to polyenes and reduced permeability to azoles may provide a common basis for double resistance.

11.2.3 Allylamines and Thiocarbamates

Drugs available for clinical use in this synthetic class are the allylamines naftifine ($C_{21}H_{21}N$) and terbinafine ($C_{21}H_{25}N$) and the thiocarbamate tolnaftate ($C_{19}H_{17}NOS$). Naftifine is applied topically, while terbinafine is taken orally. The general chemical structure consists of a naphthalene ring substituted at one position with an aliphatic chain (Figure 11.6). All three agents function as non-competitive inhibitors of squalene epoxidase (monooxygenase), an enzyme that uses NADPH to oxidize squalene to squalene epoxide as a preliminary step in membrane ergosterol synthesis. Butenafine hydrochloride ($C_{23}H_{27}N$), a synthetic benzylamine derivative that is structurally related to synthetic allylamines, can disrupt ergosterol synthesis by inhibiting squalene epoxidase as well.

Cell death is dependent on the accumulation of squalene, where high levels increase membrane permeability and disrupt cellular organization. Terbinafine inhibits the growth of dermatophytes *in vitro* at concentrations ≤ 0.01 µg/mL. Naftifine hydrochloride cream (1 %) is active against tinea cruris and tinea

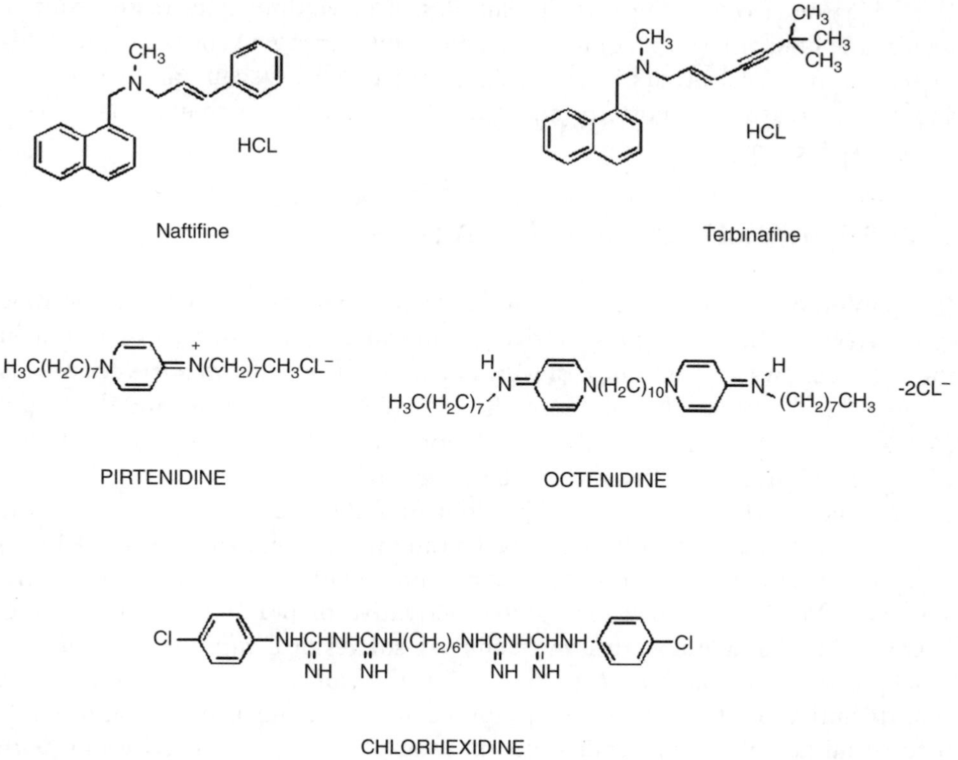

Figure 11.6 Chemical structures of allylamines (naftifine and terbinafine), pirtenidine, octenidine and chlorhexidine.

corporis, while 1 % butenafine cream is active against *Microsporum canis*, *Microsporum gypseum*, *Trichophyton mentagrophytes*, *Trichophyton rubrum* and *Epidermophyton floccosum*. Butenafine has superior fungicidal activity against many fungi compared with that of terbinafine, naftifine, clotrimazole or tolnaftate. It also displays superior activity against *C. albicans* and demonstrates low minimum inhibitory concentrations (MICs) for *C. neoformans* and *Aspergillus* spp. Although allylamine-resistant strains of *S. cerevisiae* and *Ustilago maydis* (a plant pathogen) are known, resistance to allylamines and/or thiocarbamates amongst medically important fungi is rare.

11.2.4 Octenidine and Pirtenidine

Octenidine ($C_{36}H_{64}Cl_2N_4$) and pirtenidine ($C_{21}H_{38}N_2$), which are structurally similar to chlorhexidine, were first developed as mouthwashes (Figure 11.6). They are both alkylpyridinylidine–octanamine derivatives and cause extensive leakage of cytoplasmic contents from *C. albicans* and *S. cerevisiae*. Such changes correlate with gross morphological and ultrastructural changes in treated cells. Treatment of *C. albicans* with sub-ICs of octenidine or pirtenidine alters membrane lipid and sterol contents. A significant increase in squalene and 4,14-dimethylzymosterol occurs in pirtenidine-treated cells. Octenidine-treated cells show an increase in zymosterol and obtusifoliol content, a change that affects ergosterol biosynthesis.

11.2.5 Morpholines and Other Agents

Morpholines are synthetic phenylmorpholine derivatives with amorolfine ($C_{21}H_{35}N$) as the sole representative in clinical use. The drug, discovered in the 1980s, shows a broad range of MICs for *Candida* isolates *in vitro*. It disrupts the ergosterol pathway by inhibiting the Erg24P (Δ14-reductase) and the Erg2P (Δ8-Δ7 isomerase enzyme) reactions. Amorolfine is used for topical treatment of superficial mycoses; resistance to the drug is rare.

Pradimicin A (BMY-28567) and pradimicin B (BMY-28864), described in the late 1980s, and the recently described pradimicin B analogue (BMS-181184) represent a new class of antifungal compounds with broad spectrum *in vitro* activity. BMS-181184, a water-soluble derivative of pradimicin, has an MIC between 1 and 8 µg/mL against most clinical isolates of *C. albicans*, *C. glabrata*, *C. tropicalis*, *Candida krusei*, *C. lusitaniae*, *C. neoformans* and *A. fumigatus*. The definitive mode of action of pradimicin remains unclear, but appears to involve initial calcium-dependent complexing of the free carboxyl group with the saccharide portion of cell-surface mannoproteins. This could presumably perturb the cell membrane and cause leakage of intracellular K^+ and cell death.

Quinoline nitroxoline ($C_9H_6N_2O_3$) (Figure 11.7), a urinary antiseptic, has significant activity against pathogenic *Candida* spp., with MIC ranging between 0.25 and 2 µg/mL depending on the pathogenic strain. A-39806 is another non-azole compound (Figure 11.7) with significant activity against several *Candida* spp., *Cortinarius albidus* and *Aspergillus niger*. These two compounds increase the 4,14-dimethyl sterols and concurrently decrease desmethyl sterols indicative of lanosterol 14-α-demethylase inhibition.

Ciclopirox olamine, ciclopirox and rilopirox (Figure 11.7) are hydroxypyridones with *in vitro* activity against several medically important dermatophytes, yeasts and moulds. Hydroxypyridones do not affect fungal sterol synthesis, but they inhibit cellular uptake of essential compounds and alter membrane permeability at high concentrations. Cilofungin ($C_{49}H_{71}N_7O_{17}$), the first clinically applied echinocandin drug, is an *Aspergillus*-derived antifungal that disrupt fungal cell membrane (and possibly cell wall) synthesis by specifically inhibiting the conversion of lanosterol to ergosterol.

11.3 Drugs Targeting the Cell Wall

The cell wall is the interface between cells and their environment; it protects against osmotic pressure and controls passage of molecules in and out of cells.

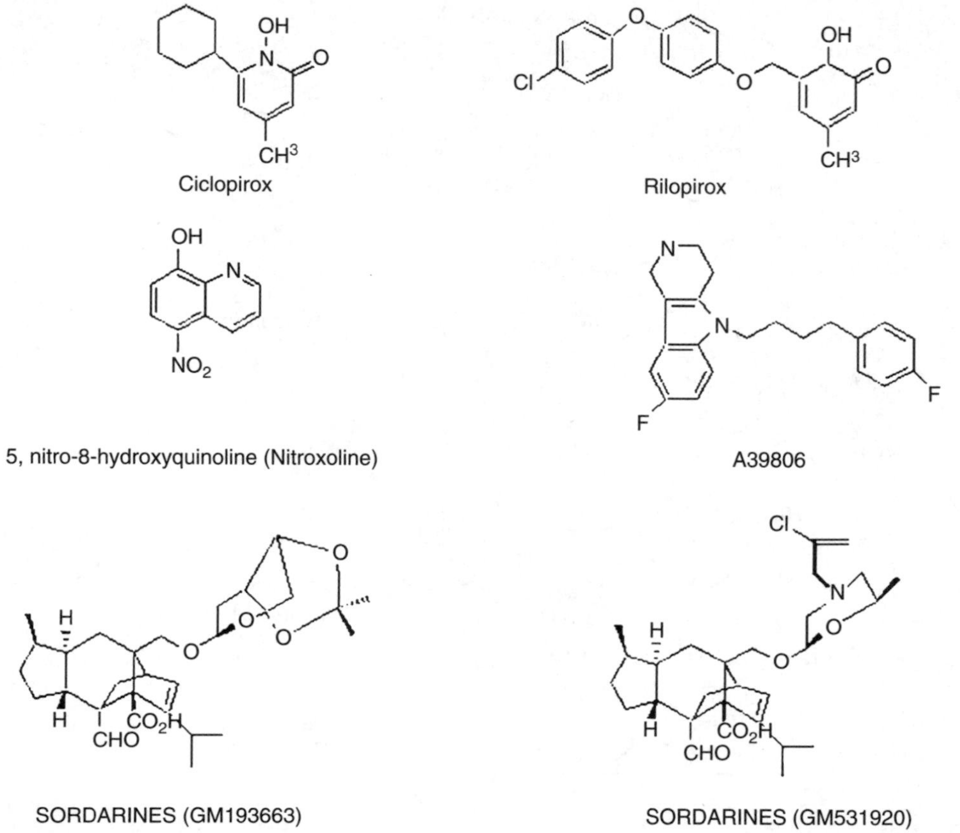

Ciclopirox

Rilopirox

5, nitro-8-hydroxyquinoline (Nitroxoline)

A39806

SORDARINES (GM193663)

SORDARINES (GM531920)

Figure 11.7 Chemical structure of two sordarins (GM 193663 and GM 531920) in addition to some promising antifungal compounds under development.

Hence, disruption of cell-wall integrity and/or synthesis is detrimental to cell viability and survival. Compounds that affect fungal cell-wall glucans are the candins (echinocandin and aminocandin) and those that target cell-wall chitins are nikkomycin and aureobasidins.

11.3.1 Echinocandins

Echinocandins in clinical use, or pneumocandins, are semisynthetic lipopeptides consisting of a cyclic hexapeptide core and a fatty acid side chain responsible for antifungal activity. Caspofungin, anidulafungin and micafungin are FDA-approved echinocandins indicated in the treatment of candidiasis and some other forms of mycosis (Figure 11.8). They are specific noncompetitive inhibitors of β-(1,3)-glucan synthase, a 210 kDa integral membrane heterodimeric protein, hence the name penicillin antifungals. Although Fks1p is the component to which echinocandins bind, it may not necessarily be the catalytic subunit that mediates the noncompetitive inhibition of glucan synthesis. Glucan synthesis

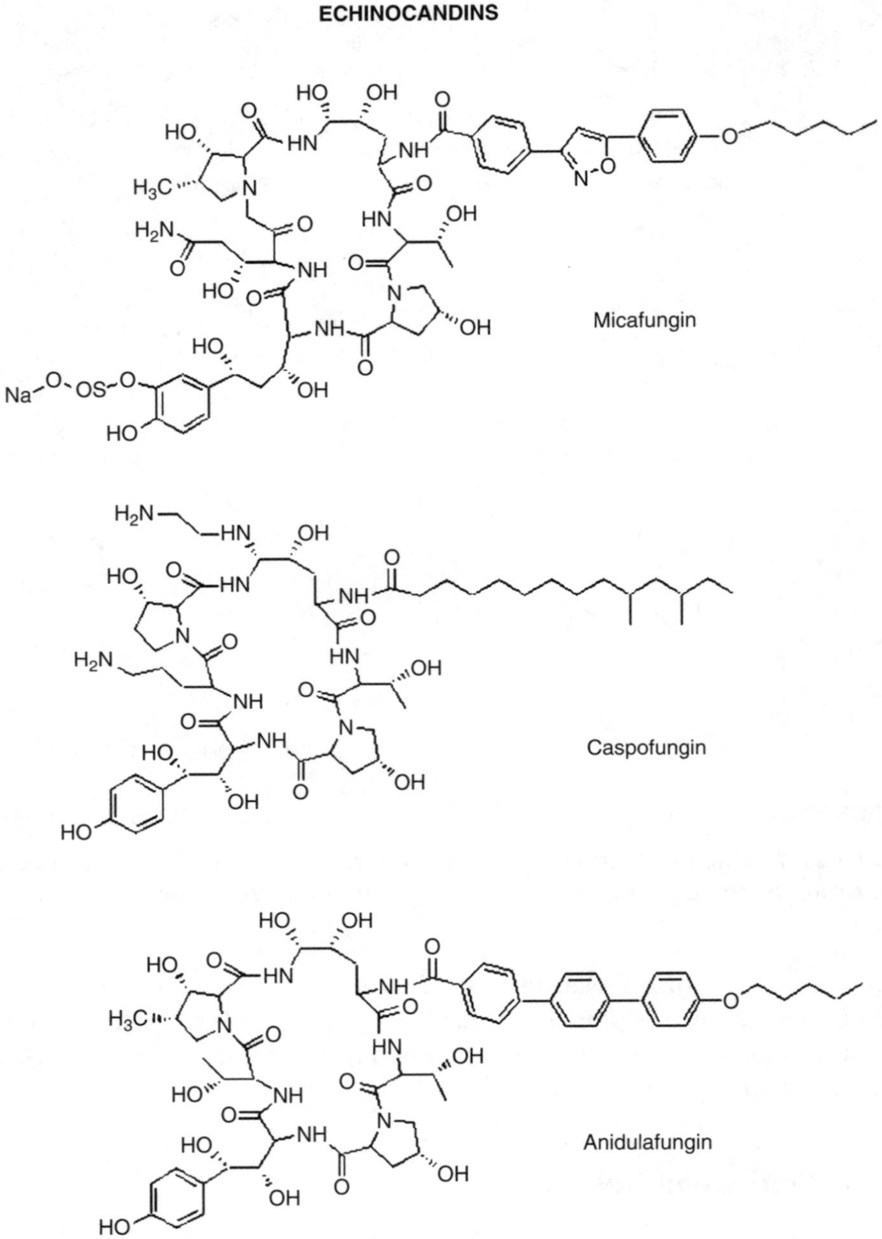

ECHINOCANDINS

Figure 11.8 Chemical structure of three echinocandins.

inhibitors have secondary effects in the form of reduced ergosterol and lanosterol content and increased cell-wall chitin content. Cytological and ultrastructural changes characterized by growth of target fungi as pseudohyphae, thickened cell wall and buds failing to separate from mother cells associate with echinocandin use. Cells also become osmotically sensitive, with lysis being restricted largely to growing tips of budding cells. Echinocandins have low oral bioavailability; therefore, they are only available as IV infusions. Nonetheless, owing to their broad spectrum of activity, minimal toxicity and negligible resistance, they are becoming the drugs of choice to treat various forms of mycosis. They are also becoming a first-line treatment for *Candida* infections before species identification and for anti-*Candida* prophylaxis in haematopoietic stem cell transplantation (HSCT) recipients.

In addition to echinocandins, genetic analysis of resistant *S. cerevisiae* mutants to the pyridobenzimidazole derivative D75-4590 suggests that this agent can specifically inhibit β-1,6 glucan synthesis (KRE6p). Interestingly, as a KRE6p homologue is present in *A. fumigatus*, partial silencing of its gene may render *A. fumigatus* susceptible to the activity of D 75-4590. D75-4590 can also inhibit hyphal elongation of *C. albicans* in a dose-dependent manner.

11.3.1.1 Caspofungin

Caspofungin (Cancidas; $C_{57}H_{88}N_{10}O_{15}$) was approved for use in 2001 for the treatment of fungal infections in febrile neutropenic adults, patients with refractory invasive aspergillosis and those with invasive aspergillosis intolerant to AMB and/or itraconazole therapy. It is also indicated for the treatment of candidemia and *Candida*-related infections (intra-abdominal abscesses, pertussis, pleural cavity infections and esophagitis). Mean duration of treatment is slightly greater than 1 month, although the range extends between 1 day and >4 months. In general terms, duration of treatment is dictated by disease severity, response to treatment and immunostatus of the patient.

So far, caspofungin seems to be associated with minimal side effects compared with AMB. Rare cases of symptomatic liver damage, peripheral oedema, swelling and hypercalcaemia have also been noted. Hypersensitivity and histamine release-related reactions, such as rashes, facial swelling, pruritus and sensation of warmth, have been reported and should be closely monitored. Embroyotoxic effects have been reported following treatment of pregnant animals with caspofungin; therefore, pregnant women should be given the drug only if benefits to them outweigh potential risks to the unborn. Adverse interactions with cyclosporin and tacrolimus are possible and deserve caution. Use of caspofungin in combination with cyclosporin may increase liver enzyme production to abnormal levels. Hence, dosage adjustment in patients with moderately impaired liver function is recommended. Although resistance in *C. albicans* has been described, it remains rare.

11.3.1.2 Anidulafungin

Formerly known as LY303366, anidulafungin (Eraxis; $C_{58}H_{73}N_7O_{17}$) is an echinocandin that received FDA approval as an antifungal in 2006. The drug has significant activity against invasive and oesophageal candidiasis and other infections caused by *Candida* and *Aspergillus* spp. The drug significantly differs from other antifungals, as it undergoes chemical degradation to inactive forms at body pH and temperature. Because it does not rely on enzymatic degradation and hepatic or renal excretion, it is safe to use in patients with hepatic or renal impairments. Levels of distribution are in the range of 30–50 L and half life is ~27 h. Close to 30 % of the drug is excreted in faeces and <1 % is excreted in urine.

11.3.1.3 Micafungin

Micafungin (Mycamine; $C_{56}H_{71}N_9O_{23}S$), a natural antifungal produced by the fungus *Coleophoma empetri*, received FDA and European Medicines Agency (EMEA) approval in 2005 and 2008 respectively. It is indicated in the treatment of candidemia and acute disseminated candidiasis as well as *Candida*-related pertussis, abscesses and oesophagitis. Recently, micafungin has received approval for use as prophylaxis in HSCT recipients at risk of developing candidiasis. The fungicidal effects of micafungin relate to its concentration-dependent inhibition of 1,3-β-D-glucan synthase, which significantly reduces 1,3-β-D-glucan formation and leads to osmotic instability and cell lysis. The drug is administered IV and dosage per day varies depending on indication: 100 mg for candidemia, 150 mg for oesophageal candidiasis and 50 mg for prophylaxis in HSCT recipients. It is metabolized in the liver as a substrate for the enzyme CYP3A4. Hypersensitivity reactions may associate with micafungin therapy.

11.3.2 Nikkomycin and Chitin Synthesis

Polyoxin ($C_{17}H_{25}N_5O_{13}$) and nikkomycin ($C_{20}H_{25}N_5O_{10}$) are streptomycete-derived nucleoside antibiotics capable of inhibiting chitin synthesis. They specifically and competitively inhibit chitin synthase by acting as decoys in place of the substrate uridine diphosphate-N-acetylglucosamine. *In vitro* testing of nikkomycins X and Z against various fungi shows moderate susceptibility of *C. albicans* and *C. neoformans*. Activity against *C. albicans* and *C. neoformans* improves significantly, however, when nikkomycin Z is used in combination with fluconazole and itraconazole. Nikkomycin is fungicidal against the dimorphic fungi *C. immitis* and *Blastomyces dermatitidis*. Aureobasidins alter chitin assembly and sphingolipid synthesis; the drug is mainly active against *Candida* spp. and *C. neoformans* but has some potency against *H. capsulatum* and *B. dermatitidis*.

11.4 Drugs Targeting Nucleic Acid and Protein Synthesis

Compounds targeting fungal protein and nucleic acid synthesis are sordarin and azosordarin, as well as 5-fluorocytosine (5-FC). 5-FC, which inhibits pyrimidine metabolism by interfering with RNA and protein synthesis, is the sole anti-mitotic antifungal in clinical use. The drug is active against *Candida*, *Aspergillus* and *Cryptococcus* spp., as well as dematiaceous fungi that cause chromomycosis (*Phialophora* and *Cladosporium*).

11.4.1 Sordarin

Sordarin inhibits protein synthesis by targeting the polypeptide elongation factor 2 (EF2) and the large ribosomal subunit during RNA translation. It disrupts the displacement of tRNA from A site to P site and blocks the movement of ribosomes along the mRNA thread. In the case of candidiasis, azasordarin targets and disrupts polypeptide chain elongation more effectively than sordarin. *C. albicans* EF2 displays more than 85 % amino acid sequence homology to the human equivalent, making the likelihood of toxicity to host cells relatively high. However, high specificity for respective fungal targets tends to minimize toxicity. Icofungin (PLD-118 or BAY-10-8888) is a synthetic derivative of the naturally occurring β-amino acid cispentacin. It targets isoleucyl-tRNA synthase and disrupts protein synthesis and cell growth. Although it has poor *in vitro* activity against *C. albicans*, it has significant activity against disseminated and invasive candidiasis in neutropenic animals.

11.4.2 5-Fluorocytosine

The oral antifungal 5-FC (Ancobon; $C_4H_4FN_3O$) is a synthetic fluorinated pyrimidine analogue. Although it lacks cytostatic and anti-neoplastic activities, for which it was first identified in the 1950s, 5-FC has noticeable antifungal activity. It is commonly used as an adjunct to polyene therapy, as AMB potentiates the uptake of 5-FC by increasing membrane permeability. The chemical formula of 5-FC is 4-amino-5-fluoro-2-pyrimidine; the compound has a molecular weight of 129 kDa (Figure 11.9). It is an odourless white crystalline substance, relatively stable at normal temperatures, and soluble in water up to 1.2 %. It tends to crystallize if kept at low temperatures and partially deaminates to 5-fluorouracil (5-FU) when stored at high temperatures or once taken by cells.

11.4.2.1 Mechanism of Action

5-FC disrupts nucleic acid and protein synthesis and alters the amino acid pool (Figure 11.9). It first enters susceptible cells by means of cytosine permease, which is usually responsible for the uptake of cytosine, adenine, guanine and hypoxanthine. Once internalized, it is converted by cytosine deaminase to 5-FU, which is then converted by uridine monophosphate pyrophosphorylase

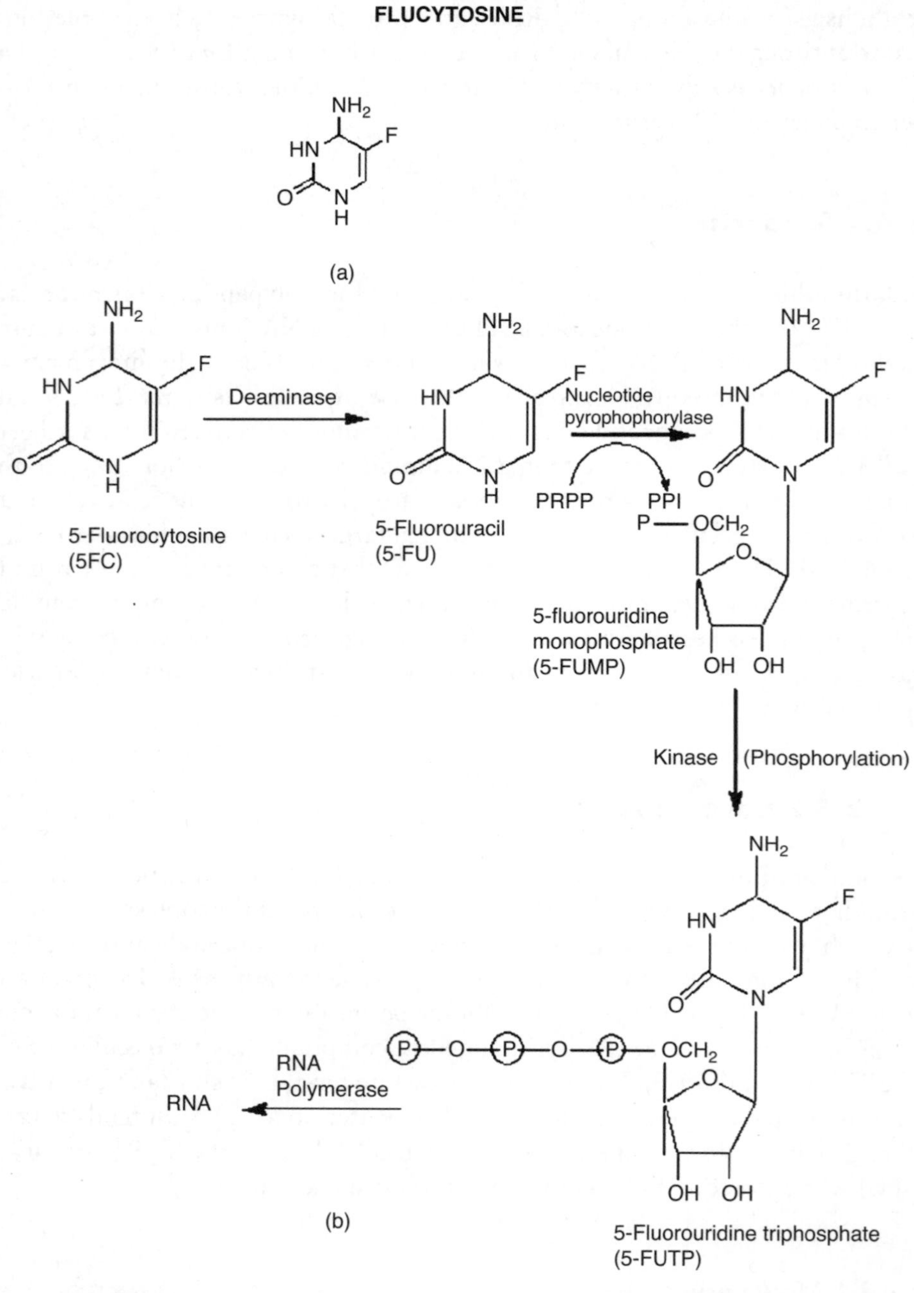

FLUCYTOSINE

(a)

(b)

Figure 11.9 Metabolic conversions and mode of action of 5-FC.

to 5-fluorouridylic acid (FUMP) that is then phosphorylated and incorporated into RNA, resulting in disruption of protein synthesis. Extensive replacement of uracil by 5-FC in fungal RNA leads to alterations in the amino acid pool. Some 5-FU can be converted to 5-fluorodeoxyuridine monophosphate, which functions as a potent inhibitor of thymidylate synthase, one of the enzymes

involved in DNA synthesis and nuclear division. Inhibition of DNA synthesis in
C. albicans can take place before 5-FU incorporation into RNA or inhibition of
protein synthesis. Resistant strains of *C. neoformans* incorporate 5-FC into RNA
at levels comparable to that in sensitive strains. This could mean that resistance
inhibition of DNA synthesis is more important than the production of aberrant
RNA in mediating the effects of 5-FC. The drug incorporates in large quantities into the 80S ribosomal subunits in *C. albicans*. The number of ribosomes
synthesized in the presence of high concentrations of 5-FC is greatly reduced.

C. *albicans* cells growing at sub-inhibitory 5-FC concentrations show increased
cell diameter due to excessive carbohydrate and protein synthesis. Additional
morphologic changes include enlarged and translucent nucleus with filamentous components, thinner cell wall and increased budding in *C. albicans* and
C. neoformans. Such changes are attributable to unbalanced growth activities
where DNA synthesis is halted while residual metabolism is retained.

11.4.2.2 Spectrum of Activity

5-FC displays significant activity against *Candida*, *Torulopsis*, *Cryptococcus* and
Geotrichum genera. It also exhibits moderate activity against *Aspergillus* spp.
and chromomycosis-causing dematiaceous fungi. However, *Coccidioides*, *Histoplasma*, dermatophytes and other medically important fungi do not respond well
to 5-FC treatment. 5-FC exerts both fungicidal and fungistatic activities against
C. albicans, *C. tropicalis*, *C. parapsilosis*, *C. glabrata* and *C. neoformans*. Fungicidal effects usually occur at relatively high concentrations and prolonged exposure. Combining 5-FC with various AMB formulations is useful in clearing the
CSF in non-HIV patients with cryptococcal meningitis, candidiasis, *Candida* en-
dophthalmitis, renal and hepatosplenic candidiasis, *Candida* thrombophlebitis of
the great veins, aspergillosis and central nervous system (CNS) phaeohyphomy-
cosis. It is also effective against large cryptococcal intra-cerebral masses, minimizing the need for surgical intervention. 5-FC activity against *C. neoformans*
and *C. albicans* is enhanced when used in combination with fluconazole.

11.4.2.3 Pharmacokinetics and Dosage

5-FC is completely absorbed from the intestines following oral intake. The drug
has excellent penetration with a volume of distribution approximating that of
total body water. The drug is available in 250 and 500 mg capsules; US manu-
facturers no longer supply it in IV form. Administration of 150 mg/(kg day) in
adults with normal renal function produces a PSC of 50–80 mg/L within 1–2 h.
CSF concentration can reach greater than 70 % that of serum concentration,
making the drug suitable to treat CNS mycoses. Initial dosage in patients with
normal renal function is 37.5 mg/kg every 6 h (150 mg/(kg day)). Elimination $t_{1/2}$
in adults with normal renal function is 3–6 h; patients on haemodialysis can

be given 37.5 mg/kg after each dialysis session as dosage adjustment. The drug accumulates in patients with azotaemia, which could be toxic unless dosage is reduced to keep blood levels within the acceptable range (50–100 μg/mL). About 80 % of 5-FC is excreted unchanged in the urine, while negligible quantities bind to serum proteins.

11.4.2.4 Adverse Effects

Gastrointestinal-tract-related symptoms like diarrhoea, nausea and vomiting are amongst the most common but least serious side effects associated with 5-FC therapy. Hepatitis can occur at concentrations >50 μg/ml or higher; it usually resolves in days to weeks. Bone marrow suppression in the form of neutropenia, thrombocytopenia and pancytopenia may develop following treatment with 5-FC. Death occurs from sepsis or intra-cerebral haemorrhage. Although human cells do not deaminate 5-FC, metabolites of 5-FU can be found in the urine and 5-FU can be found in blood at low concentrations. Colonic bacteria like *Escherichia coli* rich with cytosine deaminase could deaminate 5-FC that then gets reabsorbed. This may explain why 5-FC enteritis is largely confined to the colon. Additionally, variations in colonic mircoflora amongst patients may explain why some tolerate 5-FC blood levels of >150 μg/ml for up to 6 weeks while others show signs of toxicity much earlier.

11.4.2.5 Resistance to 5-FC

5-FC is often administered in low doses in combination with other drugs to limit secondary drug resistance, which can be profound and accompanied by clinical deterioration. Many fungi are either inherently resistant to 5-FC or develop resistance following exposure. Resistance develops by loss of cytosine permease, loss of cytosine deaminase, deficiency in uridine monophosphate pyrophosphorylase or increased *de novo* synthesis of pyrimidine via increased orotidylic acid pyrophosphorylase and orotidylic decarboxylase. Loss of feedback regulation of aspartic transcarbamylase by ATP leading to increased *de novo* synthesis of pyrimidine may also facilitate development of resistance. The frequency of *C. albicans* resistance to 5-FC at concentrations >25 μg/ml is generally greater amongst strains of serotype B than serotype A. Serotype B, though, comprising a small minority of clinical isolates, is usually responsible for the majority of primary resistance cases.

11.5 Novel Therapies

Despite the great benefits of antifungals, the incidence of opportunistic and invasive fungal infections continues to rise. Although infections caused by

C. albicans, *A. fumigatus* and *C. neoformans* remain very common, infections caused by other *Candida* and *Aspergillus* spp., opportunistic yeast-like fungi (*Trichosporon* spp., *Rhodotorula* spp., *Geotrichum capitatum*), zygomycetes, hyaline moulds (*Fusarium*, *Acremonium*, *Scedosporium*, *Paecilomyces*, *Trichoderma* spp.) and dematiaceous fungi are also on the rise. Increasing numbers of hosts with compromised immunity due to underlying diseases (cancer, AIDS and diabetes) and/or immunosuppressive therapy are partly to blame for this trend. However, toxicity, resistance, narrow spectrum of activity and inability to fully clear sites of infection, which continue to limit the clinical applicability and therapeutic efficacy of the dozen or so available antifungals, remain the chief culprits.

To overcome the limitations of antifungal chemotherapy and make it more accessible and more effective, the search for drugs with enhanced potency, broad spectrum of activity, minimal toxicity, flexible mode of administration and favourable pharmacokinetics (bioavailability and effective tissue penetration) has become a continuous work in progress. New generations of triazoles (posaconazole, ravuconazole and voriconazole), lipid formulations of AMB and echinocandins (anidulafungin, caspofungin and micafungin) go a considerable way in meeting many of these requirements. However, the emergence of resistance and the development of toxic side effects following the use of these new drugs have been reported.

New insights into fungal immunity and pathology along with several technical advances in immunology and mycology are furnishing the essential tools to search for safe and effective immune-base antifungals. This new paradigm shift has literally produced hundreds of experimental antimicrobial peptides, vaccines, monoclonal antibodies monoclonal antibody (mAbs), cell transfer procedures and cytokine regimens with potential antifungal activity (Table 11.4). Some immune-based agents and modalities are at various stages of clinical evaluation; some have even been submitted to the FDA and/or EMEA for approval.

Leukine® (sargramostim) is a granulocyte macrophage colony-stimulating factor (GM-CSF) preparation that is currently at phase III clinical trial as an antimycotic agent. A phase II clinical trial to test the safety and efficacy of granulocyte colony-stimulating factor (G-CSF)-elicited cross-matched granulocyte transfusion (GTX) therapy in neutropenic children with severe infections has been successfully concluded. Results from that study suggest that early-onset GTX therapy is effective (clears infections) and safe (minimal pulmonary reactions). A phase III clinical trial to evaluate the safety and efficacy of GTX therapy against bacterial and fungal infections during neutropenia is currently underway. Once again, high hopes for immune-based antifungal therapies have often been dashed despite early promising results. For example, a phase I clinical trial conducted several years ago on the anti-cryptococcal potential of the mAb 18B7 (specific against *C. neoformans* capsular polysaccharide) in HIV patients concluded with encouraging results. However, the agent is yet to gain approval as an agent

Table 11.4 A tentative list of promising alternative approaches to prevent and/or treat fungal infections.

	Specific examples	Antifungal activity	Mechanism(s) of action
Antimicrobial peptides	Pentraxin 3 (PTX3)	Pulmonary aspergillosis	Modulate cytokine production and complement responses
	Cathelicidin and other cationic peptides	A. fumigatus infections in animal models	Production IL-6 and TNF-α, minimize inflammation
	Kininogen-derived peptides GKH17 and HKH17	Candidiasis	Fungicidal activity
	Synthetic lactoferrin-derived peptides Lfpep and kaliocin-1	Fluconazole- and AMB-resistant strains of C. albicans	Fungicidal activity (membrane permeabilization)
	Synthetic analogues of cationic polypeptide CAP37	Fluconazole-sensitive and -resistant isolates of many Candida spp.	Fungicidal activity
	Cystatin	Azole-resistant C. albicans isolates overproducing multidrug efflux transporters Cdr1p and Cdr2p	Fungicidal activity
	Histatins 1, 3 and 5	C. albicans, C. glabrata, C. krusei, S. cerevisiae, C. neoformans and A. fumigatus	Fungistatic and fungicidal activity (target mitochondria and initiate loss of transmembrane potential)
	Trappin-2 (endogenous serine protease inhibitor)	A. fumigatus and C. albicans	Fungistatic and fungicidal activities
Monoclonal antibodies	Anti-CD40	Systemic C. neoformans infections	Production IFN-γ and TNF-α, expression of MHC-II
	Anti-β-glucan in C. albicans	Candidiasis	Inhibits synthesis of β-glucan

Table 11.4 (*Continued*)

	Specific examples	Antifungal activity	Mechanism(s) of action
	G5 (an anti-*C. albicans* IgA)	Candidiasis	Fungicidal activity
	G15 (IgM recognizing epitopes on capsular polysaccharide GXM of *C. neoformans*)	Virulent *C. neoformans* in pre-sensitized mice	Fungistatic and fungicidal activities
	A9 (IgG1 with high affinity for surface peptides	Invasive aspergillosis in mice	Inhibit hyphal development
	4F11 (an anti-*P. carinii* IgM)	Protect SCID mice against the development of pcp	Fungistatic/fungicidal activities
	Anti-gp70	*P. brasiliensis*	Abolish lung granulomas
	Anti-KT mAb KT4	KT-sensitive *C. albicans* strains	Anti-idiotype activity
	nmAb-KT (IgG1κ specific to peptide sequence 41GSTDGK46 of HM-1 KT in *Williopsis saturnus*)	Aspergillosis in neutropenic T cell-depleted bone marrow-transplanted mice	Neutralizing antibody (reduces HM-1 killing and glucan synthase activity)
Vaccines	P13 decapeptide mimotope of cryptococcal capsular polysaccharides	*C. neoformans*	Upregulate expression of IFN-γ
	HSP90-DNA vaccine	Systemic candidiasis in mice	Production of protective IgGs
	A. fumigatus-derived rAspf3	Invasive aspergillosis	T cell-dependent immunity
	N-terminus of *Candida* adhesin Als3p	Lethal candidemia and *S. aureus* infections	T cell-dependent immunity
Cytokines	IFN-γ, IL-12, anti-IL-4, G-CSF or GM-CSF	*C. neoformans, C. immitis, P. carinii* and others	Modulation of Th1-dependent immunity

(*Continued*)

Table 11.4 (*Continued*)

	Specific examples	Antifungal activity	Mechanism(s) of action
	GM-CSF and G-CSF plus conventional antifungals	Polyene- and azole-sensitive and -resistant strains of C. *albicans* and A. *fumigatus*	Modulation of Th1-dependent immunity, synergism with chemotherapy
	Human lymphocytes cultured with IL-2	C. *neoformans* *in vitro*	Conjugation and direct inhibition of fungal growth
Cell transfer therapy	Human lymphocytes cultured with encapsulated C. *neoformans*	C. *neoformans*	Fungistatic (involves NK, CD4$^+$, CD8$^+$, and CD16/56$^+$ cell activity
	Blood cells stimulated by Aspf16-pulsed DCs	*Aspergillus* spp. *in vitro*	Cytotoxic activity and IFN-γ production
	Granulocyte transfusion	In neutropenic patients	Innate immunity-dependent clearance of infecting agent
	G-CSF-activated granulocyte transfusion	Clears *Zygomycete* infections	
	CD4$^+$ or CD8$^+$ T cells from immunized animals	Immunizing pathogen in naïve mice	Modulate Th1- or Th2-dependent responses
Oligonucleotides	2′-O-methyl backbone (19-mer 2′-OMe) hairpin	C. *albicans*	Antisense oligonucleotide
	^{L}(TACCTTTC) and T^{L}CT^{L}AC^{L}GA^L CG^{L}GC^{L}C	C. *albicans*	Oligonucleotide-directed misfolding of RNA
	30 bp oligonucleotide targeting calcineurin A (CNA1)	C. *neoformans*	Antisense repression
	Olidos targeting PI3K/Akt/mTOR inflammatory pathway in lung DCs	Aspergillosis-related inflammation	Antisense small interfering RNAs (siRNA)

Table 11.4 (*Continued*)

	Specific examples	Antifungal activity	Mechanism(s) of action
Drugs and inhibitors in use for other indications	Deferasirox (iron chelator)	Mucormycosis in neutropenic mice	Fungistatic especially if combined with LAMB
	Indinavir (antiretroviral protease inhibitor)	*C. neoformans* infections	Expands splenic CD8α^+ DCs, expression of costimulatory molecules and proinflammatory cytokines
	Daucosterol (a β-sitosterol glycoside)	Disseminated candidiasis	Induction of Th2-dependent protective immunity
	Amiodarone (sodium channel blocker)	*C. albicans*, *C. neoformans*, and *S. cerevisiae* infections	Depletion of cellular Ca^{2+} stores especially if used with miconazole or fluconazole
	CPT (Camptothecin, $C_{20}H_{16}N_2O_4$)	*C. neoformans*	Topoisomerase I inhibitor
	Eflornithine ($C_6H_{12}F_2N_2O_2$)	*C. neoformans*	Inhibitor of ornithine decarboxylase
	FK506 and CsA (calcineurin inhibitors)	*C. albicans*, *C. neoformans*, *A. fumigatus*, cryptococcal meningitis	Suppress T cell-mediated immunity
Probiotic therapy	*Lactobacillus rhamnosus* GR-1 and *L. reuteri* RC-14	Mucosal and vaginal candidiasis in humans and animals	Restore and/or improve microbial balance Modulation immunity to better respond to infection
	L. casei GG	Vaginal candidiasis during human pregnancy	
	L. acidophilus (NCFM and LA-1)	Candidiasis in immunodeficient bg/bg-nu/nu mice	
	Bifidobacteria animalis	Mucosal candidiasis	

to treat cryptococcal meningitis owing to safety and efficacy concerns. A second example in this regard pertains to the anti-HSP90 mAb (Efungumab or Mycograb®), which has received a lot of attention in recent years as a candidate for the treatment of different forms of candidiasis. Although the drug entered phase III clinical trial, safety concerns prompted the FDA and EMEA to reject the drug. Novartis (the drug's developer) has recently announced that is it dropping mycograb from its list of drugs in development.

11.6 Conclusions

As discussed in the preceding sections, the complexities and intricacies involved in fungal pathogen–host interactions make the development of safe and effective antifungals, whether conventional or novel, very challenging. The rising number of hosts susceptible to invasive and opportunistic mycosis and the ever-evolving repertoire of human fungal pathogens add to the complexity and urgency of this issue. Therefore, the focus on combating fungal infections should rely on a wide array of preventive and therapeutic measures involving pathogen- as well as host-targeting agents and strategies. Short of that, the incidence of invasive and opportunistic fungal infections will continue to increase and our ability to counter them will continue to diminish precipitously.

Revision Questions

Q 11.1 Discuss the major molecular targets of antifungal chemotherapy.

Q 11.2 Discuss the mechanism of action of polyenes.

Q 11.3 Discuss the most common side effects of amphotericin B therapy, paying special attention to how polyene-induced nephrotoxicity results in anaemia.

Q 11.4 What is the general mechanism of action of azoles?

Q 11.5 What is the general mechanism of resistance to azoles?

Q 11.6 Name three specific agents that inhibit cell-wall glucan synthesis, indicating their common target enzyme.

Q 11.7 Which antifungals tend to inhibit cell-wall chitin synthesis?

Q 11.8 Describe the mechanism of action and clinical indications of caspofungin.

Q 11.9 Which antifungals target and disrupt cellular metabolism and/or mitotic activity?

Reference

Ghannoum, M.A. and Rice, L.B. (1999) Antifungal agents: mode of action, mechanisms of resistance, and correlation of these mechanisms with bacterial resistance. *Clinical Microbiology Reviews*, **12**(4), 501–517.

Further Reading

Journal Articles

Almyroudis, N.G. and Segal, B.H. (2009) Prevention and treatment of invasive fungal diseases in neutropenic patients. *Current Opinion in Infectious Diseases*, **22**, 385–393.

Fera, M.T., La Camera, E. and De Sarro, A. (2009) New triazoles and echinocandins: mode of action, *in vitro* activity and mechanisms of resistance. *Expert Review of Anti-Infective Therapy*, **7**, 981–998.

Groll, A.H. and Tragiannidis, A. (2009) Recent advances in antifungal prevention and treatment. *Seminars in Hematology*, **46**, 212–229.

Hamad, M. (2008) Antifungal immunotherapy and immunomodulation: a double-hitter approach to deal with invasive fungal infections. *Scandinavian Journal of Immunology*, **67**, 533–543.

Hwang, Y.Y. and Liang, R. (2010) Antifungal prophylaxis and treatment in patients with hematological malignancies. *Expert Review of Anti-Infective Therapy*, **8**, 397–404.

Kanafani, Z.A. and Perfect, J.R. (2008) Antimicrobial resistance: resistance to antifungal agents: mechanisms and clinical impact. *Clinical Infectious Diseases*, **46**, 120–128.

Lai, C.C., Tan, C.K., Huang, Y.T. *et al.* (2008) Current challenges in the management of invasive fungal infections. *Journal of Infection and Chemotherapy*, **14**, 77–85.

Magill, S.S., Chiller, T.M. and Warnock, D.W. (2008) Evolving strategies in the management of aspergillosis. *Expert Opinion on Pharmacotherapy*, **9**, 193–209.

Marr, K.A. (2010) Fungal infections in oncology patients: update on epidemiology, prevention, and treatment. *Current Opinion in Oncology*, **22**, 138–142.

Pasqualotto, A.C., Thiele, K.O. and Goldani, L.Z. (2010) Novel triazole antifungal drugs: focus on isavuconazole, ravuconazole and albaconazole. *Current Opinion in Investigational Drugs*, **11**, 165–174.

Pound, M.W., Townsend, M.L. and Drew, R.H. (2010) Echinocandin pharmacodynamics: review and clinical implications. *Journal of Antimicrobial Chemotherapy*, **65**, 1108–1118.

Spellberg, B., Walsh, T.J., Kontoyiannis, D.P. *et al.* (2009) Recent advances in the management of mucormycosis: from bench to bedside. *Clinical Infectious Diseases*, **48**, 1743–1751.

Staatz, C.E., Goodman, L.K. and Tett, S.E. (2009) Effect of CYP3A and ABCB1 single nucleotide polymorphisms on the pharmacokinetics and pharmacodynamics of calcineurin inhibitors: Part I. *Clinical Pharmacokinetics*, **49**, 141–175.

Books

Bryskier, A. (2005) *Antimicrobial Agents: Antibacterials and Antifungals*, 1st edn, ASM Press.

Hamad, M. and Abu-Elteen, K.H. (2010) Antifungal immunotherapy: a reality check, in *Immunotherapy: Activation, Suppression and Treatments* (ed. B.C. Facinelli), Nova Science Publishers, Inc.

Wingard, J.W. and Anaissie, E. (2005) *Fungal Infections in the Immunocompromised Patient: Infectious Disease and Therapy*, 1st edn, Informa Healthcare.

12

Fungal Pathogens of Plants

Fiona Doohan

12.1 Fungal Pathogens of Plants

A wide range of fungi cause diseases of plants (see Chapter 1 for a description of fungal taxonomy, morphology and reproduction). To be classified as a fungal plant pathogen or phytopathogen (phyto = plant), a fungus should, if possible, satisfy *Koch's postulates* or rules. Koch determined that an organism was the cause of an infectious disease if it can: (a) be isolated from a diseased host, (b) be cultured in the laboratory, (c) cause the same disease upon reintroduction into another host plant and (d) be re-isolated from that host. However, some fungal pathogens cannot be cultured or in some cases it is not easy to carry out Koch's tests.

Since early in the nineteenth century, thousands of fungi have been recognized as parasites of plants and almost all plants are hosts to particular fungal diseases. *Parasitism* occurs where one species lives off another, as distinct from *symbiosis*, where different species live in harmony with each other and the relationship is mutually beneficial, or *saprophytism*, where organisms grow on dead organic matter.

Plant-pathogenic fungi as classified as:

- **Biotrophs:** only grow and multiply when in contact with their host plants (and, therefore, cannot be cultured on nutrient media); for example, the fungi that cause rusts, powdery mildews and downy mildews.

- **Non-obligate pathogens:** grow and multiply on dead organic matter (and can, therefore, be cultured on nutrient media) as well as on living host

Fungi: Biology and Applications, Second Edition. Edited by Kevin Kavanagh.
© 2011 John Wiley & Sons, Ltd. Published 2011 by John Wiley & Sons, Ltd.

tissue. These can be further distinguished as facultative saprophytes or facultative pathogens. Facultative saprophytes complete most of their life cycle as parasites, but under certain conditions they grow on dead organic matter. Conversely, facultative saprophytes complete most of their life cycle on dead organic matter but under certain conditions they attack and parasitize living plants.

12.2 Disease Symptoms

Plant disease *symptoms* caused by fungal parasitism include:

- Spots – localized lesions on host leaves.

- Wilts – fungal colonization of root or stem vascular tissue and subsequent inhibition of translocation leading to wilting.

- Blight – browning of leaves, branches, twigs or floral organs.

- Rots – disintegration of roots, stems, fruits, tubers, fleshy leaves, corms or bulbs.

- Cankers – localized, often sunken, wounds on woody stems.

- Dieback – necrosis of twigs initiated at the tip and advancing to the twig base.

- Abnormal growth – enlarged gall-like or wart-like swelling of plant stems, roots, tubers leaves or twigs, root and shoot proliferation, etc.

- Damping off – rapid death and collapse of young seedlings.

- Decline – loss of plant vigour, retarded development.

- Anthracnose – necrotic (often sunken) ulcer-like blemishes on stem, fruit, leaf or flower.

- Scab – localized lesions of scabby appearance on host fruit, leaves, tubers.

- Rusts – many small, often rust-coloured lesions on leaves or stems.

- Mildews – chlorotic or necrotic leaves, stems and fruits covered with mycelium and fruiting bodies of the fungus, often giving a white 'woolly' appearance.

Many diseases are associated with more than one of these symptoms. Table 12.1 lists some of the economically significant fungal pathogens, associated symptoms and host plants.

Table 12.1 Examples of economically significant fungal diseases, causal organisms, hosts and associated symptoms[a].

Diseases	Fungus	Hosts	Symptoms
Vascular wilts			
Fusarium wilt	*Fusarium oxysporum*	Most vegetable and flower crops, cotton, tobacco, banana, plantain, coffee, turfgrass, ginger, soybean	Wilting, vein clearing in younger leaflets, epinasty, stunting and yellowing of older leaves
Verticillium wilt	*Verticillium dahliae* and *Verticillium albo-atrum*	Many vegetables, flowers, field crops, fruit trees, roses and forest trees	Similar to *Fusarium* wilt
Dutch elm disease	*Ophiostoma ulmi* and *Ophiostoma novo-ulmi*	Elm	Wilting, yellowing/browning of leaves, brown/green streaks in the infected sapwood underlying bark of infected branches
Oak wilt	*Ceratocystis fagacearum*	Oaks	Downward wilting and browning of leaves, defoliation. Brown discoloration of the sapwood underlying bark of infected branches
Blights			
Late blight of potato	*Phytophthora infestans*	Potato and tomato	Water-soaked lesions turning to dead brown blighted areas on lower leaves, white woolly growth on underside of infected leaves, tubers have dark surface blotches, and internal watery dark rotted tissue
Downy mildews	Several genera, e.g. *Erysiphe*	Wide range of crops, ornamentals and shrubs	Powdery growth covering shoots, leaf surfaces and sometimes flowers
Alternaria blight (early blight)	*Alternaria solani*	Potato, tomato	Dark lesions on stems, leaves, potato tuber lesions and internal dry rotting, tomato blossom blight, fruit rot and stem lesions

(*continued*)

Table 12.1 (Continued)

Diseases	Fungus	Hosts	Symptoms
Helminthosporium leaf blight	*Cochliobolus sativus*	Cereals and grasses	Dark brown spotting of leaves (also causes root rot, seedling blight and head blight)
Botrytis blossom blight (and rots)	*Botrytis* spp. (e.g. *Botrytis cinerea*)	Ornamentals and fruit trees	Water-soaked and rotting blossoms, grey/brown powdery lesions on fruit, leaves, stems and bulbs, rotting of fruit
Fusarium head blight	*Fusarium* spp.	Cereals	Premature bleaching of cereal spikelets and shrivelled pale/pink grain
Rots			
Phytophthora root rots	*Phytophthora* spp.	Fruit, forest, ornamental trees and shrubs, annual vegetables and ornamentals	Rotting of roots, plant stunting and wilting, death in severe cases
Damping off and associated rots	Several genera, e.g. *Pythium* spp.	Many hosts; broadleaved weeds and grasses are very susceptible	Seedling death, rotting of seed, roots, and fruit in contact with soil
Soft rot	*Rhizopus* spp.	Fruit and vegetables	Softening and rotting of soft fleshy organs and fruit
Brown rot of stone fruits	*Monilia* spp.	Stone fruits	Brown rotting of stone fruits
Anthracnose	Several genera, e.g. *Colletotrichum* spp.	Fruits, fruit and some forest trees, beans, cotton, ornamentals, rye, etc.	Dark sunken lesions on stems or fruits, may cause a rot of fruit
Leaf and stem spots			
Septoria leaf spot	*Mycosphaerella graminicola*	Cereals (primarily wheat)	Grey to brown water-soaked leaf lesions that, when mature, bear black visible pycnidia

Table 12.1 (*Continued*)

Diseases	Fungus	Hosts	Symptoms
Scabs			
Apple scab	*Venturia inaequalis*	Apples	Dark lesions on fruit, leaves, and sometimes on stems and bud scales
Rusts			
Black stem rust of cereals	*Puccinia graminis*	Cereals	Diamond-shaped raised orange/red powdery lesions on leaves, stems, cereal heads, that when mature, bear black teliospores
Coffee rust	*Hemileia vastatrix*	Coffee	Yellow/orange oval lesions on leaves with powdery orange/yellow lesions on the leaf undersurface, infected leaves eventually drop off
Smuts			
Covered smut of oats	*Ustilago hordei*	Oats	Grain appears black due to its replacement with black fungal spore masses
Covered smut (bunt) of wheat	*Tilletia* spp.	Wheat	As above
Powdery mildews			
Powdery mildew of cereals	*Erysiphae graminis*	Cereals	Chlorotic or necrotic leaves, stems and heads covered with mycelium and spores of the fungus, often giving a white 'woolly' appearance
Cankers, galls and malformations			
Clubroot of brassicas	*Plasmodiophora brassicae*	Brassicas	Wilting of leaves, swelling and distortion of roots, stunted growth and when severe roots rot and plant dies

[a]Many fungi cause more than one of the above diseases (e.g. *Botrytis* spp. can cause blights and rots).

12.3 Factors Influencing Disease Development

The relationship between a phytopathogen and plant and the development of disease is influenced by the nature of the pathogen and host, and the prevailing environmental conditions; all three interact and form integral parts of the *disease triangle* (Plate 12.1). Each component represents one side of the triangle. For example, if environmental conditions are unfavourable for disease development, for example wrong temperature, wind or moisture conditions, then the environment side of the triangle would be shorter and, therefore, the overall disease level would be low or nonexistent.

12.3.1 The Pathogen

In order for disease to occur, a pathogen must be virulent towards, and compatible with, its host. The aggressiveness of a pathogen will influence disease development. Some pathogens have a broad host range, while others attack relatively few plant species. Some fungal species are comprised of *formae specialis* (f.sp.) each of which parasitize and cause disease of one or a small number of host plants. For example, there are many different f.sp. of *Fusarium oxysporum* cause wilt disease of many host species. Also, races may exist within a fungal species that differ in their pathogenicity towards different genotypes (i.e. genetically distinct variants) of the same host species.

Infective propagules (e.g. fungal spores) must be present for disease to occur and, to some extent, the amount of inoculum influences disease development. The length of time that a pathogen is in contact with a host is also critical for disease development. Also, the pathogen must be able to compete with other organisms present on or in the plant.

12.3.2 The Host

Disease development requires that the host plant be at a stage of development susceptible to infection. For example, damping-off disease only affects seedlings. Infection and colonization will not occur unless the host is susceptible or tolerant to disease. Plant species and even genetic variants of the same species differ in their susceptibility to disease. For example, wheat cultivars or genotypes differ in their susceptibility to *Fusarium* head blight (FHB) disease caused by *Fusarium* spp. (Plate 12.2); this is a serious disease or cereals worldwide). Also, resistance to one pathogen does not mean immunity; that is, plants may be susceptible to other fungal diseases (e.g. the FHB-resistant wheat cultivar Sumai 3 is susceptible to the potentially serious stem rust disease caused by *Puccinia graminis*). Plants may be disease tolerant; that is, even though infected, they survive and grow and symptoms are not visible or at a nondestructive level.

12.3.3 Environmental Conditions

There is great degree of uncertainty regarding (a) what plants will be grown and what diseases will prevail under the future climatic conditions and (b) how climate change will influence disease development. Temperature, wind, moisture, sunlight, nutrition and soil quality are environmental factors that have a major impact on disease development. Fungal pathogens differ in the optimal environmental conditions required for inoculum production, dispersal and disease development. Often, disease development by fungal pathogens requires a minimum exposure time to particular temperature and moisture combinations. For example, low relative humidity can reduce the development of powdery mildew disease of tomato caused by *Erysiphe lycopersicon*.

12.4 The Disease Cycle

The disease cycle describes the events that occur from initiation of disease to the dispersal of the pathogen to a new host plant (Plate 12.3). This is distinct from the life cycle of either the pathogen or plant; life cycle describes the stage or successive stages in the growth and development of an organism that occur between the appearance and reappearance of the same state (e.g. spore or seed) of the organism. *Pathogenesis* describes events that occur in the disease cycle from infection to final host reaction.

Plate 12.4 depicts a generalized disease cycle. Inoculum is produced and disseminated and on reaching its target host plant tissue (inoculation) it penetrates the host. The type and mode of production of inoculum (e.g. sexual and asexual spores, resting spores, mycelium) and the method of dissemination (e.g. wind, water, insect) depends on the particular pathogen. For many important plant pathogens, a sexual stage has not been identified. Penetration is through wounds or natural plant pores (e.g. stomata), and some fungi produce specialized penetration structures called appressoria (singular = appressorium) (see Section 12.6.1 and Plate 12.5). Having penetrated its host, the fungus then grows within plant tissue (infection). Incubation period defines the period between inoculation and infection. As it grows within the host, it utilizes the plants' cellular resources as a nutrient source and the damage inflicted on the plant manifests as disease symptoms. Latent period defines the period between infection and symptom development. The pathogen forms survival structures such as spores that are then disseminated. During the plant pathogenesis, some diseases only involve one such disease cycle (primary cycle), while others have the potential to do more damage, as they involve a secondary or multiple cycles of disease.

Table 12.2 The gene-for-gene interaction between fungal pathogens and plants.[a]

Host →	Resistant (R)	Susceptible (r)
Pathogen ↓		
Avirulent (A)	AR (compatible)	Ar (incompatible)
Virulent (a)	aR (incompatible)	ar (incompatible)

[a]Incompatible interactions result in no disease, i.e. plant is resistant to that pathogen, while compatible interactions result in disease, i.e. plant is susceptible (Flor, 1956). Only AR interactions are resistant; in all other cases, disease will occur.

12.5 Genetics of the Plant–Fungal Pathogen Interaction

Fungal genes encode proteins that make it specific and virulent to particular hosts and, in turn, host plants have genes that make it susceptible or resistant to a particular fungus. The gene-for-gene hypothesis explains the interaction between cognate sets of pathogen and host genes. The host has dominant genes for resistance (R) and recessive genes for susceptibility (r). The pathogen has recessive genes for virulence (a) and dominant genes for avirulence or inability to infect (A). The interaction between specific sets of pathogen avirulence/virulence and host resistance/susceptibility genes determines whether disease develops. Table 12.2 explains the possible interactions between such plants and pathogens. Of the possible combinations of genes, only AR interactions are resistant and in all other cases, disease will occur. R genes encode receptors that interact with A genes. Many avirulence genes (A) act as virulence genes (a) in the absence of the corresponding host resistance gene (R).

12.6 Mechanisms of Fungal Plant Parasitism

Fungi parasitize plants using physical, chemical and biological means. In doing so, they adversely affect the photosynthesis, translocation of water and nutrients, respiration, transcription and translation in host tissue.

12.6.1 Mechanical Means of Parasitism

Mechanical means of plant parasitism include the adherence to host tissue and forceful penetration; chemical and biological molecules often facilitate such steps.

Biotrophic plant pathogens have evolved specialized structures called *appresso-ria* (singular = appresorium), which is a highly organized enlarged end of a hypha (Plate 12.4). Once the hypha senses an appropriate site, it enlarges and adheres itself to the leaf surface. This adherence is necessary to support the amount of mechanical force used to penetrate into the plant via a hypha called a penetration peg.

12.6.2 Pathogen Metabolite-Mediated Parasitism

12.6.2.1 Enzymes

Fungi produce a range of enzymes that facilitate host plant infection and colo-nization by degrading the cellular and intercellular constituents of plants (certain fungal pathogens also produce nonenzymatic proteins that inhibit the activity of plant enzymes involved in the host defence response). The cuticle forms a contin-uous layer over aerial plant parts and is an important barrier against pathogens and other stresses. Cutin is the major structural component of the cuticle bar-rier, and some fungi secrete cutinases that hydrolyse ester linkages between cutin molecules. In doing so, they, release monomers and oligomers, thus breaking the integrity of the cuticle barrier and facilitating plant parasitism. Indeed, their capacity to degrade plant cell walls makes them attractive organisms to exploit in order to degrade waste plant residue.

The plant cell wall presents a complex and important physical barrier against invading fungi. Pectinaceous substrates form part of the cell wall and are usually a major constituent of the middle lamella adhering adjacent plant cells. They are polysaccharides consisting mostly of galacturonan molecules interspersed with rhamnose molecules and side chains of galacturonan and other sugars. There are many pectin-degrading enzymes that attack different parts of the polysaccharide, including pectin esterases, polygalgacturonases and pectate lyases, resulting in a general disintegration of host tissue.

Cellulose, the most abundant natural polymer and a component of plant cell walls, consists of repeating units of glucose molecules and these chains cross-link to form fibrils embedded in a matrix of other polymers, such as pectin and lignin. Fungi often produce cellulose-degrading enzymes, or cellulases, in the latter stages of plant parasitism. Different cellulases attack the cross-linkages, and the polymeric, or degraded oligomeric or dimeric cellulose chains in a series of steps, resulting in the degradation of cellulose into small oligomers or the monomer glucose. Hemicellulose represents a complex mix of polysaccharides that forms part of the plant cell-wall matrix and is also present in the middle lamella. The primary polysaccharide component is xyloglucan, and others include glu-comannan, galactomannans and arabinogalactans. Fungi produce an array of hemicellulose-degrading enzymes necessary for complete hydrolysis of the sub-strate. Thus, cellulases and hemicellulases reduce the host cell-wall integrity.

Lignin is the second most abundant natural polymer on Earth, being ubiquitous in monocots and dicots and a major component of woody plants. The basic lignin polymer consists of chains of substituted phenylpropanoid molecules, and this polymer is substituted with a variety of side chains. Certain fungi, mostly basidiomycetes, can degrade ligin. The white-rot fungi produce liginase enzymes that degrade most of the lignin in the world.

Various other fungal enzymes facilitate plant pathogenesis. These include protein-degrading proteases, chitinases and lipid and starch-degrading enzymes. Proteases can interfere with membrane integrity by degrading the protein component of the plasma membrane. Chitinases (that degrade the chitin component of fungal cell walls) are recognized as important pathogenicity determinants in some plant diseases. Some enzymes degrade plant compounds involved in the host defence response (e.g. β-glucosidases cleave and detoxify glucoside residues).

12.6.2.2 Polysaccharides, Toxins and Growth Regulators

Certain fungi also produce polysaccharides that facilitate host colonization or deactivation of the host defence response. Some fungi produce toxic compounds that seriously damage or kill the host cells. Toxins vary in composition from low molecular weight metabolites to proteins. Examples of fungal toxins, their producers and the associated disease and hosts are listed in Table 12.3. Fungal toxins may be host or non-host specific. Host-specific toxins are usually required for pathogenicity, but are only toxic to the host plants of a pathogen and show no or low toxicity to other plants. In contrast, non-host-specific toxins affect a wide range of host plants and may act as virulence factors, increasing the extent of the disease, but are not essential for disease development. In addition to being phytotoxic, some non-host-specific toxins are also classified as mycotoxins; that is, they are harmful to human and animal health (e.g. trichothecenes, fumonisins and fusaric acid). Fungal toxins may cause visible symptoms, such as wilting, chlorosis, stunting (Plate 12.5) and so on, but may operate at the biochemical level and not cause visible effects.

Increased levels of growth regulators or hormones are often associated with diseases of plants; whether these increases originate from the fungus or the plant is often unclear. Growth regulators affect plant growth in many ways. Increased cytokinin production is associated with tumour formation (e.g. clubroot of brassicas caused by *Plasmodiophora brassicae*; Plate 12.6) and peach leaf curl caused by *Taphrina deformans*. The pathogen *Gibberella fujikuroi* secretes gibberellins, causing 'foolish seedling' disease of rice characterized by overgrown, weak spindly plants. Ethylene is a volatile hormone, and increased levels are sometimes associated with fungal diseases, but its origin is often unclear; that is, fungus or plant.

Table 12.3 Examples of fungal toxins, their specificity, producer fungi and associated diseases.

Toxin	Producer fungi	Associated diseases
Host specific		
Victorin	*Cochliobolus victoriae*	Victoria blight of oats
HS toxin	*Bipolaris sacchari*	Eyespot disease of sugar cane
HC toxin	*Cochliobolus carbonum* race 1	Northern leaf spot and ear rot of maize
T-toxin	*Cochliobolus heterostrophus* race T	Southern corn (maize) leaf blight
AK-toxin, AF-toxin, ACT-toxin, AM-toxin, AAL-toxin, ACR(L) toxin	F.sp. of *Alternaria alternata*	Diseases of Japanese pear, strawberry, tangerine, apple, tomato, rough lemon respectively
PM-toxin	*Mycosphaerella zeae-maydis*	Yellow leaf blight of corn (maize)
Peritoxin (PC-toxin)	*Periconia circinata*	Sorghum root rot
Ptr ToxA, Ptr ToxB	*Pyrenophora tritici-repentis*	Tan spot of wheat
Non-host-specific		
Tentoxin	*Alternaria* spp.	Various
Solanopyrones	*Alternaria solani*	Early blight of potato
	Didmella rabiei (anamorph: *Ascochyta rabiei*)	Blight of chickpea
Trichothecenes	*Fusarium* spp.	Head blight, seedling blight and root rot of wheat
Fumonisins	*Gibberella fujikuroi* (anamorph: *Fusarium moniliforme*)	Ear rot of maize
	Anternaria alternata	
Enniatins	*Fusarium* spp., e.g. *F. avenaceum*	Dry rot of potato
Fusicoccin	*Fusicoccum amygdali*	*Fusicoccum* canker of peaches, nectarine and almonds
Sirodesmin PL, depsipeptide HST, phomalide	*Leptosphaeria maculans*	Blackleg of crucifers
Oxalic acid	*Sclerotium, Sclerotinia*	
Fomannoxin	*Heterobasidion annosum*	Root and butt rot of conifers
Cerato-ulmin	*Ophiostoma novo-ulmi*	Dutch elm disease
Cercosporin	*Cercospora* spp.	Various

12.7 Mechanisms of Host Defence

Different host species and genotypes of a host plant species vary in their susceptibility to fungal diseases and non-hosts are disease resistant. This variation in disease resistance is due to structural and/or metabolic differences amongst plants. Table 12.4 highlights some of the preformed and induced structural and metabolic defence strategies used by different plants or plant cultivars to combat fungal diseases. The host defence response may be localized or systemic.

Plant disease resistance (R) genes encode receptors that interact with fungal avirulence (A) genes. The protein products of R genes have to both perceive and activate signals; that is, to recognize the pathogen signal (A gene) and activate a plant defence response. Many R genes encode structurally related proteins with motifs that target their intracellular or extracellular localization (leucine-rich repeat or LRR motifs) and nucleotide binding (NB) motifs. Having perceived the pathogen, R genes' products in disease-resistant hosts induce an active host defence response.

Plant defences vary in their time of activation in response to fungal attack (for a review of rapid and slow plant defence responses, see Strange (2003)). Rapid defence responses include the oxidative burst, callose synthesis and deposition, generation of nitric oxide (NO) and cross-linking of cell wall proteins. NO and oxidative burst-generated O_2^- and H_2O_2 possess antimicrobial activity. Callose, a β-1,3-linked glucan that is deposited as papillae (localized wall appositions), together with the cross-linked cell-wall proteins, increases the cellular resistance to fungal penetration.

Slower responses include the production of phytoalexins, pathogenesis-related proteins (PRPs) and hydroxyproline-rich glycoproteins (HGRPs), and the induction of lignification, suberization, the hypersensitive response (HR) and induced systemic resistance (ISR) or systemic acquired resistance (SAR). Phytoalexins are low molecular weight compounds that, nonspecifically, inhibit a wide range of organisms. HGRPs are a constitutive part of cells, but can also be induced by wounding and infection. Although it is known that they play a role in cross-linking of cell-wall proteins, much remains to be discovered as to their role in plant defence. Lignin and suberin are both constitutive components of healthy plant cell walls; enhanced synthesis and deposition of these molecules in response to infection is thought to increase the resistance of cell walls to penetration and degradation. HR involves a genetically programmed suicide of the cells surrounding the infecting fungus, and both O_2^- and H_2O_2 may participate in this response. The HR is particularly debilitating to biotrophic and hemibiotrophic pathogens that require living plant tissue to survive; its role in response to necrotrophic pathogen invasion is not so clear. Also, HR is not an obligatory component of disease resistance and is absent, or occurs as a late event, in some resistant responses of plants to pathogens.

Table 12.4 Preformed and induced defence strategies used by plants to combat fungal diseases.

Type	Time of formation	Examples
Structural	Preformed	Thick wax or cuticle layer covering the epidermis
		Thickness and/or tough cell walls (e.g. highly lignified walls impede pathogen advancement)
		Size, shape and location of stomata
	Induced	Cork cells impeding pathogen advancement
		Isolation of infected area by formation of abscission layers
		Tylose formation in xylem vessels (blocking pathogen advancement through the vascular system)
		Gums deposited in intra- and inter-cellular spaces forming an impenetrable barrier to the pathogen
		Morphological cell-wall alterations (e.g. lignification)
		Hypersensitive response (destroying cells in contact with the pathogen and thus starving the fungus of nutrients)
		Fungitoxic exudates
Metabolic	Preformed	Lack of elicitor production
		Lack of host receptor production
		Low content of essential pathogen nutrients
		Antifungal phenolics, tannins, saponins or antifungal enzymes (e.g. glucanases and chitinases)
		Plant defence proteins
	Induced	Formation of callus and cork cells around infection
		Production of phenolics and phytoalexins (that may be hypersensitive reaction-associated metabolites)
		Transformation of non-toxic to toxic compounds, e.g. phenolics, cyanides
		Production of fungal-degrading enzymes
		Production of compounds that inhibit pathogen enzymes
		Detoxification of pathogen toxins
		Complexing of plant substrates to resist fungal enzyme attack
		Metabolites involved in localized and systemic induced resistance

SAR describes the resistance that develops in plants at a distance from the initial infection point; that is, a process that is induced by infection and immunizes the plant against future pathogen attack. A salicyclic acid-dependent process mediates the SAR response and mitogen-activated protein kinase (MAPK) signal transduction cascades have been implicated as negative regulators of salicyclic acid accumulation and inducers of resistance. PRPs are acidic, protease-resistant proteins, some of which possess antifungal activity (e.g. glucanases and chitinases that degrade the chitin component of fungal cell walls). Some increase the permeability of fungal cell membranes (thionins), or affect membrane transport (defensins) or inactivate fungal ribosomes (ribosome-inactivating proteins). SAR results in the formation of plant PRPs. In response to infection, PRPs are known to accumulate in resistant plants in tissue distant from the infection point and their accumulation is correlated with the development of SAR. ISR differs from SAR, at least in some systems, in that it is mediated by a jasmonic acid/ethylene-dependent process and is not accompanied by the accumulation of PR proteins.

12.8 Disease Control

12.8.1 Cultural Practices that Aid Disease Control

Removal of crop debris, stubble or alternative host plants such as weeds helps control fungal diseases, as many causal organisms survive and overwinter on these materials. The most famous example of a fungus with an alternative host was that of *Puccinia graminis* f.sp. *tritici*, the cause of the devastating black stem rust disease of wheat. This fungus completes its life cycle on barberry bush. For this reason, barberry bushes are eradicated in the USA. Plant rotation in cropping systems will help prevent inoculum build-up provided that all hosts are not susceptible or do act as alternative hosts for prevalent diseases. Good sanitary practices will help prevent dissemination of inoculum by mechanical means, such as contaminated equipment. The date and method of sowing and the quality of the seedbed can influence disease development. As a general rule (but dependent on the pathogen, host and environmental conditions), earlier-sown seed tends to suffer more disease problems than does late-sown seed. Planting seed too deep will increase the time to emergence and increase the susceptibility to root rot and damping-off diseases.

12.8.2 Fungicidal Control of Plant Pathogens

For a more complete overview of modern fungicides and their modes of action, please refer to Kramer and Schirmer (2007). There is a huge array of fungicides on the market for controlling fungal diseases, many being variants of the same active ingredient(s). Depending on the disease and the host, fungicides are applied as a fumigant, spray, dust, paint, soil treatment and so on. Fungicides are classified in several different ways according to how they penetrate plant tissue, what effect they have and according to their mode of action. Fungicides may act as:

- **Protectant:** They do not penetrate the plant, but affect pathogen viability and germination on the surface of the host plant. Such fungicides must adhere to the plant surface and resist weathering and are most effective when applied as a preventative measure; that is, before plant inoculation with the fungus.

- **Systemic:** They can act on the plant surface and be translocated throughout the plant vascular system to kill the fungus. Systemic fungicides may also exhibit translaminar movement within the leaves.

- **Eradacative:** They are applied post-infection and act on contact by killing the organism or by preventing its further growth and reproduction.

Strobilurins represent a relatively new class of broad-spectrum fungicides that exhibit translaminar properties. Some are systemic and some are mesostemic; that is, act on plant surface, penetrate plant tissue, show translaminar movement, are absorbed by the waxy layer and exhibit vapour movement and redistribution within the crop canopy.

Fungicides are further classified into numerous groups according to their target sites. For example, systemic sterol biosynthesis inhibiting fungicides (SBIs) interfere with ergosterol production in fungi and, hence, with the integrity of the cell membrane. Strobilurins disrupt the production of ATP. Some chemicals induce systemic disease resistance in the plant (defence activators).

Many fungal pathogens have evolved resistance to fungicides by changing the target site of the fungicide. For example, many target pathogens have evolved resistance to the broad-spectrum methyl benzimidazole carbamate (MBC) fungicides. These fungicides inhibit β-tubulin assembly (and hence mitosis) and a point mutation in the β-tubulin gene of some fungi renders them unaffected by such fungicides. More recently, strains of *Mycosphaerella graminicola* that cause *Septoria tritici* blotch (STB) disease of wheat have developed tolerance to SBI fungicides and resistance to strobilurin fungicides. These are the major classes of fungicide used to control this pathogen, and because STB is one of the most economically important diseases of wheat, this poses a significant threat for the future security of wheat grain supply.

12.8.3 Host Resistance to Disease

If screening studies identify genotypes of a host that are disease resistant, the resistance can be introgressed into host cultivars that possess other desirable characteristics, such as high yield. If the resistance is already in a cultivar of the required plant species, or in closely related species, this may be achieved by conventional plant breeding techniques. Also, identified gene(s) associated with resistance can be incorporated into the required plant via genetic engineering. Ongoing research attempts to identify such resistance genes using a variety of molecular biology approaches. Candidate genes for enhancing disease resistance include those that code for proteins that are either antifungal or regulate pathways involved in the host defence response (see Section 12.7). New molecular technologies, such as enabling the plant to turn off or silence genes involved in host susceptibility or pathogen virulence (RNA silencing), will increase the number of potential target genes for genetic engineering.

The decline in the effectiveness of genes to confer resistance over time to one or more races of a pathogen in the field led to the quest for durable (long-lasting) non-race-specific resistance. Vertical resistance describes resistance to particular races of a fungal pathogen (i.e. race specific) and is usually conferred by one (monogenic) or a few genes. Horizontal resistance describes intermediate resistance to all races of the pathogen (i.e. non-race-specific) and is usually conferred by many genes (polygenic). This is achieved by pyramiding R genes into a single or multiple isogenic plant lines (genotypes that are genetically identical except for a single character, e.g. the R genes), or by growing mixes of genetically distinct genotypes that each contains one or more different R genes.

12.8.4 Biological Control of Fungal Pathogens

Biological control uses organisms (i.e. biocontrol agents (BCAs)) to directly or indirectly control fungal diseases of plants. Most BCAs used to control fungal diseases of plants are environmentally acceptable and non-plant-pathogenic bacteria or fungi. Biological control is not a new phenomenon, as it occurs naturally. For example, the natural microbial population of certain soils suppresses the development of wilt diseases caused by vascular wilt fungi (e.g. *F. oxysporum*). Also, many organic matter soil amendments (e.g. compost) provide nutrients for, and therefore increase, the resident microbial population, some of which may be disease suppressive. Biological control can, therefore, be attempted either by the introduction of foreign organisms or, less commonly, through the manipulation of the natural microbial population to enhance the activity of resident disease-suppressive organisms.

There are three means by which microbes can inhibit the development of a fungal disease (Plate 12.7):

- **Direct:** The BCA colonizes directly, suppressing the pathogen by one or more of the following means: hyphal interference, secretion of antifungal compounds (e.g. antibiotics) or enzymes. For example, the fungus *Trichoderma harzianum* is used for the biocontrol of various fungi, which it achieves by means of hyphal interference and the secretion of hydrolytic enzymes that attack the causal fungi.

- **Indirect:** The BCA physically excludes the pathogen from contact with its host or out-competes the pathogen for nutrients in a particular niche. For example, in the rhizosphere, hypovirulent (non-virulent) strains of *F. oxysporum* can out-compete some virulent strains associated with certain vascular wilt diseases.

- **Induction of host defence:** The BCA induces the host defence mechanisms. For example, *Pseudomonas* spp. induce SAR to vascular wilt fungi in various hosts.

An effective BCA must be produced, formulated and delivered in such a way that it can reach its target ecological niche in a viable and active form. Although many potential BCAs are effective under experimental conditions, many prove erratically effective under field conditions. Many potential BCAs are isolated from a habitat similar to that in which they will be applied (e.g. Plate 12.8). Table 12.5 lists some of the commercially available BCAs for the control of fungal diseases.

12.9 Disease Detection and Diagnosis

Host disease symptoms may enable the assessor to make a preliminary disease diagnosis and give the first clue as to the causal organism. Isolation and purification of the pathogen on culture media and subsequent macroscopic and microscopic analysis of fungal structures (e.g. hyphae, spores, resting structures) may confirm its identity. For new pathogens, their ability to fulfil Koch's postulates should be determined.

These traditional diagnostic techniques, although invaluable, often require a considerable amount of time and plant pathology expertise. Also, the disease will have advanced to the stage where symptoms are expressed before detection and diagnosis can occur, and at this stage the control measures adopted may be reduced in their efficacy. These disadvantages have led to the development of more rapid diagnostic techniques for many diseases, which can often detect as little as a few propagules of the pathogen in asymptomatic plant material. Such techniques include the commonly used pathogen-specific polymerase chain reaction (PCR) test that detects DNA sequences specific to the pathogen. Future developments are likely to include biosensors and microchips for disease diagnosis. Biosensors convert a biological reaction (for example, between a pathogen cell-wall component and an enzyme) into a detectable, quantifiable signal. Microchips contain nucleic acid, protein or antibody probes for a range of pathogens on a single micro-slide. A reaction between a cellular extract from a plant suspected to be diseased and an entity on the probe indicates a positive reaction. A positive reaction can be detectable via several methods (most often via fluorescence chemistry).

Table 12.5 Examples of commercially available biocontrol agents for the control of fungal diseases.

Biocontrol agent	Commercial product trademarks	Disease
Bacteria		
Bacillus spp.	Companion	Soil-borne diseases of greenhouse and nursery crops
Bacillus subtilis QWT713	Serenade	Various diseases including powdery and downy mildews of vegetables, grapes, hops and other crops
Burkholderia cepacia	Intercept, Deny	Damping off and rot diseases caused by soil-borne *Rhizoctonia*, *Pythium* and *Fusarium* spp.
Candida oleophila I-182	Aspire	Diseases of citrus and pome fruit caused by *Botrytis* and *Penicillium* spp.
Pseudomonas aureofaciens	BioJect Spot-Less	Anthracnose, damping-off and pink now mould diseases of turfgrass and other hosts
Pseudomonas chlororaphis	Cedomon	Seed-borne and foliar diseases of cereals caused by various pathogens
Fungi		
Ampelomyces quisqualis isolate M-10	AQ10 Biofungicide	Powdery mildew of apples, cucurbits, grapes, ornamentals, strawberries and tomatoes
Fusarium oxysporum (non-pathogenic)	Biofox C, Fusaclean	*Fusarium* wilt of basil, carnation, cyclamen, tomato
Gliocladium virens GL-21	Soilgard	Damping off and root rot diseases of ornamentals and greenhouse and nursery food crop plants, e.g. caused by *Rhizoctonia solani* and *Pythium* spp.
Trichoderma spp.	Various, e.g. Binab, Supresivit PlantShield, T-22 Planter box	Soil-borne fungi causing rots, damping off and wilts of various hosts, including fruit, ornamentals, turf, and vegetables

12.10 Vascular Wilt Diseases

Wilt diseases arise from a water deficiency in plant foliage. This wilting results because fungus infection and colonization of the host results in a blocking of the xylem vessels and consequent inhibition of water translocation. The fungus may directly or indirectly mediate the blocking of the xylem vessels:

- **Direct:** The fungus colonizes the xylem vessels and physically blocks water movement within the vessel.

- **Indirect:** The host responds to fungal infection by secreting substances that block its xylem vessels in order to limit the spread of the pathogen.

Fusarium and *Verticillium* species cause vascular wilt diseases of a range of host plants; other fungal wilts of economic significance include Dutch elm disease and oak wilt (Table 12.1).

12.10.1 *Fusarium* Wilts

Vascular wilt caused by soil-borne *Fusarium* species, especially *F. oxysporum*, is common in many parts of the world. It produces asexual macroconidia (characteristic of *Fusarium* species) (Plate 12.9a) and resting spores called chlamydospores (hyphal swellings) and smaller asexual microconidia characteristic of select *Fusarium* species. Mycelia are yellow/red pigmented when grown in culture on potato dextrose agar (Plate 12.9b). The sexual state of this fungus, if it exists, has not yet been found.

F. oxysporum attacks and causes serious economic losses of most vegetable and flower crops, cotton, tobacco, banana, plantain, coffee, turfgrass, ginger and soybean in warm and temperate climates. More than 120 f.sp. have been identified (e.g. wilt disease of tomato is caused by *F. oxysporum* f.sp. *lycopersici*, while wilt disease of banana is caused by *F. oxysporum* f.sp *cubense*). Some f.sp. are primarily associated with root rots, foot rots or bulb rots of plants, rather than with vascular wilt disease. Within many f.sp., *F. oxysporum* isolates are further classified into races according to their virulence against different host genotypes. Non-pathogenic populations of *F. oxysporum* also exist in soils, some of which have, or show potential to provide, biological control of wilts and other fungal diseases of plants.

Plate 12.9b depicts the typical disease symptoms associated with vascular wilt disease of tomato plants caused by *F. oxysporum* f.sp. *lycopersici*. In addition to wilting, other symptoms associated with this disease include vein clearing in younger leaflets in the early stages of disease, epinasty, stunting and yellowing of older leaves.

Plate 12.10 depicts the typical wilt disease cycle for *F. oxysporum*. Conidia of the fungus germinate in response to root exudates, producing penetration hyphae that attach to the root surface and penetrate it directly. Mycelium then advances intercellularly through the root cortex and enters the xylem vessels through the pits. Within the xylem vessels, the fungus produces microconidia that are carried upwards in the sap stream. Upon germination, microconidia penetrate the upper wall of the vessels, producing more microconidia in the next vessel. The plant is subjected to water stress and subsequently wilts due to both the accumulation of fungal mycelium and/or toxin production, and the host defence responses induced as a result of pathogen attack (e.g. production of gels, gums, tyloses and physical crushing of the vessels due to induced proliferation of adjacent parenchyma cells). Once the plant is killed, the fungus spreads to invade the parenchymatous tissue and sporulates profusely on the plant surface. These conidia can be returned to the soil, where they can survive for extended periods. Conidia are disseminated to new plants or areas by wind, water and so on.

Plant infection and colonization by *F. oxysporum* is facilitated by the pathogen's perception of signals present in root exudates, hyphal adherence to the root surface, secretion of a battery of cell-wall-degrading enzymes and the pathogen's possession of multiple mechanisms of overcoming host defence mechanisms. Root exudates elicit a fungal response and evidence suggests that cellular signal transduction cascades mediate this response. Host molecules that activate such cascades and the mechanisms by which the fungus perceives such signals (i.e. the signal receptors) are not yet known. *F. oxysporum* secretes an array of cell-wall-degrading enzymes that may contribute to infection and colonization (e.g. polygalacturonases, pectate lyases, xylanases and proteases). Pectate lyases are thought to play an important role in the virulence of *Fusarium* wilt fungi. Plants have evolved a number of physical, chemical and biological defence strategies against *Fusarium* wilt disease, including the deposition of callose adjacent to infected cells, the 'blocking off' of colonized vessels by gels, gums and tyloses, the synthesis of fungitoxic compounds and antifungal enzymes.

Control of *Fusarium* wilt is difficult due to the soil-borne nature of the pathogen and its ability to persist for extended periods in the absence of a host. Disease control is usually achieved via an integrated pest management system; that is, the combined use of one or more of the following: cultural practices, resistant host genotypes, fungicides and biological control agents. Cultural practices that help control this disease include using rotating land with non-hosts such as cereals (avoiding solanaceous crops such as potato and tomato), removal of crop debris, non-excessive irrigation and the maintenance of vigorous plants via fertilizer application. Using *Fusarium* wilt-resistant host cultivars is an option: many host cultivars differ in their susceptibility to this disease. Resistance to this disease is thought to be mediated by a gene-for-gene relationship and Takken and Rep (2010) review the arms race that is ongoing between *F. oxysporum* and tomato. Wilt-resistance genes can be introgressed into cultivars with other desirable characteristics (e.g. carrying resistance genes for other races, good yield, etc.).

Traditionally, chemical control of *Fusarium* wilt was often attempted by soil fumigation with methyl bromide to kill the soil-borne conidial inoculum. In more recent years, environmentally friendly alternatives have been revisited, such as soil steam sterilization of limited amounts of substrates (e.g. for glasshouse use) and reduction of seed-borne inoculum using hot water treatment. Fungicide seed treatment can protect the plant during the early stages of establishment. Certain systemic fungicides can protect the plant at later developmental stages, but their use is not encouraged for food plants. Biological control of *Fusarium* wilt has received a lot of attention in recent years as an attractive means of disease control. Currently, potential biocontrol organisms for controlling *Fusarium* wilt are on the market (Table 12.5), and others are being developed (e.g. non-pathogenic strains of *F. oxysporum*).

12.10.2 Other Wilts of Economic Significance

Other wilts of economic significance include Verticillium wilt, Dutch elm disease and oak wilt. Verticillium wilt is commonly caused by the soil-borne *Verticillium albo-atrum* and *Verticillium dahliae*. These lack known sexual reproductive structures (deuteromycetes) and the asexual conidia are produced. *V. albo-atrum* grows best at 20–25 °C, while *V. dahliae* grows better at 25–28 °C. *Verticillium* wilt affects a wide array of vegetables, flowers, field crops, fruit trees, roses and forest trees. Both *V. albo-atrum* and *V. dahliae* exist as various strains or races whose host range, virulence and other characteristics vary considerably. The symptoms and disease cycle of *Verticillium* wilt are similar to those of *Fusarium* wilt. Symptoms, often not obvious until either dry weather or later in the growing season, include wilting, stunting, vascular discoloration, leaf epinasty, and chlorosis and necrosis of leaves. Compared with *Fusarium* wilt, *Verticillium* wilt symptoms develop more slowly, at lower temperatures, often only develop on one side of a plant and are usually confined to lower plant parts. Both *Verticillium* species can be spread in contaminated seed, propagative plant parts, by wind, water and soil and can overwinter and persist as mycelium in perennial hosts, plant debris and vegetative propagative parts. As with *Fusarium* wilt, control is usually based on an integrated pest management strategy similar to that described above for *Fusarium* wilt. Resistant or partially resistant cultivars of some susceptible plant species are available, and durable resistance has been incorporated into many commercial tomato cultivars.

Dutch elm disease has devastated elm populations worldwide over the last 100 years; the devastation continues, particularly in the USA. Two species of the ascomycetous heterothallic *Ophiostoma* genus are responsible for the disease, *Ophiostoma ulmi* and the more aggressive *Ophiostoma novo-ulmi* (the causal organism was traditionally called *Ceratocystis ulmi*). Disease symptoms include wilting, yellowing/browning of leaves of individual branches or of the entire tree (Plate 12.11) and the formation of brown or green streaks in the infected sapwood underlying the bark. The disease is most destructive when trees are infected early in the growing season. The fungus overwinters as a saprophyte in dying or dead elm trees or logs as mycelium or conidia-bearing coremia. The fungus is spread by emerging adult elm bark beetles that become coated with sticky conidia. These beetles wound and colonize other elms, to which they transfer their conidial loads (the fungus can also spread through grafted root systems). The conidia germinate and rapidly parasitize and colonize the wood until it reaches the large xylem vessels of the spring wood, where it may produce more conidia that are carried up in the sap stream and initiate new infections. Control of Dutch elm disease spread may be facilitated by proper sanitation, removal of localized branch infections by pruning, avoidance of contact between healthy and infected roots and immediate Elms. Root injection with systemic fungicides may give protective or short-term curative control, but this may be expensive. Elimination of the disease vector, the elm bark beetles, can be attempted (using pesticides, pathogenic nematodes or trapping using pheromones), but, in the past, this has not proved to be very effective in controlling Dutch elm disease. Control via innate host resistance is unlikely, as most elm species are susceptible to this disease.

The ascomycetous fungus *Ceratocystis fagacearum* causes oak wilt disease. Infected trees rapidly wilt from the top of the canopy downward and trees usually die within 2 months. Leaf symptoms of oak wilt include wilting, clearly delineated browning and defoliation. These symptoms are sometimes accompanied by a brown discoloration of the sapwood underlying the bark of symptomatic branches. Red oaks are very susceptible to the disease; members of the white oak family are generally more resistant. The fungus overwinters as mycelium in still-living, infested trees and as fungus pads on dead trees. Insects such as sap- and bark-feeding beetles can spread the fungus. Control of oak wilt disease spread may be facilitated by means similar to those described above for Dutch elm disease.

12.11 Blights

Fungal blight diseases of plants are characterized by a rapid browning and death of the plant leaves, branches, twigs and floral organs. Table 12.1 lists some of the economically important blights caused by plant-pathogenic fungi (many of these fungi cause additional disease symptoms, such as rots, etc., on these and other hosts). The most famous example is late blight of potatoes.

12.11.1 Late Blight of Potatoes

Phytophthora comprises a genus of fungi that cause a range of disease symptoms on different host plants. Of the different species, *Phytophthora infestans* is best known as the causal agent of late blight of potatoes and tomatoes. This is a ubiquitous disease in most potato-growing regions of the world and is very destructive in the potato-growing regions of Europe and the USA. This disease precipitated the Irish potato famine in the nineteenth century.

P. infestans is a heterothallic oomycete that produces motile zoospores (that have two flagella) in lemon-shaped sporangia that form at the top of sporangiophore branches (Plate 12.12). At higher temperatures ($>15\,°C$), sporangia can germinate directly by producing a germ tube, while at lower temperatures (>12–$15\,°C$) sporangia germinate almost entirely by means of zoospores. Sexual reproduction requires the mating of an A1 and A2 type of the fungus whose hypha, on contact, differentiate to form an antheridium and oogonium. The antheridium fertilizes the oogonium (a process called karyogamy) that develops into a thick-walled oospore that germinates by means of a germ tube (Plate 12.12b). This germ tube then forms a sporangium or, occasionally, directly forms a mycelium. Until recently, mating type A1 was rarely found outside Central and South America and, therefore, sexual reproduction did not play an important part in the disease cycle in many parts of the world. Several races of *P. infestans* exist that vary in their pathogenicity towards different host cultivars.

This disease affects stem, leaf and tuber tissue. Leaf symptoms of late blight of potato include water-soaked spots on lower leafs that, under moist conditions, subsequently enlarge to form brown areas (Plate 12.12c). On the underside of these infected leaves a white woolly fungal growth appears, and soon all the leaflets on the leaf die. Under prolonged periods of wet or damp weather, all aerial plant parts blight and rot, giving a foul odour. Affected tubers have brown/black blotches on their surface, and internally they exhibit water-soaked dark-brown rotted tissue (Plate 12.12d). Such tissue may be colonized by secondary soft-rotting invaders (bacteria and fungi), resulting in a foul odour.

The pathogen overwinters as mycelium in infected potato tubers. Overwintering mycelium infects and spreads though the cortical region of the stems of plants arising from infected tuber seed, and later colonizes the pith cells of the stem. Once the mycelium reaches the aerial plant parts it produces sporangiophores that protrude through the leaf and stem stomata and project into the air. Sporangia are produced on the sporangiophores and, when mature, they are released into the air or are dispersed by rain. These sporangia land on potato leaves or stems where they form appresoria, and under moist conditions they germinate and the arising penetration peg penetrates the plant and causes new infections. The penetrating hypha grows intercellularly, and the resulting mycelium penetrates and forms intracellular haustoria; that is, hyphae that enter and draw nutrients from the host cells. These colonized cells die and the disease spreads to fresh tissue. A large number of these asexual generations and new infections can occur in a short time period. These infections result in the blighting and premature death of foliage and a reduction in tuber yield. Tuber infection begins when, under wet conditions, sporangia are washed downwards from infected foliage into the soil, releasing zoospores that reach the tubers near the surface of the soil. Zoospores germinate and the germ tube penetrates the lenticels or wounds of tubers. The fungus colonizes the tuber cells by growing intercellularly and producing haustoria within the cells. Infected and colonized tuber tissue decays and rots, either in the field or in storage. Also, seemingly healthy tubers inoculated with sporangia from soil can become infected and rot during storage.

Relatively warm days and early evenings promote infection and sporulation; lowering of temperature during the night induces dew formation and zoospore production; gradually increasing temperature during the day promotes zoospore encystment and penetration. Infection and colonization of potato by *P. infestans* is facilitated by adhesion of the pathogen to the host surface, penetration and infection via physical force, degradative enzymes and other virulence molecules, and by suppression of host defence responses. Zoospores of other *Phytophthora* species exhibit electrotactic swimming towards weak electric fields generated by plant roots, chemotaxis (i.e. attraction towards plant-derived compounds) and chemotrophic and contact-induced responses.

Control of late blight of potato has received much attention. It is generally attempted by an integrated approach involving cultural, chemical and disease forecasting control strategies. Proper sanitation, including the use of disease-free seed and removal of plant debris and volunteer potato plants, helps control late blight disease. Regular fungicide applications are often used as a preventative measure under late-blight-favourable conditions; that is, wet or humid weather and cool nights. Disease forecasting systems are widely used to indicate environmental conditions conducive to blight. Advanced systems take into account the resistance of the potato cultivar to late blight, the effectiveness of the fungicide and in some cases also the local disease risk. Most commercial cultivars are susceptible to late blight disease of potato. However, some do offer partial resistance to the disease, and research is continually ongoing to develop new cultivars with enhanced disease resistance via both traditional breeding and transgenic approaches. *R* genes (for single-gene resistance and genes for quantitative resistance) to late blight are known to be present in both wild and cultivated potato germplasm. The potential of controlling late blight using biological control methods is being extensively investigated. Isolates of several fungal species (e.g. *Aspergillus, Fusarium, Mucor, Penicillium, Rhizoctonia* and *Trichoderma*) and bacterial species (e.g. *Bacillus subtilus*) have shown promise for disease control and some have been commercialized.

12.12 Rots and Damping-Off Diseases

Many fungi cause rots of a wide variety of plants. These fungi vary greatly in terms of their taxonomic classification, their host and tissue specifics and mechanisms of parasitism. Different fungi cause soft rots of fruits, tubers, corms bulbs, root rots, stem rots and so on, and Table 12.1 lists some of the economically important rots of plants caused by fungal diseases. Rots can occur in the field, or in stored fruits, tubers and bulbs. In addition to causing a rotting of roots and stems of young plants, some fungi cause a rapid death and collapse of young seedlings; that is, damping-off disease.

12.12.1 Pythium Damping-Off Disease

Species of the genus *Pythium* include *Pythium irregulare*, *Pythium aphanidermatum*, *Pythium debarynum* and *Pythium ultimum*. *P. ultimum* can cause root rot, seedling damping off and seed rot of many plants and soft rot of fleshy fruits that are in contact with the soil. Broadleaf and grass plant species are particularly susceptible to the disease. *P. ultimum* prevails in cooler to cold soils, while *P. irregulare* and *P. aphanidermatum* are adapted to higher soil temperatures. Like *Phytophthora*, *Pythium* species are soil-borne oomycetous fungi that produce asexual motile sporangia-derived zoospores (Plate 12.13a) and sexual oospores (Plate 12.13b), although some pathogenic isolates have no known sexual reproduction stage, and most species are homothallic.

The symptoms caused by damping-off fungi such as *Pythium* species depend on the plant age and developmental stage. Seeds sown into infested soil can become infected and disintegrate. Post-germination seedling infections manifest as either a poor rate of emergence above the soil line (i.e. pre-emergence damping off), or the collapse of emerged seedlings (i.e. post-emergence damping off) (Plate 12.13c). *Pythium* species can attack and cause lesions of the stems or root rot of older plants that, if severe, can cause plant stunting, leaf yellowing, wilting and death. *Pythium* soft rots of fleshy vegetable organs manifest as woolly fungus growth on the surface (often referred to as 'cottony leak'), accompanied by internal soft rotting of the organ. These symptoms are also be commonly caused by species of the genera *Rhizoctonia*, *Fusarium* and *Phytophthora*, other fungi and some bacteria.

Sporangia and thick-walled oospores constitute important soil-borne survival structures *Pythium* species. Seed, root and stem infection (at the soil line) occurs when, under moist conditions, soil-borne mycelium or released zoospore germ tubes come into contact with, and penetrate the tissue (either directly or through cracks, wounds or natural openings). Exudates from seeds and roots can induce fungal spore germination, hyphal growth and penetration. The fungus then uses physical force and an enzymatic degradation of middle lamellae, cell walls and protoplasts to colonize the host, leading to pre-emergence death or to post-emergence damping off of seedlings. Once the tissue is colonized by mycelium, sporangia and oospores are produced inside or outside the host tissue and these return to the soil with decaying plant material, providing a new source of inoculum for future infections. The disease is spread through soil water and through the movement of infected plant debris.

Good resistance to *Pythium* damping off and associated diseases is virtually unknown amongst cultivars of many hosts. Cultural practices, including good sanitary practice, shallow seed planting, provision of adequate plant nutrition and maintaining well-drained plots or beds, help prevent or reduce the risk of *Pythium* diseases. Because many plants are susceptible to these diseases, crop rotation may not prevent disease and inoculum build-up in soil. Pre-plant soil fumigation was commonly used in contained conditions (e.g. glasshouses) to kill soil-borne inoculum. Soil sterilization, and soil solarization in warmer countries, may reduce inoculum build-up. Systemic fungicides can be applied as a preventative seed dressing or as a water-soluble preventative or curative soil drench. However, as with many diseases, when applied post-infection, the fungicide efficacy in controlling disease will be reduced. The success of biological control agents in controlling *Pythium* diseases has been very variable, especially under field conditions. Bacteria that have shown potential to control *Pythium* diseases include species of *Pseudomonas*, *Burkholderia*, *Streptomyces* and *Bacillus*; fungal antagonists include species of *Trichoderma*, *Gliocladium* and non-pathogenic *Pythium* and *F. oxysporum* isolates.

12.13 Leaf and Stem Spots, Anthracnose and Scabs

Leaf and stem spots are characterized by the occurrence of localized lesions on either the host leaves or stems. Several fungi cause such foliar and stem diseases of cereals that manifest as localized spots (Table 12.1). *Septoria* leaf blotch (Plate 12.14) (and glume blotch) are considered amongst the most serious diseases of cereals, especially in maritime climates. Anthracnoses are diseases in which spots or sunken lesions with a slightly raised rim occur on the stems, foliage or fruits of plants; other symptoms of this disease may include dieback of twigs or branches and infected fruit may drop and rot (Table 12.1). Anthracnose diseases of curcurbits and common beans caused by *Colletotrichum* species cause significant yield losses worldwide; the latter is particularly serious in Africa, Central and South America and Asia (Plate 12.15). Scabs are characterized by localized lesions of scabby appearance on host fruit, leaves and/or tubers. Fungal scab is the most important disease of apples, reducing quality and size of infected fruit (causal organism = *Venturia inaequalis*).

12.13.1 Spot and Blotch Diseases Caused by *Mycosphaerella* Species

Mycosphaerella comprises an ascomycetous genus of fungi with anamorphs of various types; for example, *Cercospora*, *Septoria*, *Ascochyta*, *Ramularia* and *Didymella*. Associated disease names often incorporate the anamorph state. Between them, members of this genus, or their anamorph state, cause serious leaf spots of cereals, banana, pea and other vegetable crops. Economically important diseases caused by fungi from this genera include leaf spot diseases of banana caused by *Mycosphaerella* species, *Ramularia* blotch of barley caused by *Ramularia collo-cygni* (telomorph not known) and *Septoria* blotch of wheat caused by *Mycosphaerella graminicola* (anamorph: *Septoria tritici*).

Septoria blotch of wheat is amongst the most serious diseases of cereals, particularly in maritime climates. This fungus produces sexual ascospores and asexual conidia called pycnospores (produced in a spore case called a pycnidia). Disease symptoms include elongate oval lesions on leaves, running parallel to the leaf. These become water soaked and brown, and a chlorotic halo may develop around the lesion. Black pycnidia develop on mature lesions and cirri may form on the lesions if the weather is dry for prolonged periods. The fungus overwinters on seed, stubble, debris and overwintering cereal crops. In the spring, it reproduces sexually to form ascospores that are wind dispersed to wheat leaves. But asexual pcynospores that are water splashed to host leaves provide the major source of inoculum for disease development. Spore germ tubes directly penetrate the plant or enter via stomata where they colonize host tissue and produce pycnidia. Moisture in the form of rain, dew or irrigation stimulates pycnospore release from conidia that are water splashed to leaves to cause new infections. Control of this disease is attempted by good sanitary practices, such as removal of crop debris, crop rotation, using resistant wheat cultivars and chemical control. Many isolates of this fungus are resistant to MBC fungicides. Systemic fungicides are used to control this disease, but recent resistance to the newer strobilurin fungicides has been reported amongst strains of the fungus. Some wheat cultivars show more resistance to the disease than others.

12.14 Rusts, Smuts and Powdery Mildew Diseases

Rusts, smuts and powdery mildew diseases are amongst the most common and devastating fungal diseases of plants. Control of rusts, smuts and powdery mildews is achieved by good clean cultural practices, using resistant hosts and fungicide application. Rust diseases have caused devastating losses on coffee and grain crops, but also of pine, apple, other field crops and ornamentals. There are thousands of rust fungi that attack different plants. They are basidiomycetes and most are obligate parasites. The most destructive diseases caused by rusts include stem rusts of wheat and other cereals caused by *Puccinia graminis* (Plate 12.16), coffee rust caused by *Hemileia vastatrix* and *Puccinia* rusts of field crops, vegetables and ornamentals caused by *Puccinia* species. Rust diseases appear as yellow to brown, white or black rusty spots; the rusty appearance results from epidermal rupture by the emerging spore masses that often have a powdery appearance.

Smut diseases, also caused by basidiomycetes, were a serious problem prior to the development of contemporary fungicides. These fungi are not obligate parasites (i.e. they can be cultured on media), but in nature they exist almost exclusively as parasites. Most smut fungi attack and devastate attack the ovaries of cereal grains and grasses, turning seed to black powdery masses (e.g. covered smut of oats caused by *Ustilago hordei* and covered smut or 'bunt' of wheat caused by *Tilletia* species; Plate 12.17), but some attack leaves, stems or seeds. Smut diseases either destroy the affected tissue or replace it with black spores.

Powdery mildew diseases are amongst the most conspicuous plant diseases affecting all kinds of plants. They are caused by obligate parasites that are members of the fungal family Erysiphaceae. Plate 12.18 depicts the symptoms of powdery mildew disease of wheat caused by *Blumeria* (= *Erysiphae*) *graminis*. Grasses and cereals are amongst the plants severely affected by this disease, as are ornamentals such as roses, field crops and trees. Symptoms of powdery mildew disease include chlorotic or necrotic leaves, stems and fruits covered with mycelium and fruiting bodies of the fungus, often giving a white 'woolly' appearance.

12.15 Global Repercussions of Fungal Diseases of Plants

The economic implications of plant disease are a reduction in crop or harvest value due to decreased quality and/or quantity of produce. Plant disease epidemics can directly or indirectly impact on the health of humans and animals. Direct impacts include mycotoxicosis associated with the consumption of mycotoxin-contaminated foods. Mycotoxins have also been investigated as potential biological warfare agents; but, conversely, some mycotoxin derivatives have provided medicinal compounds beneficial to human health. Historic examples that clearly illustrate the far-reaching and indirect repercussions of fungal diseases of plants include ergot disease of rye and other cereals and the Irish potato famine. A more recent example is sudden oak death caused by *Phythophthora ramorum*.

Ergot is a disease of cereal heads caused by *Claviceps* species, including *Claviceps purpurea* and *Claviceps fusiformis*. In infected grain, *C. purpurea* produces ergotamine ergocristine alkaloid mycotoxins and *C. fusiformis* produces clavine alkaloid mycotoxins; these are responsible for gangrenous and convulsive ergotism diseases respectively of humans and animals. Gangrenous ergotism is characterized by a loss of extremities and convulsive ergotism normally manifests as a nervous dysfunction of the victim. In the Middle Ages these compounds caused many disease epidemics, and outbreaks of ergotism have been documented throughout history. They were responsible for many thousands of deaths and social upheaval in Europe. In France in 800–900 AD, the consumption of bread contaminated with toxins (and the ergot alkaloid derivative lysergic acid diethylamide (LSD) produced during the baking of contaminated wheat) caused what was termed ignis sacer (sacred fire) or St Anthony's fire disease of humans. This ergotism epidemic was so called because it was believed that a pilgrimage to St Anthony's shrine would alleviate the intense burning sensation associated with ergotism disease. It is theorized that women accused of witchcraft and sentenced to death in European and in the Salem trials of 1692 were suffering from ergotism. An outbreak of gangrenous ergotism occurred in Ethiopia in 1977–1978; convulsive ergotism occurred in areas of India in 1975. But contemporary grain processing and cooking processes employed in the western world eliminate or detoxify most ergot alkaloids present in grain. Contrastingly, ergot alkaloids have yielded beneficial medicinal compounds; the commonly used derivative ergotamine is often prescribed for patients suffering from vascular headaches.

The Irish potato famine (1845–1850) was the consequence of *P. infestans* late blight of potato. Although this epidemic of late blight was not unique to Ireland, the consequences for Ireland were more severe because, being nutritious, easy to grow and to store, the potato then constituted the bulk of the diet of the impoverished Irish. This dependence on potato, particularly on the blight-susceptible cultivar lumper, together with the blight-favourable weather conditions in Ireland, rendered the population at the mercy of potato blight disease that arrived from North America in 1845. More than 1 million died as a consequence of the famine, and many people tried to escape from hunger by emigrating to the USA, Canada and Britain (reducing Ireland's population from 8 million to 5 million in a matter of years). The famine highlighted the need for cultural change in Ireland and heralded the end of the traditional practice of dividing family estates in small plots amongst descendants. It also heralded the beginning of political upheaval in the then British-ruled Ireland.

The advent and development of fungicides over the last century means that, today, the repercussions of fungal diseases are not as widespread as those of the Irish potato famine or ergotism in the Middle Ages. However, such compounds do not easily control some pathogens, and others evolve resistance to fungicides. Also, new pathogens and associated disease outbreaks are continuously emerging. Sudden oak death caused by *Phytophthora ramorum* is an example of such a disease. This has emerged over the last decade as a devastating disease of oak and other forest trees in cooler, wet climatic regions and has reached epidemic proportions along the Californian coast. It has spread to Europe, and more recently this pathogen has been identified as the causal agent of disease of other broad-leaf trees and larch. In 2010 it devastated larch plantations in Britain.

Plant pathogens and their toxic metabolites have been investigated as terrorism and military weapons and, thus, have raised public and security concerns. But the usefulness of such organisms as biological weapons is open to debate. Some suggest that yellow rain caused thousands of deaths in South East Asia (1974–1981) and that T-2 toxin, which belongs to the trichothecene group of mycotoxins produced by some *Fusarium* species, was the agent of disease, but this is highly controversial. Strains of the non-trichothecene-producing *Fusarium* species *F. oxysporum* and the fungus *Pleospora papaveracea* that are natural pathogens of poppies are under investigation for their potential control of narcotic-yielding poppy fields to aid in the 'war on drugs'. However, the

consequences of the deliberate release of such organisms or derivative compounds is highly controversial and a subject of world debate. In the future it
must be considered that genetic engineering provides the potential for improved
virulence or altering the metabolite profile of a fungus, thus increasing its usefulness as a biological weapon.

12.16 Conclusions

This chapter is too concise to cover all fungal plant diseases of economic significance, and some important diseases (e.g. those that cause cankers and galls)
did not even feature. From this very shallow insight into plant pathology, it
is very obvious that fungal diseases of plants have, and will in the future, put
tremendous stress on world food production. From the wilts to the gall diseases,
these fungi have evolved, and are continually evolving, diverse mechanisms for
infecting and colonizing the world plant population. Plant pathology researchers
have made a tremendous contribution to the science of understanding these diseases. The 'global economy' and changing cultural and breeding practices (e.g.
organic farming and genetic engineering) are now placing further challenges on
plant pathologists to determine how these virulent pathogens are moving around
the world (gene flow), if host or non-host disease resistance exists and how it
can be introgressed into agronomically desirable crops and if new environmentally friendly and durable chemical and biological disease control methods can
be developed. Both traditional and more novel molecular biology techniques
are being adopted to answer these questions in order to safeguard the world
plant population. But pressures on food and feed supply are ever increasing due
to both a growing population and climate change. There is an urgent need to
test models in order to get a clearer picture as to how climate change will impact upon agronomic practices, pathogen dynamics and population biology, and
disease epidemiology.

Acknowledgements

Thanks to Seamus Kennedy for help with graphical illustrations and to Brian
Fagan and Gerard Leonard for supplying diseased material.

Revision Questions

Q 12.1 Define the terms parasitism, symbiosis and saprophytism.

Q 12.2 What is the disease triangle?

Q 12.3 Describe a generalized disease cycle.

Q 12.4 Describe how enzymes produced by fungi are critical to pathogenesis.

Q 12.5 Define the two types of fungal toxin.

Q 12.6 Define protectant, systemic and eradactive in relation to fungicides.

Q 12.7 Give an example of the use of direct biological control for the control of fungal pathogens.

Q 12.8 What plants are affected by *Verticillium* wilts?

Q 12.9 List some of the global impacts of fungal diseases of plants.

References

Agrios, G.N. (ed.) (1998) *Plant Pathology*, 1st edn, Academic Press, California.

Flor, H.H. (1956) The complementary genic systems in flax and flax rust. *Advances in Genetics*, 8, 29–54.

Kramer, W. and Schirmer, U. (eds) (2007) *Modern Crop Protection Compounds*, vol II, Wiley–VCH Verlag GmBH & Co., Weinheim, Germany.

Strange, R.N. (ed.) (2003) *Introduction to Plant Pathology*, John Wiley and Sons, Inc., New York.

Takken, F. and Rep, M. (2010) The arms race between tomato and *Fusarium oxysporum*. *Molecular Plant Pathology*, 11, 309–314.

Further Reading

Journal Articles

Brasier, C. (2010) Plant pathology: sudden larch death. *Nature*, 466, 824–825.

Di Pietro, A., Madrid, M.P., Caracuel, Z. *et al.* (2003) *Fusarium oxysporum*: exploring the molecular arsenal of a vascular wilt fungus. *Molecular Plant Pathology* 4(5), 315–325.

McDowell, J.M. and Woffenden, B.J. (2003) Plant disease resistance genes: recent insights and potential applications. *Trends in Biotechnology*, 21, 178–183.

Books

Agrios, G.N. (ed.) (1998/2005) *Plant Pathology*, 1st and 5th edn, Academic Press, California.

Beckman, C.H. (1987) *The Nature of Wilt Diseases of Plants*, American Phytopathological Society Press, St. Paul, MN.

Bélanger, R.R., Bushnell, W.R., Dik, A.J. and Carver, T.L.W. (eds) (2002) *The Powdery Mildews: A Comprehensive Treatise*, American Phytopathological Society Press, St. Paul, MN.

Dickinson, M.D. and Beynon, J.L. (eds) (2004) *Molecular Plant Pathology*, Blackwell Publishing, Oxford.

Kramer, W. and Schirmer, U. (eds) (2007) *Modern Crop Protection Compounds*, vol **II**, Wiley–VCH Verlag GmBH & Co., Weinheim, Germany.

Lucas, J.A., Bowyer, P. and Anderson, H.M. (eds) (1999) *Septoria on Cereals: A Study of Pathosystems*, CABI Publishing, Oxford.

Lucas, J.A. and Dickinson, C.H. (eds) (1998) *Plant Pathology and Plant Pathogens*, Blackwell Scientific Publishing, Oxford.

Murray, T.D., Parry, D.W. and Cattlin, N.D. (eds) (2008) *A Colour Handbook of Diseases of Small Grain Cereals*, Manson Publishing, London.

Parker, J. (ed.) (2008) *Molecular Aspects of Plant Disease Resistance. Annual Plant Reviews*, vol. **34**, Wiley–Blackwell Publishing.

Slusarenko, A.J., Fraser, R.S.S. and van Loon, L.C. (eds) (2000) *Mechanism of Resistance to Plant Diseases*, Kluwer Academic Press, Dordrecht.

Strange, R.N. (ed.) (2003) *Introduction to Plant Pathology*, John Wiley and Sons, Inc., New York.

Answers to Revision Questions

Chapter 1

A 1.1 Principle macromolecular components of both yeast and fungal cells: glucans, chitin, mannoproteins arranged in microfibrillar network. Major difference between yeasts and fungi being the predominance of chitin in the latter. Major physiological roles: cell structure stability, stress protection and cell–cell interactions.

A 1.2 Main nutrients are: sugars for carbon and energy; ammonium or amino acids for protein synthesis; oxygen for respiration and fatty acid/sterol synthesis; minerals for ionic balance and enzyme activity; vitamins for coenzymes/growth factors. Nutrients transported, depending on the type of nutrient, by active (energy-mediated) transport or by diffusion (free, facilitated or channels).

A 1.3 (a) Oxygen used as terminal electron acceptor in respiratory metabolism of glucose (via glycolysis, then mitochondrial oxidative phosphorylation) to water and carbon dioxide. (b) Fermentative metabolism of glucose by glycolytic enzymes in the cytoplasm to ethanol and carbon dioxide. Much more energy in the form of ATP generated by (a) than by (b).

A 1.4 Temperature, pH, oxygen and osmotic pressure. Concept of minimum, optimum and maximum ranges for these factors and some typical values for fungal growth.

A 1.5 Diagrammatic description of budding cycle in yeast cells, typified by *Saccharomyces cerevisiae*. Description of hyphal elongation, septum formation in filamentous fungi.

Fungi: Biology and Applications, Second Edition. Edited by Kevin Kavanagh.
© 2011 John Wiley & Sons, Ltd. Published 2011 by John Wiley & Sons, Ltd.

A 1.6 Logarithmic doubling, for a limited time period (and after recovery from a lag phase), of cells as nutrients are plentiful and if physical growth conditions are conducive. Calculations of doubling time given on p. 30.

Chapter 2

A 2.1 Fungi are very amenable to molecular/genetic analysis, as they grow very quickly, are easy to culture, many have a haploid genome which aids mutagenesis studies, many genomes have been sequenced and advanced tools, such as a genome-wide knockout set in *S. cerevisiae*, are available.

A 2.2 Yeast cells grow vegetatively by budding, and daughter cells completely separate. Filamentous fungi grow vegetatively through highly polarized cells called hyphae. Hyphae do not separate into different cells, but lay down septa cross-walls which separate the hypha into distinct compartments.

A 2.3 An ascus of *S. cerevisiae* contains four ascospores which can be isolated. The individual products of meiosis can therefore be tested. In *N. crassa*, the four products of meiosis go through an additional round of mitosis, creating eight ascospores. The asci are linear, which allows ordered arrangement of the ascospores and mapping of genes in relation to the centromere.

A 2.4 Metabolic mutants were identified in *N. crassa*. Conidia were irradiated to induce mutagenesis, and then maintained on rich media to allow growth. Conidia were then tested for their ability to grow on minimal media.

A 2.5 Temperature sensitivity was used in a screen by Morris to identify cell cycle and nuclear distribution mutants in *A. nidulans*. Drug resistance was used to identify tubulin genes in this organism as well. Sensitivity to mating pheromone was used to identify components of the mating pathway in *S. cerevisiae*.

A 2.6 *C. albicans* is diploid and does not undergo meiosis.

A 2.7 Tetratypes are rare. Tetratypes require that the two genes are on the same chromosome and crossover takes place, or there is crossover between at least one gene and the centromere. Both genes are close to the centromere, and on different chromosomes. This would not change if the starting parents were the different genotype indicated.

A 2.8 *N. crassa* mating loci are called idiomorphs because the sequences vary significantly between opposite mating types, whereas the mating types are referred to as alleles in *S. cerevisiae*. In *N. crassa*, the mating loci also regulate heterokaryon compatibility, and mating type switching does not occur. In *C. cinereus*, there are more than 12 000 mating types, as opposed to two in *S. cerevisiae*.

A 2.9 Parasexual genetic analysis involves recombination in the absence of meiosis, or mitotic crossing-over. This has been used in *A. nidulans* to help map genes to chromosomes. In *C. albicans*, meiosis does not occur, but tetraploid nuclei formed through mating can break down into diploid products which demonstrate different genotypes from the parents. Diploids form from tetraploids through chromosome loss, indicating that mitotic crossing-over gave rise to the progeny.

A 2.10 Plasmids in *S. cerevisiae* can be maintained as extrachromosomal elements, as they contain an autonomous replication sequence (ARS) and centromere sequence (CEN). These allow the plasmids to be stable and to segregate. Genes can be introduced and expressed without integrating into the genome, and plamid loss can be induced.

Chapter 3

A 3.1 Chytridiomycota, Zygomycota, Basidiomycota and Ascomycota.

A 3.2 Sanger sequencing utilizes the replication enzyme DNA polymerase to synthesize a DNA strand. A typical reaction tube will contain single-stranded DNA, DNA polymerase, DNA primers, the four nucleotide triphosphates and a specific dideoxynucleotide triphosphate. Dideoxynucleotide triphosphates contain no reactive 3'-OH and, therefore, DNA synthesis is terminated once they have been incorporated into a growing DNA strand. Newly synthesized DNA strands are separated based on size using a polyacrylamide gel. Specific nucleotide triphosphates are radiolabelled to aid visualization.

A 3.3 Next-generation sequencing refers to commercial sequencing techniques that enable us to generate millions of sequence reads in a single sequencing reaction. Some of the reads can be over 500 nucleotides in length and are ideal for the *de novo* sequencing of bacterial and small eukaryotic genomes. They have also been used extensively in resequencing strains, isolates and individuals where a reference strain is already available.

A 3.4 SGD, CGD and AspGD are composite databases. As well as containing all DNA and protein sequence information for species of interest, they are also highly annotated with supporting literature linked to each gene.

A 3.5 A global alignment aligns two sequences over their entire length and generally works best when the sequences are highly conserved and approximately the same length. On the other hand, a local alignment aligns two sequences at regions of high similarity.

A 3.6 A phylogenetic tree displays the evolutionary relationships amongst species and is usually reconstructed using a single phylogenetic marker. Instead of using a single marker, supertree methods enable us to reconstruct a representative species phylogeny from multiple phyloegentic trees derived from different phylogenetic markers.

A 3.7 The Gene Ontology (GO) is a universal vocabulary that describes the roles of genes and gene products in an organism. It was introduced in 2000 and aims to facilitate cross-comparisons of species by allowing users query different databases/genomes using the same search terms. The three main branches of the GO are cellular component, biological process and molecular function.

A 3.8 Comparative genomics involves doing a direct comparison between two or more closely related species. For example, by comparing two species, one pathogenic and one non-pathogenic, we can locate difference in the genetic repertoires of each species which may account for increased virulence in one species over another.

A 3.9 Gene synteny describes the relative order of genes on chromosomes that share a common ancestor. Chromosomal regions are considered syntenic if multiple consecutive genes are found in a conserved order between the two species under consideration.

A 3.10 Horizontal gene transfer is the acquisition and incorporation of DNA by one species from an unrelated species. It is termed non-vertical transmission of DNA as the recipient species acquires DNA from a non-parental source.

Chapter 4

A 4.1 Because not only was it the first eukaryotic organism to have its genome sequenced, its homologous recombination system allowed for the systematic deletion of each ORF in search of phenotypes; microarray technology allowed its global mRNA profiles to be identified; and two-hybrid technology looked at all possible protein–protein interactions.

A 4.2 Pre-genomic research was characterized by hypothesis-driven sequential experiments; post-genomic research is driven by the massively parallel analysis of biological information, followed by pattern recognition within datasets.

A 4.3 This is the name given to a cross-disciplinary approach to developing computer models of how molecules interact to generate biological phenomena. It includes contributions from biology, chemistry, physics, mathematics, computer science and engineering.

A 4.4 This is a sequence of DNA that starts with a methionine codon and runs for another 99 codons, without hitting a nonsense one.

A 4.5 Approximately 5000 new ORFs were discovered. Approximately 3000 ORFs were annotated with functions. A further 1000 have had functions attributed to them since then.

A 4.6 That *S. cerevisiae* underwent *genome* duplication, *S. pombe* underwent many *gene* duplications and that both have active transposable elements. On the other hand, there is no evidence of genome or multiple gene duplication in *N. crassa*. Moreover, the latter also displays a paucity of transposable elements.

A 4.7 *In silico* analysis uses BLAST searches of previously annotated DNA sequences to provide insight into an ORF function depending on how well matched the new sequence is to ones already in the databases. *Reverse genetics* associates phenotypes with deliberately disrupted known ORFs. *Identification by association* requires a comparative analysis of genomes from closely and distantly related organisms, thereby identifying genes that are held in common by biologically related species. For example, a gene which is found in yeasts but not in filamentous fungi probably encodes information that defines some specific aspect of yeast biology and vice versa.

A 4.8 DeRisi *et al.* PCR-amplify each of the 6400 distinct ORFs described in the first edition of the yeast genome and printed these DNA molecules at unique addresses on glass slides using a simple robotic printing device. This was then used simultaneously to explore the expression profile of mRNA as yeast cells underwent a diauxic shift from fermentative growth to aerobic respiration and to explore gene expression patterns in regulatory mutants of yeast. Red fluorescently-labelled cDNA was synthesized from cells grown under different culture conditions. Green fluorescent cDNA was also prepared using mRNA from control cells. The expression level of each gene was measured by hybridizing fluorescently labelled cDNA to the microarray probe, visualizing the fluorescence pattern using a confocal microscope and using a computer to analyse the image the relative intensity of the spots. Yellow spots indicate that the experimental and control cells express that particular ORF equally. Red colour indicates gene expression increased relative to the reference, while green colour indicates gene expression decreased relative to the reference. The analysis revealed that 28 % of all yeast genes underwent a significant alteration in gene expression level as a result of a diauxic shift and identified groups of genes whose pattern of expression changed in association with one another. It also detected co-ordinately regulated genes in the regulatory mutants.

A 4.9 *BLAST (basic local alignment sequence tool):* An algorithm that searches a sequence database for sequences that are similar to the query sequence. There are several variations for searching nucleotide or protein databases using nucleotide or protein query sequences. *Complementary DNA library:* A collection of clones each containing a cDNA derived from the RNAs isolated from a specific tissue or cell and inserted into a suitable cloning vector. *Expressed sequence tags:* Short cDNA sequences that are derived from sequencing of all the mRNAs present in a cell. ESTs represent the expression profile of the cell at the time point of RNA isolation. *SAGE (serial analysis of gene expression):* Extremely short ESTs that are linked together as DNA chimeras consisting of 15 base pair sequences from 40 different mRNAs. The sequence analysis of thousands of these 40 X15bp chimeras permits a quantitative estimation of the mRNAs in the original sample. *Sequence alignment:* A linear comparison of nucleotide or protein (amino acid) sequences. Alignments are the basis of sequence analysis methods and are used to identify conserved regions (motifs).

A 4.10 The challenges facing proteomics in defining 4-D biology include establishing the amino acid sequence and 3-D structure of every protein encoded by the genome. It also requires information on how the location and concentration of each protein change with alterations in the cell. It also needs to establish when and where proteins are post-translationally modified, and finally how proteins interact with other proteins in different contexts.

A 4.11 Fusion proteins are made by using recombinant DNA techniques to splice, in frame with a yeast promoter, the DNA sequence for each yeast gene to the DNA sequence encoding the other member of the fusion protein. Fusion proteins find two major uses in post-genomic yeast analysis: (a) fluorescent protein fusions are used to visualize where each yeast protein is located in the cell; (b) the two-hybrid system exploits the fact that transcription factors consist of two separable protein domains – a DNA binding domain and an activation domain. If the DNA binding domain is fused to one yeast protein and the appropriate activation domain is fused to another yeast protein, then the two domains will only be close enough to drive expression from a reporter gene if the two yeast proteins interact closely in the cell. The yeast protein X is fused to a DNA binding domain protein (DBD) and yeast protein Y is fused to the appropriate activation domain (AD). Specific interaction of proteins X and Y brings the DBD and AD together and when the DBD binds to the promoter the AD can drive reporter gene transcription. If X and Y do not interact, then the DNA binding domain can still bind to the promoter, but, in the absence of the activation domain, cannot activate expression.

A 4.12 In *step 1* of the systems approach all available information is used to define an initial model of the phenomenon. In *step 2* each known component is systematically perturbed, yielding numerous separate cellular conditions. The global cellular response to *each perturbation* is detected and quantified using array technology. Also, in at least some of these cases large-scale protein expression analysis is performed. In *step 3* all of this new information is integrated with the initially defined model and also with the information currently available on the global network of protein–protein and protein–DNA interactions in yeast. In *step 4* the researchers formulated new hypotheses to explain the observations not predicted by the model and then designed and executed additional perturbation experiments to test the new 'improved' model. This is then re-iterated until the all of the observations can be accommodated within one coherent model.

Chapter 5

A 5.1 Solid fermentation is used in the production of edible mushrooms and involves growing mycelium over a solid substrate. Batch fermentation is employed in the production of beers and wines and involves growing yeast in a liquid medium derived from barley and grapes respectively. Fed batch is used for the production of baker's yeast and involves adding substrate to the fermentation at predetermined times to increase the cell density and uses the cells from the end of one fermentation to start the next fermentation. Continuous fermentation is used in the manufacture of Quorn mycoprotein and involves keeping the fungus at a steady state to maximize the production of biomass.

A 5.2 Cells can be entrapped within a membrane, attached to a solid substrate, entrapped in a bead composed of alginate, for example, or induced to flocculate to give higher densities than might normally be expected. Immobilization ensures that a high cell density is maintained and removes growth inhibition due to the production of a toxic metabolite. The main disadvantage of immobilization is that fungal cell viability decreases over time and that the immobilization system may degrade with continual usage.

A 5.3 Productivity can be affected by a variety of factors. The nature of the fermentation system is a major consideration, since a particular system may be good for producing one product but bad for another. For example, batch fermentation is used for ethanol production but not usually for antibiotic production by fungi. The physical nature of the fungal cell is important – yeast cells can be agitated with spargers (paddles) but fungal mycelium will be disrupted if spargers are employed. The level of aeration will influence productivity, particularly when ethanol is being produced. Too high a concentration of sugar can adversely affect productivity.

A 5.4 Novel means of genetically manipulating yeast have allowed the creation of brewing strains with altered characteristics. For example, yeasts have been produced which (a) do not produce the organoleptic diacetyl, (b) contain genes for amylase and glucanse production and (c) flocculate at the end of the fermentation. Although not used commercially, genetically modified brewing strains should allow beers to be produced faster, cheaper and with new flavours/aromas.

A 5.5 Quorn mycoprotein is produced under continuous conditions in an air-lift fermenter. Fungal mycelium is fed a medium composed of glucose, biotin and mineral salts at a constant rate. Culture is harvested at the same rate as new medium is added to the fermenter and thermally treated to reduce the RNA content.

A 5.6 The synthesis of antibiotics by fungi can be affected by the nature of the fermentation system, the level of nitrogen or phosphate in the culture medium, and feedback inhibition where the antibiotic inhibits its own synthesis.

A 5.7 Historically, fungi were cultivated under solid fermentation conditions for the isolation of enzymes. More recently, submerged, fed-batch and continuous systems have been utilized.

A 5.8 The compost is spawned with a culture of *Agaricus bisporus* (edible mushroom) grown on sterilized cereal grains and cultured at 25–28 °C and high humidity for 14 days to allow the mushroom mycelium to colonize the compost. After this period the compost is 'cased' with a layer of neutralized peat, which has the effect of stimulating the formation of large primordia that will later develop into mushrooms.

Chapter 6

A 6.1 Primary metabolites include, fats, alcohol and organic acids; secondary metabolite examples are compounds such as antibiotics, statins, cyclosporins and ergot alkaloids, strobilirubin and giberellic acid.

A 6.2 Penicillin-binding proteins.

A 6.3 The compound inhibits the production of interleukin-2 by T-lymphocytes.

A 6.4 They act as potent competitive inhibitors of 3-HMG-CoA-reductase.

A 6.5 *Cephalosporium acremonium.*

A 6.6 *S. cerevisiae* can utilize selenium as it does sulfur, resulting in the biosynthesis of various organic selenium compounds.

A 6.7 *Phaffia rhodozyma.*

A 6.8 *Ashbya gossypii.*

A 6.9 It can be used as a vasoconstrictor to control excessive bleeding after childbirth.

A 6.10 Antibacterial, antifungal, anti-cancer, anti-tumour, imunnosuppressive.

Chapter 7

A 7.1 Examples of the use of enzymes include cheese making, brewing, baking, the production of antibiotics, and the manufacture of commodities such as leather, indigo and linen. They are also used in areas such as detergent and paper production, the textile industry and in the food and drinks industry in products ranging from fruit juice, to coffee, tea and wine.

A 7.2 Examples include *Aspergillus* spp., *Trichoderma* spp., *Rhizopus* spp., *Mucor* spp., *Trametes* spp., *Saccharomyces* spp.

A 7.3 Protease enzymes hydrolyse the peptide bond (CO-NH) in a protein molecule.

A 7.4 Endopeptidase and exopeptidase.

A 7.5 The three main cellulolytic enzymes include endoglucanases, exoglucanases and β-glucosidases.

A 7.6 The two main fungal organisms responsible for xylanase production are *Trichoderma* sp. and *Rhizopus* sp.

A 7.7 Xylanase degrades xylan into xylooligosaccharides and xylose.

A 7.8 The three main types of amylase include α-amylase, β-amylase and glucoamylase.

A 7.9 The principal function of lipase is to hydrolyse triglycerides into diglycerides, monoglycerides, fatty acids and glycerol, but it can also catalyse esterification, interesterification and transesterification reactions in nonaqueous media.

A 7.10 The principal benefit of phytase inclusion into monogastric diets is that the antinutritional properties of phytic acid are destroyed.

Chapter 8

A 8.1 *Recombinant DNA* simply requires the cutting and joining of DNA molecules from two different sources. However, the production of *recombinant proteins* requires DNA engineering and also the information has to be in the correct context to be transcribed and then translated. This means the careful insertion of the DNA downstream of an appropriate promoter so that it is in the correct reading frame. It often also requires a terminator sequence to be present in the vector. Once transcribed, the mRNA must carry the appropriate translation signals. Finally, the protein needs to fold properly (and often requires further processing by addition of sugars, etc.) before it has the correct structure and, hence, biological function.

A 8.2 A *cDNA* sequence is derived by reverse transcription of the heterologous mRNA. It therefore lacks introns and regulatory sequences and is, therefore, easier to have expressed in the majority of heterologous expression systems.

A 8.3 A backbone of bacterial plasmid DNA carrying a bacterial origin of replication and a selectable marker for use in *E. coli*. A selectable marker for the intended fungal host. A strong promoter to drive the production of the heterologous mRNA. Appropriate DNA sequences to ensure efficient termination of transcription and polyadenylation of the mRNA. Appropriate sequences to ensure the correct initiation and termination of translation.

A 8.4 Enzymatically removing the cell walls and exposing the resulting sphaeroplasts to the DNA in the presence of calcium ions and polyethylene glycol. Electroporation of yeast cells and fungal sphaeroplasts. Transformation of yeast cells by treating them with alkali cations (usually lithium) in a procedure analogous to *E. coli* transformation.

A 8.5 *Saccharomyces cerevisiae* grows rapidly by cell division. It has its own autonomously replicating plasmid. It can be transformed as intact cells. It forms discrete colonies on simple defined media. It can carry out post-translational modifications of expressed proteins. It secretes a small number of proteins into the growth medium, which can be exploited to simplify the purification of heterologous proteins. It has got a long, safe history of use in commercial fermentation processes. It does not produce pyrogens or endotoxins.

A 8.6 *Comparison:* Both have a backbone of bacterial plasmid DNA carrying a bacterial origin of replication and a selectable marker for use in *E. coli*. Both carry appropriate promoter and terminator sequences. Both carry appropriate selectable markers for use in their respective host cells.

Contrast: The yeast plasmid is an autonomously replicating one maintained at high copy number in the yeast cells. The *Pichia* vector cannot replicate autonomously in the fungal cells but integrates into the AOX1 locus by homologous replication. Also, the copy number will be very low – unlike the yeast one.

A 8.7 The copy number of the expression vector. The strength of the promoter. The presence of a terminator. The presence of appropriate translation signals in the mRNA. Low-level protease activity. The production of an authentic protein product (glycosylation and/or processing by proteases).

A 8.8 Although the viral protein was expressed, it was not glycosylated. Furthermore, the host cells failed to produce the 22 nm phospholipid–protein particles. The *E. coli* recombinant protein, therefore, failed to elicit an appropriate immune response in animals.

A 8.9 *Comparison:* Both have a backbone of bacterial plasmid DNA carrying a bacterial origin of replication and a selectable marker for use in *E. coli*. Both are autonomously replicating plasmids.

Contrast: The *Trp1* selectable marker in the proof of principle plasmid was replaced by the poorly expressed *Leu-2d* gene. This caused an increase in plasmid copy number. A terminator sequence was inserted after the gene in the scale-up plasmid to ensure that stable mRNAs were transcribed. The promoter was changed from *ADH1* in the proof of principle plasmid to the more powerful *GPDH* one in the scale-up version.

A 8.10 Use homologous recombination to insert a reporter gene construct into the yeast chromosome. This consists of a selectable marker (*URA1*) and a disabled *cyc1* promoter, containing oestrogen response elements, fused to a *LACZ* gene. Arrange to constitutively express the receptor protein from an autonomously replicating expression vector. In the presence of oestrogen the receptor protein binds to the oestrogen response elements in the promoter of the *LACZ* gene. The level of β-galactosidase activity is therefore a measure of how successfully the receptor–oestrogen interaction is in terms of ERE-mediated gene activation. Therefore, it can be used to assay the effect of molecules that interfere with/accentuate the hormone–receptor interaction.

A 8.11 Under normal conditions, when the mating pheromone a-factor binds to the a-factor receptor in Mat a cells it triggers a pathway which activates a number of genes (e.g. *FUS1*) through a promoter element and causes the cells to arrest their growth in the G1 phase of the cell cycle through *Far1*.

The yeast mating pathway was re-engineered so that the a-factor receptor is re-placed by a heterologous human receptor protein (SST_2). Binding of the appro-priate human ligand (somatostatin) results in activation of the mating pathway cascade. However, by deleting *Far1* the cells fail to arrest their cell cycle, and by fusing the *HIS3* gene to the *FUS1* promoter the cascade triggers the expression of *HIS3*. The ligand–receptor interactions can therefore be assayed by the number of cells with an $HIS3^+$ phenotype.

A 8.12 The two-hybrid system exploits the fact that transcription factors consist of two separable protein domains: a DNA binding domain and an activation domain. If two proteins interact closely in the cell, then if one is attached to a DNA binding domain and the other protein is attached to the appropriate activation domain, the two domains will be close enough to drive expression. Therefore, proteins are genetically engineered so that protein X is fused to a DNA binding domain protein (DBD) and expressed from one vector whereas protein Y is fused to the appropriate activation domain (AD) which is expressed from vector. Specific interaction of proteins X and Y brings the DBD and AD together and when the DBD binds to the promoter the AD can drive reporter gene transcription. If X and Y do not interact, then the DNA binding domain can still bind to the promoter but, in the absence of the activation domain, cannot activate expression.

Chapter 9

A 9.1 In general, proteins must be isolated, either in pure form or free from contaminants, which would prevent detection by protein MS, to enable inves-tigation of fungal proteomes. Access to extensive fungal genome or transcrip-tome sequence is preferable, if not essential, for your fungus of interest. SDS- or

2D-PAGE can be used for protein separation prior to MALDI-ToF MS or LC–MS protein mass spectrometric analysis. Comparative 2D-PAGE and image analysis allows identification of altered protein expression under different experimental conditions. Individual proteins or shotgun proteomics can be used to identify proteins by LC–MS/MS.

A 9.2 Intracellular protein isolation from fungi is difficult due to the rigid nature of the fungal cell wall and the presence of large amounts of interfering carbohydrate polymers. The level of proteins secreted by fungi is low in many wild-type organisms and is dependent on the culture medium and conditions used for growth. No filamentous fungal genome sequences were available until the 2000s, which meant that protein MS was of limited use in facilitating protein identification. Filamentous fungal genes were known to contain multiple introns, unlike bacterial, yeast and viral genes. This meant that interrogation of available genomic DNA databases (following *in silico* translation) with peptide mass data was often of limited value for protein identification, since inadequate bioinformatic tools were available for intron–exon splice site identification in genes.

A 9.3 The primary structure of proteins not only consists of the peptide bonds which result in the linear sequence of amino acids in a protein, but also the disulfide bridges between cysteine residues (intra- or inter-chain). Disulfide bridges are cleaved using reducing agents (e.g. dithiothreitol, β-mercaptoethanol or sodium borohydride) prior to trypsin cleavage to prevent the presence of tryptic peptides linked by disulfide bridges and to ensure complete release of the maximum amount of tryptic peptides. Since disulfide bridges can reform, post-cleavage, alkylation agents are used to covalently react with the free thiols and prevent disulfide bridge reformation. They can also react with cysteine residues not originally found in disulfide bridges. The mass of alkylation agents can be automatically taken into account during subsequent protein mass spectMSrometry.

A 9.4 2D-PAGE analysis yields purified proteins for trypsin digestion and protein mass spectrometric analysis. It is compatible with many protein dye reagents. Combined with image analysis techniques, 2D-PAGE enables identification of altered protein expression levels in response to altered environment or genotype. Although significant sample pretreatment is required prior to 2D-PAGE, it yields much information on protein charge and mass which aids confident identification. 2D-PAGE analysis is not always compatible with high molecular mass and hydrophobic (membrane) protein separation. It requires competent skills to perform reproducibly.

A 9.5 Owing to improvements in LC fractionation for the separation of peptides and in the algorithms used to assign detected peptides to a corresponding protein in a database, it is now possible to simultaneously identify many of the proteins present in a complex mixture by protein MS. This approach is known as 'shotgun proteomics' and bypasses the need to purify the protein of interest prior to trypsinization and MS analysis. The resultant peptide mixture, after trypsin digestion, contains peptides derived from all proteins present in the initial sample, whereby hundreds of proteins may yield many thousands of distinct peptides. Following LC–MS/MS analysis, a list of proteins is obtained, which have been identified by comparison against the appropriate *in silico* database. Some proteins may be identified from as little as a single peptide, while in most cases the detection of two peptides per protein, or sequence coverage of 5–10 %, is necessary for confident identification of constituent proteins. Shotgun proteomics yields large amounts of data from a single experiment, but it can be time consuming in terms of LC run time.

A 9.6 Proteomics enables protein identification in fungi, determination of altered protein levels in response to environmental conditions, phenotypic analysis of gene deletion mutants, identification of protein–protein interactions, confirmation of *in silico* intron–exon splice site identification, the demonstration of hypothetical gene calling and discovery of novel proteins with biotechnological or biomedical potential.

A 9.7 In most cases, proteins facilitate the actual adaptation to new environmental conditions, although this may have been preceded by an alteration in gene expression. Therefore, changes in protein levels may be more precisely related to observed phenotype changes in the cell. Moreover, transcriptome analysis cannot reveal information on protein modification-induced alterations in cell status. Proteomics can reveal these types of change. Protein–protein interactions can really only be studied at the protein, not the nucleic acid level.

A 9.8 Glycosylation, phosphorylation, acetylation, phosphopantetheinylation, ubiquitination and methylation. The masses of individual types of PTM can be pre-programmed into LC–MS software. Subsequent identification of any peptide which results in two ions, differing precisely in the PTM mass, indicates occurrence of that group on the relevant amino acid (e.g. phosphate group on tyrosine). LS–MS/MS analysis, via alterations in ionization strength, can sub-fragment PTM groups to yield specific fragmentation patterns characteristic of that modification.

A 9.9 MALDI-ToF MS is used to study whole conidia proteomics. It is suitable because conidia could block the very narrow capillaries present in LC–MS instrumentation and the biomolecules released are readily detectable by MALDI-ToF MS. Also, the minimum specimen pretreatment used is not compatible with LC–MS analysis.

A 9.10 α-Cyano-4-hydroxycinnamic acid (HCCA), 2′,5′-dihydroxybenzoic acid or sinapinic acid. HCCA is best for peptide mass fingerprinting, while the other matrix materials can be used for whole-protein mass detection by MALDI-ToF MS.

Chapter 10

A 10.1 Ringworm (tinea) and athlete's foot (tinea pedis). These infections are caused by dermatophytes, such as *Trichophyton* spp. and *Microsporon* spp.

A 10.2 *Candida albicans*, *Candida glabrata*, *Candida parapsilosis* and *Candida tropicalis*. *C. albicans* is the most pathogenic yeast species. However, non-*C. albicans Candida* species are increasingly diagnosed in cases of superficial and systemic infection, particularly in patients with severely compromised immune systems.

A 10.3 HIV infection and poor denture hygiene (for oral candidosis). Solid organ and bone marrow transplantation, Anti-neoplastic therapy, premature birth, catheterization, broad-spectrum antibiotic and steroid treatment, prior colonization with *Candida* (for invasive candidosis).

A 10.4 (i) The ability to grow in yeast and hyphal form (i.e. dimorphism). The yeast form allows dissemination, while the hyphal form may facilitate tissue invasion. (ii) Phenotypic switching. (iii) The ability to produce adhesins that allow *Candida* cells to attach to host tissues (e.g. Als3 and Hwp1). (iv) The ability to produce extracellular hydrolytic enzymes, such as aspartyl proteinases (Saps) and phospholipases.

A 10.5 Invasive aspergillosis, aspergilloma and allergic bronchopulmonary aspergillosis.

A 10.6 *Pneumocystis jiroveci*, which is the causative agent of pneumocystis pneumonia (PCP).

A 10.7 *Histoplasma capsulatum* exists in the hyphal form in soil at 25 °C; however, in the human body at 37 °C, the pathogenic form of the fungus is small yeast like cells, known as conidia. This is in contrast to *C. albicans*, which can exist as yeasts and hyphae in the host, both of which are believed to be important for pathogenesis.

A 10.8 Aflatoxin B_1 is the most clinically important form of this family of mycotoxins. It is believed to be carcinogenic; in particular, ingestion of food contaminated with aflatoxin B_1 may be a risk factor for hepatic cancer.

Chapter 11

A 11.1 Current antifungals can inhibit the synthesis or disrupt the integrity of fungal cell walls, plasma membranes, nucleic acid and protein synthesis, and mitotic activity. Some novel (experimental) antifungals can alter cytoplasmic and nuclear processes and signals.

A 11.2 The interaction between polyenes and membrane sterols generates aqueous pores that leak cytoplasmic components. This results in inhibition of aerobic and anaerobic respiration, cell lysis and death. Polyenes tend also to induce oxidative damage to fungal plasmalemma, which may contribute to fungicidal activity.

A 11.3 Nephrotoxicity is the most significant side effect of AMB treatment. Manifestations may include decreased glomerular filtration rate, decreased renal blood flow, casts in urine, hypokalaemia, hypomagnesaemia, renal tubular acidosis and nephrocalcinosis.

A 11.4 Azoles inhibit cytochrome P450 (CYP)-dependent membrane ergosterol synthesis. The principle molecular target of azoles is CYP-Erg 11P, which catalyses the oxidative removal of the 14 α-methyl group in lanosterol and/or eburicol. Inhibition of 14 α-demethylase leads to depletion of ergosterol and accumulation of sterol precursors, which alters membrane permeability and fluidity. They can also inhibit the membrane ATPase in *C. albicans* and other yeasts. Azoles can inhibit cytochrome C oxidative/peroxidative enzymes, alter cell membrane fatty acids (leakage of proteins and amino acids), inhibit catalase systems and decrease fungal adherence. At fungicidal concentrations, azole treatment causes disintegration of mitochondrial internal structures and nuclear membranes.

A 11.5 In general terms, resistance to azoles may involve reduced permeability to azoles and possible alterations in the cell membrane rather than perturbations in cytoplasmic enzymes.

A 11.6 Echinocandins (caspofungin, anidulafungin and micafungin) are the main inhibitors of glucan synthesis through blocking the activity of glucan synthase. Other agents, like the pyridobenzimidazole derivative D75-4590, can also disrupt glucan synthesis by blocking the activity of β-1,6 glucan synthase.

A 11.7 Polyoxin, nikkomycin and aureobasidin.

A 11.8 It is a noncompetitive inhibitor of the β-(1,3)-glucan synthase, the enzyme involved in glucan synthesis. It also has secondary effects in the form of reduced ergosterol and lanosterol content and increased chitin content. Caspofungin activity leads to cytological and ultrastructural changes characterized by growth of target fungi as pseudohyphae; thickened cell wall and buds failing to separate from mother cells associate with echinocandin use. Cells also become osmotically sensitive, with lysis being restricted largely to growing tips of budding cells. Caspofungin therapy is indicated in the treatment of different forms of candidiasis and aspergillosis.

A 11.9 Sordarin, azosordarin and 5-fluorocytosine.

Chapter 12

A 12.1 *Parasitism* occurs where one species lives off another, as distinct from *symbiosis*, where different species live in harmony with each other and the relationship is mutually beneficial, or *saprophytism*, where organisms grow on dead organic matter.

A 12.2 The disease triangle may be defined as the relationship between a phytopathogen and plant and the development of disease which is influenced by the nature of the pathogen and host, and the prevailing environmental conditions.

A 12.3 Inoculum is produced and disseminated, and on reaching its target host plant tissue (inoculation) it penetrates the host. The type and mode of production of inoculum (e.g. sexual and asexual spores, resting spores, mycelium) and the method of dissemination (e.g. wind, water, insect) depends on the particular pathogen. For many important plant pathogens, a sexual stage has not been identified. Penetration is through wounds or natural plant pores (e.g. stomata), and some fungi produce specialized penetration structures called appressoria (singular = appressorium). Having penetrated its host, the fungus then grows within plant tissue (infection). Incubation period defines the period between inoculation and infection. As it grows within the host, it utilizes the plants cellular resources as a nutrient source and the damage inflicted on the plant manifests as disease symptoms. Latent period defines the period between infection and symptom development. The pathogen forms survival structures such as spores that are then disseminated. During the plant pathogenesis, some diseases only involve one such disease cycle (primary cycle), while others have the potential to do more damage as they involve secondary or multiple cycles of disease.

A 12.4 Fungi produce a range of enzymes that facilitate host plant infection and colonization by degrading the cellular and intercellular constituents of plants (certain fungal pathogens also produce non-enzymatic proteins that inhibit the activity of plant enzymes involved in the host defence response).

A 12.5 Fungal toxins may be host specific or non-host specific. Host-specific toxins are usually required for pathogenicity, but are only toxic to the host plants of a pathogen and show no or low toxicity to other plants. In contrast, non-host-specific toxins affect a wide range of host plants and may act as virulence factors, increasing the extent of the disease, but are not essential for disease development.

A 12.6 *Protectant:* Do not penetrate the plant, but affect pathogen viability and germination on the surface of the host plant. Such fungicides must adhere to the plant surface and resist weathering and are most effective when applied as a preventative measure; that is, before plant inoculation with the fungus. *Systemic:* Can act on the plant surface and be translocated throughout the plant vascular system to kill the fungus. Systemic fungicides may also exhibit translaminar movement within the leaves. *Eradactive:* They are applied post-infection and act on contact by killing the organism or by preventing its further growth and reproduction.

A 12.7 The biological control agent suppresses the pathogen by one or more of the following means: hyphal interference, secretion of antifungal compounds (e.g. antibiotics) or enzymes. The fungus *Trichoderma harzianum* is used for the biocontrol of various fungi, which it achieves by means of hyphal interference and the secretion of hydrolytic enzymes that attack the causal fungi.

A 12.8 Vegetables, flowers, field crops, fruit trees, roses and forestry trees.

A 12.9 Famine (Irish potato famine 1845–1849), ergotism (Ethiopia in 1977–1978), Disease due to mycotoxins; potential use of fungal toxins as bio-terrorism agents.

Index

Fungi: Biology and Applications, Second Edition. Edited by Kevin Kavanagh.
© 2011 John Wiley & Sons, Ltd. Published 2011 by John Wiley & Sons, Ltd.